Gastroenterology and Hepatology

Gourdas Choudhuri • Anil C Anand
P Piramanayagam
Editors

Gastroenterology and Hepatology

Bench to Bedside

Springer

Editors
Gourdas Choudhuri
Department of Gastroenterology and
Hepato-biliary Sciences
Fortis Memorial Research Institute
Gurgaon, Haryana, Haryana, India

Anil C Anand
Department of Gastroenterology and
Hepatology
KIIT University
Bhubneshwar, Odisha, India

P Piramanayagam
Department of Gatroenterology
Apollo Hospitals
Chennai, Tamil Nadu, India

ISBN 978-981-99-9286-7 ISBN 978-981-99-9287-4 (eBook)
https://doi.org/10.1007/978-981-99-9287-4

This Springer imprint is published by the registered company Springer Nature Singapore Pte Ltd.
The registered company address is: 152 Beach Road, #21-01/04 Gateway East, Singapore 189721, Singapore

Foreword

As the world grappled with one of the most catastrophic pandemics in living memory, the critical relevance of "basic science" to clinical medicine and, indeed, human health, had never been more obvious. Our ability to deal with this new coronavirus ultimately depended not only on a complete knowledge of its virology but also on the ingenuity of those who, employing various molecular strategies, worked on the development of an effective vaccine. COVID-19 has taught us, in a most dramatic fashion, how important our "basic science" colleagues are to those of us whose primary preoccupation is caring for our patients.

You will have noted that I enclose "basic science" in quotations—this is deliberate as I feel that this term should be abandoned and the separation between the laboratory and the hospital ward must end. I, therefore, applaud Dr Choudhuri's bench-to-bedside approach—it is the only way forward! Gone are the days when any practicing doctor can dismiss biochemistry, physiology, anatomy, microbiology, or molecular biology as something tolerated to get through medical school and pass the requisite but much hated examinations; a necessary but transient evil, but which can be safely ignored now that one has graduated and is in the "real world." Unfortunately, an anti-science stance has become a virtual badge of honor for some, including some of our political leaders. This is a dangerous and potentially fatal approach. How can we help our patients if we do not know how to optimize testing that is now so frequently based on molecular techniques? Referring back to the pandemic, an inability to differentiate between antigen vs antibody tests or to understand why tests based on polymerase chain reaction (PCR) are still the most accurate can have enormous implications for entire populations. In gastroenterology, the evaluation of dysphagia, acid reflux disease, dyspepsia, and constipation are based on fundamental concepts of anatomy and physiology. How can one manage hepatitis without understanding the virology or optimally treat the GI cancer patient and counsel their families without an understanding of cancer biology and genetics? I make an earnest plea, therefore, to all involved in medical education at all levels, and including continuing medical education/continuing professional development, to eschew the temptation to place "basic science" and "clinical practice" in separate silos and base their teaching in the complete integration of the biological sciences with patient care. Medical science is a continuum, not a compendium of distinct disciplines. I have been most fortunate in spending several wonderful years in a totally integrated research institute, The Alimentary Pharmabiotic Center (APC—now APC Microbiome Ireland), in Cork, Ireland, where gastroenterologists (and many other clinical specialists) and molecular microbiologists worked together on clinical problems without giving thought to whether one was from "basic science" or the hospital—such distinctions were not even considered.

This, as Dr Choudhuri shows us in this volume, is the only way forward. We, as clinicians, must stand up for science, confer with all our colleagues, and keep abreast of advances relevant to our practice—if we fall behind we will soon be no better than the charlatans who peddle false hope, dodgy tests, and worthless "cures" based on mythology, conspiracy theories, and sheer greed to prey on vulnerable sufferers and their families.

I congratulate Dr Choudhuri on his foresight in adopting the bench-to-bedside (and back to bench!) approach as a fundamental basis for his definitive textbook. I know that you will find it invaluable in the daily care of your patients and in directing you to new avenues of clinical research.

August 2020

Eamonn M.M. Quigley
Gastroenterology and Hepatology
Lynda K and David M Underwood Center for Digestive Disorders
Institute of Academic Medicine, Department of Internal Medicine
Houston Methodist Hospital, Weill Cornell Medical College
Houston, TX, USA

Texas A and M Health Sciences Center College of Medicine
Bryan, TX, USA

School of Medicine, University College Cork
Cork, Ireland

Acknowledgments

This book in its present form has had a long history in making in which several people have played important roles.

I am grateful to Prof Eamonn Quigley, who, despite his very busy schedule and commitments, agreed to write the Preface and contribute four key chapters.

Dr P Piramanayagam, an eminent gastroenterologist and teacher, has helped me through several stages of the book: from preparing the initial road map to writing and compiling several chapters and sections. Without his dedicated commitment and unflinching support, this book would not have been possible. I am honored to have him join me as an "editor."

The last few years have seen rapid expansion in our knowledge in the field of "liver diseases." Prof Anil C Anand, an eminent hepatologist and researcher, has taken up the daunting task of writing and compiling the section on liver.

Many of my colleagues and fellows have contributed chapters either for this updated book or its earlier version that appeared in 2014. Some who deserve special mention and thanks are Dr Lakshmi CP, Dr Abhai Verma, Dr Govind Makharia, Dr Vineet Ahuja, Dr Uday C Ghoshal, Dr Gursimran Kaur, Dr Karthik Natarajan, Dr Kushan Sengupta, Dr Bhavin, Dr Vivek Singh, Dr Neha Choudhary, and Ms Seema Kaleel.

My special thanks to Mr Gaurav Singh, Ms Raman Shukla, Dr Naren Agarwals, Ms Saanthi Shankhararaman, Ms Sukanya, and Ms Sangamithirai of Springer Nature for putting it all together and helping us to bring this edition to you.

My last and most special acknowledgment is to the students and fellows across the country whom I had the privilege of interacting over the last 35 years…their constant quest for updated knowledge and untiring aspiration for learning how to apply them in the bedside provided me the stimulus to undertake this task.

Gourdas Choudhuri

Introduction

Modern medicine distinguishes itself from all other forms of healing by being rooted in the knowledge of structure and functions of the human body. It has in fact penetrated deep in recent times from the level of organs to that of cells, sub-cellular mechanisms, and molecules.

The explosion of knowledge in recent times has however been so exponential and enormous in recent times that the modern physician often finds himself seriously challenged to keep himself updated with what he needs to know of this "bench knowledge" and to be able to apply it on the patient's bedside.

Volume 1 of *Gastroenterology and Hepatology: Bench to Bedside*, which deals with the GI tract, made its appearance in 2014, and the second volume dealing with the liver, biliary tree, and pancreas was significantly delayed. We have therefore compiled a single volume incorporating all sections this time and updated the contents to make it relevant to the present time.

This book aims to bridge this widening gap so that trainees, fellows, and practitioners get a whiff of "basic" science knowledge that serves as a foundation for our clinical decision-making. This approach can sharpen our perspective as we practice clinical gastroenterology and hepatology.

This book is not meant to replace the existing standard textbooks of Gastroenterology and Hepatology but is aimed at becoming a handy compendium of few select topics of contemporary importance. We have therefore intentionally kept it short and crisp.

Given the very short shelf life of medical knowledge, we are mindful that the life of this book is likely to be short, and that it will require frequent updates and revisions to remain in step with the times. The editors are cognizant of this challenge and are planning measures to keep up.

We shall be grateful for your feedback, comments, suggestions, and criticisms in our future endeavors.

Gourdas Choudhuri
Anil C. Anand
P. Piramanayagam

Contents

Chapter 1
Esophagus

Gourdas Choudhuri

1.1 Lower Esophageal Sphincter

1.1.1 Anatomy

An elongated tube-like structure, the esophagus is muscular, extending over an average length of 25 cm connecting the hypopharynx to the stomach. At either end lie the esophageal sphincters. The lower esophageal sphincter (LES) is a small stretch (3–4 cm) of raised pressure which marks the end of the esophagus and approaches the stomach. While the proximal and distal LES borders are similar in length (1.5–2 cm), the former lies above the squamocolumnar junction, whereas the latter is positioned inside the abdominal cavity. Extrinsically, the pharyngoesophageal ligament that arises from the diaphragm, specifically its transversalis fascia, connects and fixes the esophagus within its hiatus. Short clasp muscles encircle the medial side of the distal esophagus, and gastric sling fibers laterally on the left side result in a high-pressure zone.

1.1.2 Embryology

The esophagus develops at 4 weeks from the common tube endoderm as a diverticulum, and the musculature of the esophagus develops at 11 weeks.

G. Choudhuri (✉)
Department of Gastroenterology and Hepatobiliary Sciences, Fortis Memorial Research Institute, Gurugram, India

G. Choudhuri et al. (eds.), *Gastroenterology and Hepatology*,
https://doi.org/10.1007/978-981-99-9287-4_1

1.1.3 Physiology

When swallowing occurs, the LES pressure decreases within 1.5–2.5 s, and this continues for about 7 s, during which the peristaltic contracting wave transverses the esophageal body.

The lower esophageal sphincter relaxes completely during swallows, and the swallow-induced inhibition probably includes both active inhibitions of the muscle by NANC inhibitory neurons releasing NO and cession of tonic neural excitation to the sphincter.

The lower esophageal sphincter functions as an anti-reflux barrier. It is tonically contracted and maintains an average resting pressure of 20 mmHg. Other components of the antireflux barrier are clasp fiber, gastric sling fiber, phreno-esophageal ligament, intra-abdominal esophagus, and angle of His.

Transient LES relaxation is also mediated by nitric oxide and occurs independently of swallows, persists for a longer duration, and is associated with reflux disease.

Every 90–120 min, the LES pressure oscillates along with the migrating motor complex, with pressures recorded highest in association with the vigorous motor activity during phase III in the stomach. Lower esophageal sphincter tone is modulated following hormones, food, and drugs.

1.2 Wall Layer of the Esophagus

The histology of the esophagus is similar to the digestive tract consisting of mucosa, submucosa, muscularis, and adventitia; however, the serosa is absent. The muscularis propria, which is the last layer, is, in turn, made up of three zones of muscles with a mixed skeletal and smooth muscle sandwiched between a proximal striated or skeletal and a distal smooth muscle in a 1/3 ratio across. This layer predominantly performs the peristaltic function to propel the esophageal content towards the distal end.

1.3 Esophageal Constrictions

There are three areas of esophageal constrictions:

1. At the junction of the cricopharynx with the proximal esophagus.
2. Mid esophagus, where the arch of the aorta and left main bronchus indent the anterior wall of the esophagus.
3. At the site where the esophagus passes through the diaphragm into the abdominal cavity.

Importance:

The site of corrosive strictures/pill-related esophagitis is due to temporary stasis at these sites.

1.4 Blood Supply of the Esophagus

The blood supply to the esophagus can be divided into three sections, with each being vascularized by different arteries. The proximal third part of the esophagus is innervated by the inferior thyroid artery, branches from the descending thoracic aorta innervate the central third portion, and the distal third by branches from the left gastric artery. The veins from the proximal, central, and distal sections drain into the inferior thyroid veins, the azygos veins, and the left gastric vein, respectively.

1.5 Portosystemic Anastomosis

The portal-systemic veno-venous, or the portocaval anastomosis, positioned at the distal end of the esophagus consists of tributaries of the systemic azygos veins and the left gastric vein of the portal system.

1.6 Lymphatic Drainage of Esophagus

The lymphatic fluid from the esophagus drains from the proximal, central, and distal sections into the deep cervical nodes, the superior and posterior mediastinal nodes, and the celiac nodes, including those lying lateral to the left gastric blood vessels, respectively.

1.7 Esophageal Rings

The Schatzki rings (A and B) define the limits of the proximal and distal margins of the esophageal vestibule.

1.7.1 A Ring

The A ring positioned at the proximal margin is a broad 4–5 mm congruous spread of muscle that has undergone hypertrophy enclosed by squamous epithelium. It narrows the esophageal lumen at its intersection with the vestibule corresponding to the proximal margin of the LES. Making a seldom appearance with the usual asymptomatic nature, it differs in caliber of the esophagography based on the level of esophageal distension. Dysphagia for food is infrequently linked with the occurrence of an A ring. The esophageal A rings that manifest symptoms can be managed with a 50-Fr mercury-weighted dilator or with botulinum toxin injection.

1.7.2 B Ring

Unlike the rarely found A ring, the mucosal B ring is frequently reported in patients undergoing routine GI series. When assessed using barium, it is consistently found to be coexistent with a hiatal hernia. It is a fine membrane (2 mm) that narrows the esophageal lumen at the vestibulo-gastric cardia intersection. The proximal squamocolumnar junction of the ring is lined by squamous epithelium and columnar epithelium on the upper and the lower surface, respectively. The ring per se comprises solely of mucosa and lacks the muscularis propria. Schatzki rings have been associated with GERD and are also argued to be congenital. B rings are usually asymptomatic. However, if luminal diameter becomes less than 12 mm, patients may present with irregular dysphagia or steakhouse syndrome. Although asymptomatic B rings need not be treated, symptomatic rings may need to be managed either with a large (≥50-Fr) mercury-weighted dilator either once or sequentially to achieve adequate patency. However, in instances where these rings are resistant to dilatory methods, endoscopy with electrocautery may be required.

1.7.3 Corrugated Ringed Esophagus

Apart from the solitary esophageal ring abnormalities, sometimes other anomalies include multiple rings or corrugated-ringed esophagus. In this condition, the esophagus manifests centrally aligned muscular structures which are parallel to each other and sometimes extend the entire length of the esophagus or are limited to the proximal and distal ends. These rings may be detected on endoscopy since these persist even in the presence of the highest air insufflation, but barium studies may not recognize these. These rings may be of congenital origin with a relevant history of dysphagia, frequently accompanied by impactions, experienced in late infancy and primarily found in males. These rings may also occur in patients with GERD and eosinophilic esophagitis. Dilation is restricted to 40-Fr bougie because it is frequently associated with a risk of mucosal tears and sometimes perforations in these patients.

1.8 Esophageal Webs

Unlike the rings, the esophageal webs do not appear as concentric circles but form planar membranes comprising of stratified squamous epithelium in the proximal and central sections of the esophagus. These webs protract from the anterior section and extend laterally but do not encircle the esophagus. Frequently found in the cervical esophagus, these are easily diagnosed on the lateral view of an esophagogram. Although predominantly esophageal webs manifest as dysphagia for solids, these membranes are easily treated with bougienage with mercury-weighted dilators due to their fragile nature. Cervical esophageal webs manifesting as dysphagia along with iron deficiency anemia are typical findings of a patient with Plummer–Vinson or Paterson–Kelly syndrome. Its identification is beneficial since this group of patients is particularly susceptible to squamous cell carcinoma of the pharynx and esophagus. Treatment of IDA in this syndrome may provide relief from the symptoms as well as eliminate the esophageal webs.

1.9 Esophagus in Portal Hypertension

The portal vein is formed from vitelline veins. There are two vitelline veins—right and left each of which forms a respective tributary of the portal vein; however, the principal portal vein is formed primarily from the left vitelline vein.

Venous supply from the lower esophagus and OG junction.

The venous drainage system can be divided into four different bands (Fig. 1.1):

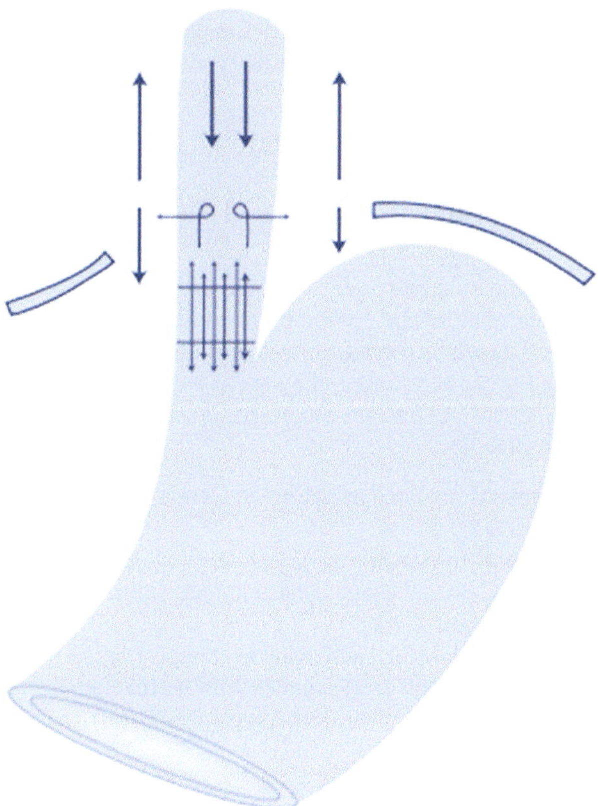

Fig. 1.1 Venous anatomy at gastroesophageal junction showing perforating zone

1. Gastric zone: extends 2–5 cm below esophagogastric junctions and veins are longitudinal and drain into left and short gastric veins.
2. Palisading zone: extends 2–3 cm above the esophagogastric junction, vein runs longitudinally in lamina propria, along four corresponding mucosal folds, no communication with veins surrounding the esophagus, and commonest site of rupture of varices in portal hypertension.
3. Perforating zone: extends 2 cm above into the esophagus and site of communication between internal and external veins with veins passing across the muscular wall of the esophagus. Veins form loops and networks in this layer-site of systemic and portal circulation communication.
4. Truncal zone: extends 8–10 cm proximally into the esophagus and veins run longitudinally in lamina propria.

1.9.1 Changes in Portal Hypertension

In portal hypertension, most of the portal venous flow is diverted through left and short gastric veins to form gastroesophageal varices and fundic varices. In the absence of perforating veins in the palisading zone, there is marked dilatation and tortuosity of veins, and it is the commonest site of rupture of lower esophageal varices.

1.10 Tracheo-Esophageal Fistula

When the lung bud does not isolate from the foregut, a trachea-esophageal fistula is formed. One of the common anomalies of esophageal development is esophageal atresia, in which the primitive foregut fails to recanalize. Isolated esophageal atresia is rare ~7%. Most often, they are associated with trachea-esophageal fistula.

The types of trachea-esophageal fistulas as described by Gross et al. are (Fig. 1.2) the following:

- Type A: Isolated esophageal atresia without trachea-esophageal communication (8%).
- Type B: Proximal esophageal atretic segment communicates with trachea.
- Type C: Distal esophageal segment communicates with the trachea, which is the commonest (87%).
- Type D: Proximal and distal esophageal segments communicate with the trachea.
- Type E: Esophageal stenosis without trachea-esophageal communication.
- H-type fistula where the esophagus is not atretic and communicates with the trachea.

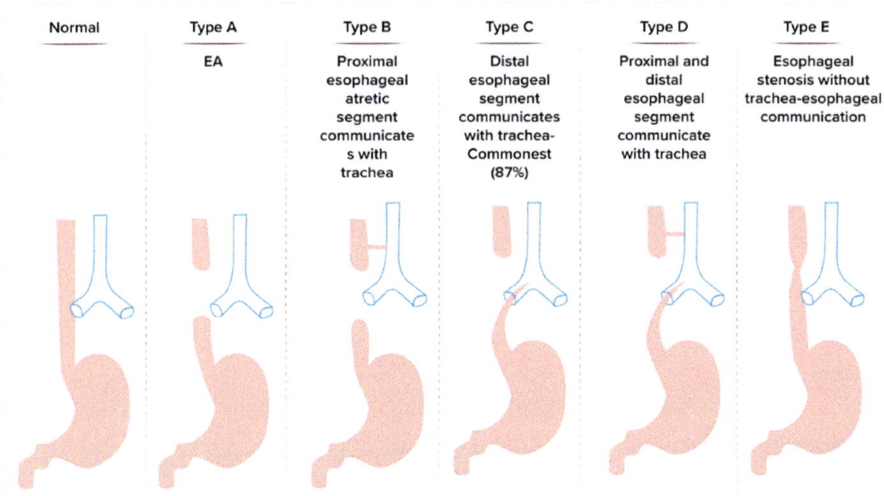

Fig. 1.2 Classification of trachea-esophageal fistula (GROSS)

1.10.1 Diagnosis

- In fetuses with esophageal atresia, polyhydramnios occurs during gestation since the fetus is unable to swallow amniotic fluid. Neonates present with choking and distress soon after birth after the first feed.
- The neonates with trachea-esophageal fistula (communication with the proximal esophagus) have a history of recurrent aspiration pneumonia. If a distal esophageal segment is in continuity with the trachea, the abdomen will be filled with gas. If only the proximal esophageal segment is in continuity with the trachea and the distal segment is not communicating, a gasless scaphoid abdomen is seen. X-ray of the chest with the abdomen helps in diagnosing the type of trachea-esophageal fistula. Failure to pass a nasogastric tube often confirms the diagnosis.
- Neonates with esophageal stenosis alone may grow into childhood and may have delayed presentation.

1.10.2 Treatment

Esophageal atresia and trachea-esophageal fistula are preferentially resolved with surgery and isolated esophageal stenosis. Post-surgery, on follow-up, there is a high incidence of GERD due to accompanying esophageal motility disorder.

1.11 Transient Lower Esophageal Sphincter Relaxation

Temporary LES relaxation initially observed in the absence of swallowing action is an underlying phenomenon of belching which was later termed transient lower esophageal sphincter relaxation (TLESR). It occurs along with concurrent inhibition of the crural diaphragm to prevent reflux.

1.11.1 Definition

TLESRs are sudden decrements in LES pressure to the same degree as the intragastric pressure but in the absence of the swallowing trigger. These transient relaxation periods are longer (10–45 s) than those that occur during swallowing. A prominent after-contraction at the end of the relaxation period is also a typical feature of TLESR.

1.11.2 Physiology

The TLESR occurs in two settings based on the manner of LES pressure before its onset. It can occur either spontaneously when LES pressure is stable or instantly post-end of an LES relaxation induced due to swallow.

The challenge of recognizing TLESRs is differentiating these from the usual swallow-induced ones. The optimal means to identify relies on the temporal association of swallow and initiation of LES relaxation, which requires a precise record of swallow either using pharyngeal manometry or mylohyoid electromyography. Similarly, based on the duration of LES relaxation, TLESR can be distinguished from swallow-induced relaxation, which is shorter than the former (>10 s versus 6–8 s). The TLESRs due to GERD are generally complete (<2 mm of intragastric pressure).

Thus, the basis of precise TLESRs may be listed as follows:

1. No pharyngeal swallow signal either 4 s pre- and 2 s post the beginning of LES relaxation or absence of mylohyoid electromyogram complex for 3 s before the onset of LES relaxation.
2. Drop of >1 mmHg/s in the LES pressure.
3. Time-lapse of <10 s for one complete LES relaxation.
4. The lowest pressure of –2 mmHg.

1.11.3 Events Associated with TLESRs

Pharynx: The contractions arise in the pharynx and mylohyoid muscle at the onset of ESR in about 30% of cases.

Esophageal body: Distal esophageal pressure waves not associated with swallowing frequently occur at the onset of TLESRs. These waves at multiple sites typically occur simultaneously.

Stomach: A minor drop (2–4 mmHg) in intragastric pressure may occur during TLESR.

Diaphragm: During TLESR, partial or complete inhibition of the diaphragm may occur, notwithstanding the sustained activity of the costal diaphragm.

1.11.4 Control of TLESRs

1. Gastric distension.
2. Pharyngeal intubation.
3. Posture—supine position is known to completely attenuate the stimulation of TLESRs triggered by gaseous gastric distention.
4. Sleep—the reflux instances (~10 s) are restricted to bouts of arousal during sleep and do not occur during in stable sleep period.
5. Stress—cold stress may decrease the recurrence of TLESRs.
6. Drugs—drugs that may attenuate TLESRs consist of CCKA-1 antagonists, antimuscarinic agents, GABA-B agonists, 5-HT_3 antagonists, somatostatin, nitric oxide inhibitors, and morphine.

1.11.5 Neural Pathway-Mediating TLESRs

The TLESR is primarily a vagal reflex in response to the stimulation of muscarinic receptors located in the gastric fundus and pharynx and, to some extent, the larynx. The relationship of the events that comprise the LES, stomach, esophageal body, pharynx, and crural diaphragm during TLESRs aligned to their pattern in the brainstem is time related. In otherwise healthy subjects, two-thirds of their TLESRs are associated with refluxes, while in those with GERD, nearly three-fourths experience refluxes.

1.12 Refractory GERD

Refractory GERD, defined as persistent symptoms to 1–2 months of twice-daily PPI pharmacotherapy, may occur in more than a quarter of patients diagnosed with it. Non-adherence to the medication or posology may be the underlying cause of non-responsiveness to pharmacotherapy. It can be diagnosed with upper GI endoscopy essentially to determine the presence of esophagitis, peptic ulcer, or sometimes malignancy. Current evidence shows that only 10% of patients with refractory GERD display abnormalities in the upper GI endoscopy.

1.12.1 Refractory Gastroesophageal Reflux Disease Without Esophagitis

Nocturnal Gastric Acid Breakthrough (NAB)

- Nocturnal sustained gastric acid secretion normally plays an important role in safeguarding the Gastrointestinal system (GIT) from bacterial infection.
- In patients with GERD, a nocturnal gastric acid breakthrough may result in symptoms refractory to PPI therapy.
- Histamine-2-receptor antagonists (H_2RA) such as ranitidine (150–300 mg at nighttime) have been proven efficacy in attenuating NAB. However, tolerance develops after 1–2 weeks with a recurrence of NAB.
- Intermittent H_2RA use after heavy meals or drinking may be an optional approach for minimizing tolerance.

Non-acid Gastroesophageal Reflux

- Non-acid reflux, which is mildly acidic (pH > 4.0), is reported in 20–40% of patients to occur during the day. It can be detected using impedance testing along with the determination of pH.
- The other non-acid reflux is bile reflux, also known as duodenal gastroesophageal reflux, which occurs in 10–15% of cases.
- Baclofen (5–20 mg thrice a day) may decrease duodenal gastroesophageal reflux, but long-term efficacy is unknown.

Missed Gastroesophageal Reflux Disease

- Traditional pH catheters may miss acid reflux as it does not occur every day.
- False negative as eating and activity may be altered due to nasal catheter.
- Acid reflux can be monitored over 48 h, even in unusual situations (heavy exercise, swimming, etc.), using the wireless pH monitor (Bravo TM capsule).

- Missed acid reflux as a cause of unyielding symptoms in refractory GERD is not known.

Functional Heartburn

- There is intermittent retrosternal burning with no identified GER pathology, motility, or anatomical disorder.
- They were frequently reported in females who are not obese. In this subgroup, the underlying mechanism is abdominal hyperalgesia.
- The normal pH test result is seen in 30–50% of patients with heartburn and normal upper endoscopy.
- Given the mechanism of functional heartburn, tricyclic antidepressants and selective serotonin reuptake inhibitors may be effective for pain modulation rather than PPI.

Wrong Diagnosis

- Achalasia with minimal concurrent esophageal dilation may resemble GERD. However, it can be easily diagnosed using esophageal manometry and barium studies (Fig. 1.3).
- Gastroparesis, in which regurgitation is the predominant symptom than heartburn, may occur in 10–20% of patients with refractory reflux symptoms.

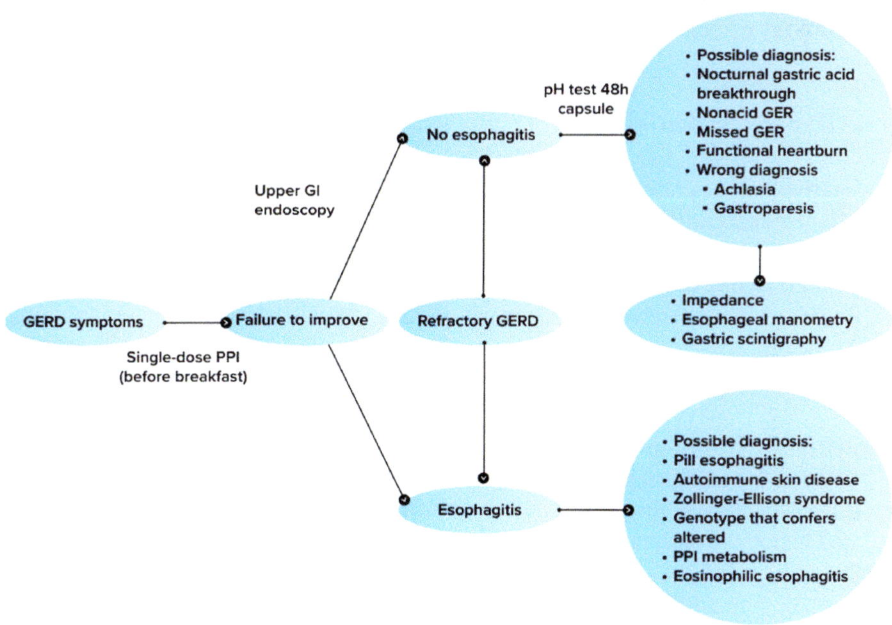

Fig. 1.3 Refractory GERD. *GERD* gastroesophageal reflux disease, *PPI* proton pump inhibitor, *GI* gastrointestinal, *GER* gastroesophageal reflux

1.12.2 Refractory Gastroesophageal Reflux Disease with Esophagitis

Pill-Induced Esophagitis

- Medicinal agents can induce esophagitis in both elderly and young patients.
- Odynophagia, chest pain, and dysphagia accompany heartburn.
- Esophageal motility disorder or stricture is commonly associated with the elderly.
- Doxycycline, tetracycline, alendronic acid, quinolone, ferrous sufhate, naproxen, aspirin, and ascorbic acid are the common culprits.
- The presence of ≥1 discrete esophageal ulceration circumscribed by healthy mucosa at the intersection of the proximal and central sections is a peculiar finding.
- Sparing of squamocolumnar junction helps to differentiate from reflux esophagitis.

Autoimmune Skin Diseases

- Epidermolysis bullosa acquisita, lichen planus, cicatricial pemphigoid, and pemphigus vulgaris can cause intractable esophagitis.
- It is commonly seen in middle-aged females who complain of dysphagia or odynophagia.
- Skin lesions, oral mucosal, or anogenital lesions might not be obvious.
- Histology of dense histiocytic infiltrates is suggestive of lichen planus.

Acid Hypersecretion

- Nearly 45% of patients with Zollinger–Ellison syndrome display esophagitis or strictures on endoscopy.
- Twice-a-day PPI therapy with lowering gastric acid secretion to less than 10 mEq/h may result in symptom relief in two-thirds of patients.
- Recurrent esophageal stricture may need gastric acid secretion reduction to <1 mEq/h.

Eosinophilic Esophagitis

- It is suspected in young men with intermittent solid food impaction and associated history of asthma or food allergy.
- Endoscopy helps in the identification of multiple esophageal rings, linear furrows, or tiny white exudates.

- Histology showing >15–20 eosinophiles per high-powerfield is suggestive of eosinophilic esophagitis.
- Response with inhaled fluticasone propionate (440 μg b.d.) occurs in 2–3 months; however, some may require additional therapies like leukotriene D4 antagonist and oral prednisolone.

Genotypic Differences

- Proton pump inhibitors are mainly metabolized by hepatic cytochromes (P4502C enzymes).
- Considerable genetic variations occur among patients in their hepatic cytochromes (P4502C enzymes) to metabolize PPIs.
- Rapid PPI metabolism is seen in 12–20% of the Asian population as compared to 3–6% of the white population.
- Esomeprazole is metabolized by CYP3A4, so it can overcome the problem with other PPIs.

1.12.3 Newer Treatment in Gastroesophageal Reflux Disease

Extended Release Proton Pump Inhibitors

- Dexlansoprazole MR releases two different granules at distinct pH (5.5 and 6.8), hence, termed dual-release formulation. Single dosing and oral intake, irrespective of meals, may result in better compliance of patients.
- Ilaprazole is a benzimidazole compound with a greater half-life and safety profile than other PPI. A minimal dose of 10 mg is more efficacious than omeprazole.
- Esomeprazole strontium delayed release is an incrementally modified drug where strontium is used instead of magnesium salt. Clinical trial results are awaited, although recently, it has been tentatively approved by FDA.

Potassium-Competitive Acid Blocker

- Vonaprazan is a premier reversible competitive K^+ acid blocker with a quick onset of action and swift inhibition of acid secretion. Its efficacy in alleviating erosive esophagitis is comparable to esomeprazole.
- Soraprazan has also shown quick inhibition of acid secretion in experimental studies.
- Revaprazan is a recently developed acid blocker with equivalent acid suppression as compared to PIs.

TLESR Reducers

- Delta-9-tetrahydrocannabinol, which is a dual cannabinoid receptor 1 and 2 antagonists, suppresses the frequency of TLSERs. However, clinical trials were prematurely terminated due to major adverse events such as tachycardia, hypotension, and emesis.
- Rimonabant, a CB1 receptor antagonist, also decreased the TLESRs and consequently post-prandial reflux—nevertheless, increased risk of neuropsychological adverse events led to its withdrawal.
- Cholecystokinin/gastrin receptor antagonists have been tested in clinical trials for decreasing the TLESRs (loxiglumide, itriglumide, and spiroglumide). Currently, evidence is inadequate to confirm whether its efficacy in physiologic post-meal reflux can be mirrored in its efficacy in pathologic reflux.

Mucosal Protectant

Rebamipide is an anti-inflammatory agent which may therefore demonstrate efficacy in healing esophageal mucosal ulcers along with a PPI.

1.13 Achalasia Cardia

1.13.1 Normal Esophageal Motor Innervation

Smooth muscles of the distal esophagus and LES are innervated by postganglionic vagus nerve fibers via the myenteric plexus (Fig. 1.4).

1.13.2 Pathogenesis of Primary Achalasia

1. Various factors have been implicated in triggering inflammation leading to neuronal loss
   - Infectious (Fig. 1.5)
   - Familial
   - Autoimmune
   - Neurodegenerative
2. The decline of inhibitory ganglion cells in myenteric plexus
3. Loss of inhibitory neuro-transmitter (NO)
4. Intact cholinergic (Ach) transmission

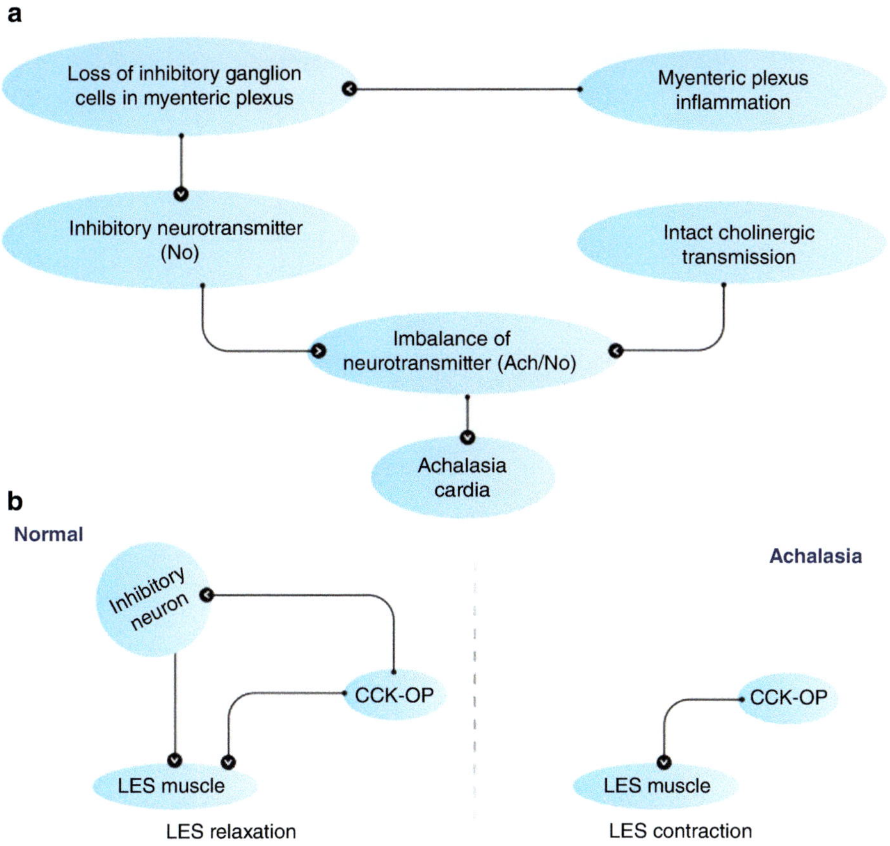

Fig. 1.4 Achalasia: (**a**) pathophysiology and (**b**) pathogenesis of primary achalasia. *LES* lower esophageal sphincter, *CCK-OP* cholecystokinin octapeptide

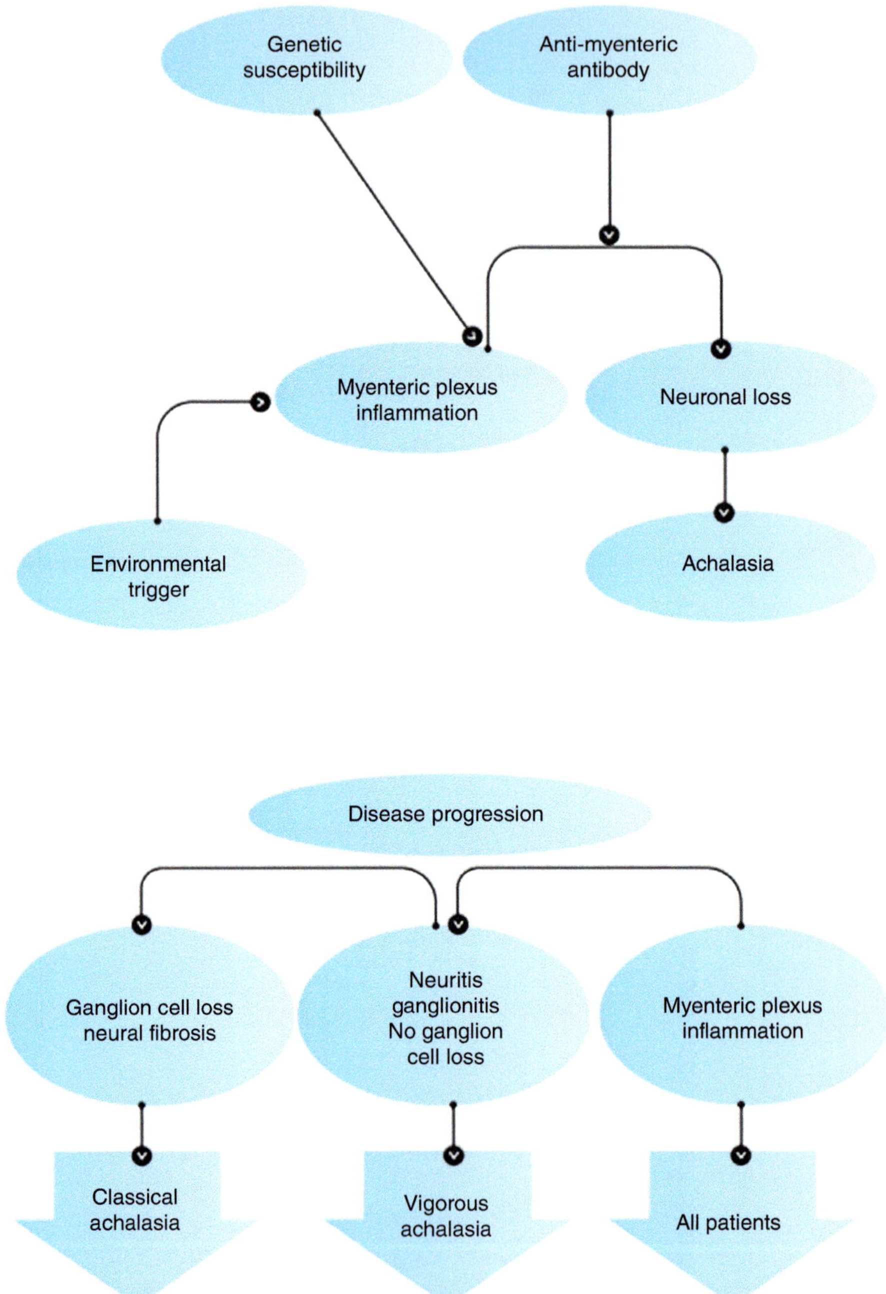

Fig. 1.5 Primary achalasia: disease progression

1.14 Eosinophilic Esophagitis

Eosinophilic esophagitis is a disorder which is diagnosed by clinicians based on both clinical and pathologic investigations described as the following:

- Manifestations associated with esophageal dysfunction.
- Esophageal biopsy showing excessive eosinophil presence indicating inflammation (value ≥15 eosinophils/hpf).
- Mucosal eosinophilia restricted to the esophagus is maintained even after treatment with PPI.
- Any other etiology of esophageal eosinophilia should be excluded.

Clinical characteristics are as follows:

- Atopy is common, including atopic diathesis.
- Occurs predominantly in males (male-to-female ratio 3:1).
- Most common in racial and ethnic groups (non-Hispanic Whites).
- Manifestations begin in infants and toddlers as feeding difficulties or in school-going children as emesis, food impaction, abdominal pain, and/or failure to thrive.
- The disorder may also present in adults between the age of 30 and 40 years with solid-food dysphagia and impaction of food, along with regurgitation, abdominal pain, and chest pain in some.
- There are no physical abnormalities particular to the diagnosis of eosinophilic esophagitis.

1.14.1 Endoscopic Findings

The ACG Clinical Guidelines proposed classification and grading for endoscopic assessment of eosinophilic esophagitis using four major and two minor criteria (adapted from ACG Clinical Guidelines 2013). The ACG major features include edema, furrows, and strictures with grades 0 or 1 for their absence or presence, fixed rings with grades 0 through 3 for absence and progressive difficulty in passage of the endoscopic tube, exudates grade 0 through 2 for absence and presence varying severity of lesions on the esophageal surface, furrows, and strictures each with grades 0 and 1 for their absence and presence. The minor features include two grades, 0 and 1 each, for the absence or presence of crepe paper and narrow caliber esophagus.

While eosinophilic esophagitis (EoE) Endoscopic Reference Score (EREFS) classification is another recently validated method to recognize and record the endoscopic findings in eosinophilic esophagitis with similarities to the one recommended by ACG, except that minor features are not included.

Table 1.1 Differences between eosinophilic esophagitis and gastroesophageal reflux disease

	Gastroesophageal reflux disease	Eosinophilic esophagitis
Number of eosinophils/hpf	<15	>15
Location of eosinophils	Distal esophagus	Proximal + distal esophagus
Thickened mucosa with basal layer hyperplasia	–	+
Mast cells and dendritic cells	+	+++
Papillary lengthening	+	+++
Eosinophil surface layering	–	+
Eosinophil microabscesses	–	+

1.14.2 Histologic Characteristics

Apart from the excessive number of eosinophils, the other histologic characteristics that differentiate EE from GERD include the presence of extracellular eosinophil granules, excess lymphocytes, hyperplasia of the basal cell layer, microabscess, rete-ridges penetrating the underlying layers and fibrosis of the lamina propria (Table 1.1).

1.14.3 Treatment Options

Eosinophilic esophagitis can be managed with non-pharmacological, pharmacological, and surgical methods.

1. Dietary treatments

   The three dietary recommendations include complete exclusion of probable food allergens commonly associated with atopic diathesis (soy-based, nuts, mild, eggs, seafood, and wheat) and, second, excluding dietary components confirmed as an allergen based on the results of allergy tests such as skin prick or patch test, and, lastly, substituting the eliminated components with amino acid-containing formulae. These dietary interventions have proven to be effective not only in resolving symptoms but also in improving the histological abnormalities across age groups in patients with eosinophilic esophagitis.
2. Pharmacologic treatments

   The initial pharmacotherapy includes a 2-month treatment with topical steroids such as fluticasone or budesonide, which should be swallowed (Table 1.2). In patients necessitating quick symptomatic relief or non-responsiveness to initial steroids, prednisone may be prescribed. Patients who were refractory to either may be initially treated with higher doses for an extended duration of topical steroids and subsequently switched to oral corticosteroids. Currently, clinical evidence to advocate pharmacotherapy containing histamine or leukotriene antagonists and biologics remains inconclusive.

Table 1.2 Pharmacological treatment of eosinophilic esophagitis

Medication	Age group	Dosing
Fluticasone	Pediatric	88–440 μg/day in a divided dose
	Adult	880–1769 μg/day in a divided dose
Budesonide	Pediatric	1 mg/day in a divided dose
	Adult	2 mg/day in a divided dose

3. Endoscopic treatment

   In patients who remain symptomatic and show no histological improvement to dietary and pharmacotherapy, and in those who present with severe grade eosinophilic esophagitis, dilation may be performed. It is associated with risks of an invasive procedure, such as bleeding and post-procedure chest pain, both of which are common and rarely perforation.

1.15 Genetic Polymorphism in GERD

1.15.1 Introduction

- Considered a complex disease, GERD varies in prevalence and sequelae. Along with various other extrinsic factors such as environment and food habits, intrinsic factors such as physiology and genetics have been shown to play important etiological or prognostic roles.
- Some of the genetic polymorphisms which increase or decrease the susceptibility of a person to GERD are described below.

1.15.2 Various Types of Polymorphism in Gastroesophageal Reflux Disease

1. **CYP2C19 polymorphism**: Polymorphic CYP2C19 metabolizes PPIs and manifests as different phenotypes, including.

   (a) extensive metabolizer
   (b) poor metabolizer
   (c) presence of one wild type and one mutant allele

   The presence of this polymorphism affects the healing rate of GERD, which is probably greater in those with poor metabolizers and the combined wild and mutant polymorphism compared to the extensive metabolizers. Among the various PPIs, only rabeprazole escapes hepatic metabolism and, therefore, may provide uniform efficacy regardless of the type of the CYP2C19 polymorphism, unlike omeprazole or lansoprazole.

2. **Glutathione S-transferase (GST) polymorphism**: The enzyme GST is associated with essential processes such as the elimination of xenobiotics and carcinogens in the GIT. However, some polymorphic variants may be pathogenic or oncogenic, e.g. GSTM1, GSTT1, and GSTP1, with the latter increasing susceptibility to GERD due to its role in pathogenesis.
3. **IL-1β, IL-1RN, and IL-10 gene polymorphism**: Interleukins are associated with inflammatory and anti-inflammatory pathways. Polymorphisms in IL-1β and IL-1RN, namely IL1β-31 and ILRN-2 polymorphic alleles, decrease the risk of GERD. Further, the IL1RN + 2018 and IL-101082G/A genotypes are more commonly associated with Barrett's esophagus than GERD. IL-101082G/A genotype is also linked to adenocarcinoma.
4. **CCND1 polymorphism**: Cyclin D1 gene codes regulatory protein in a cell cycle. Polymorphism of CCND1 increases the risk not only for GERD but also Barrett's esophagus and esophageal adenocarcinoma.
5. **DNA repair gene polymorphism**: The presence of the XPC PAT homozygous variant genotype increases the risk of esophageal adenocarcinoma, while *XRCC1* has a protective effect in GERD and Barrett's esophagus. The protective effect of such polymorphisms may explain the progression of only a small subset of patients with GERD to Barrett esophagus or adenocarcinoma.

1.16 Biomarkers for Barrett's Esophagus

Biomarkers are indicators of normal biologic or pathologic processes or may act as prognostic markers of treatment responses. The level of biomarkers should be specific such that it should be minimal in premalignant lesions and elevated in early phases of malignancy such as dysplasia. This underlies the basis of the application of various genetic pathways as biomarkers for malignancy.

1.16.1 Diagnosis

In most cases, Barrett's esophagus precedes adenocarcinoma; however, only a fraction of patients are diagnosed with it before the detection of malignancy.

The objective of the Early Detection Research Network is to innovate biomarkers for early detection of malignancy.

The currently identified biomarkers have not demonstrated satisfactory sensitivity and specificity. Among the various biomarkers identified, the relative risk of progression to adenocarcinoma after a decade is listed below in descending order

- 17p loss of heterozygosity (relative risk 10.6)
- Tetraploidy (relative risk 8.8)
- Aneuploidy (relative risk 8.5)

– TP53 mutation (relative risk 7.3)
– 9p loss of heterozygosity (relative risk 2.3)

- The combined use of biomarkers may enhance the prediction of the relative risk, such as 17p and 9p loss of heterozygosity, along with DNA-content anomaly, demonstrating a cumulative 10-year AC risk of 38.7, which is higher than any individual biomarkers.
- Hypomethylation and hypermethylation of A-kinase anchoring protein 12 (AKAP12) and tachykinin1 (TAC1) and modifications to histone complexes may result in worsening of BE to AC. Inactivation of a tumor suppressor gene *CDKN2A* due to methylation is linked with twice the risk of developing AC in the subsequent decade.
- Although the loss of Y-chromosome, anomalous or decreased expression of β-catenin, E-cadherin, and α-catenin is also found in patients with AC along with the abnormalities listed above, the evidence to confirm this association is inadequate, and these findings need to be confirmed in larger studies specifically designed for biomarker evaluation.

1.17 Esophageal High-Resolution Manometry: Recent Updates on Chicago Classification 4.0

1.17.1 Procedure

CC4.0 working group recommendations:

- The patients should fast for ≥4 h before the procedure and patient consent is necessary.
- A solid-state high-resolution manometry catheter with <2 cm sensor spacing with combined impedance sensors should be used. Nevertheless, water-perfused catheters and the application of normative values are also acceptable, if available (Table 1.3).

1.17.2 Classification of Esophageal Motility Disorders as per Chicago Classification 4.0 (Fig. 1.6)

Table 1.3 Study procedure for esophageal high-resolution manometry as recommended in Chicago 4 update

1. **Study begins in supine position** [use supine normative values]
• ≥60 seconds adaptation period
• Document position with at least three deep inspirations.
• ≥30 seconds baseline period
• 10 supine wet (5 mL) swallows
• One multiple rapid swallow sequence (multiple rapid sequences may be repeated up to three sequences if failed attempt or abnormal response)
2. **Change position to upright** [use upright normative values]
• ≥60 seconds adaptation period
• Document position with at least three deep inspirations.
• >30 seconds baseline period
• ≥5 upright wet (5 mL) swallows
• One rapid drink challenge
3. **If no clinically relevant motility disorder is found consider the following manometric tests:**
• In a patient with a high probability of a missed diagnosis, especially EGJ outflow obstruction: Solid test swallows, solid test meal, and/or pharmacologic provocation (i.e., amyl nitrite and cholecystokinin) in the upright position to assess for obstruction
• For suspected rumination/belching disorder: post-prandial high-resolution impedance observation
4. **If equivocal results are found and/or there is suspicion of an EGJ outflow obstruction that does not fulfill the criteria for achalasia, consider the following supportive tests:**
• Timed barium esophagram, preferably with tablet
• Endoluminal functional lumen imaging planimetry (FLIP)

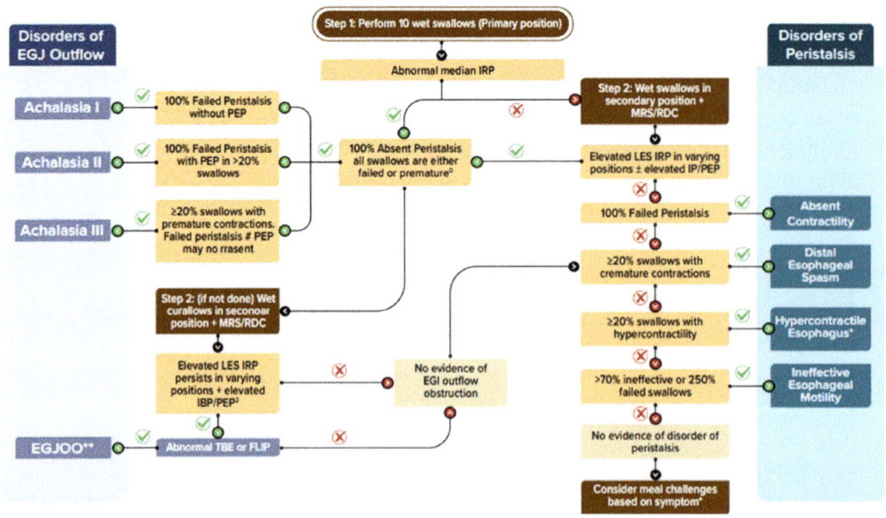

Fig. 1.6 Chicago Classification 4.0 scheme

1.18 24-Hour pH Impedance: Lyon Consensus and What Has Changed

Using multichannel intraluminal impedance (MII) along with pH testing can aid in the identification of both acid and non-acid GER.

1.18.1 Indications

- This approach is useful to estimate and typify GER, particularly in patients with suboptimal or lack of response to PPI therapy and in patients with the absence of unusual, abnormal endoscopic findings.
- It is also useful in the assessment of patients with non-specific GER symptoms such as chronic cough, dysphonia/hoarseness, and painless feeling of a lump in the throat or globus pharyngeus, leading to constant non-productive throat clearing.

No definite guidelines exist on whether dietary restrictions should be advised before the test, and, therefore, the testing centers have their practices.

1.18.2 Testing On or Off Proton Pump Inhibitors

- Monitoring of the reflux can be performed either “on” or “off” PPI therapy.
- However, the Lyon Consensus has argued that testing must be performed off PPI therapy:
  - to find baseline acid exposure time (AET) in unconfirmed GERD (absence or low-grade esophagitis on endoscopy and absence of positive pH testing)
  - to assess the condition for anti-reflux surgery
- Patients with confirmed GERD (prior LA grade C or D esophagitis, long segment Barrett’s esophagus, or prior abnormal pH meter) should be assessed on double-dose PPI therapy to distinguish between true refractoriness of GERD and inadequate response due to non-adherence to PPI therapy. pH impedance is particularly useful compared to pH monitoring since the use of PPI weakens the acidity (pHs 4–7).

1.18.3 Interpretation of pH and pH-Impedance Monitoring

- Acid exposure time (AET) is the most consistent parameter which can be duplicated and can estimate the response to pharmacotherapy as well as surgery. Lyon Consensus suggested that.

- AET <4% should be regarded as definitively normal (physiological)
- AET >6% should be regarded as definitively abnormal
- AET 4–6%: should be regarded as inconclusive

- pH-impedance monitoring is the frequency of reflux episodes (acidic, weakly acidic, or weakly alkaline). However, it is limited by the overestimation due to automated analysis, which necessitates manual inspection of the output. The Lyon GERD Consensus has proposed that,
  - >80 reflux episodes per 24 h should be regarded as conclusively abnormal
  - <40 reflux episodes per 24 h should be regarded as conclusively physiological
  - intermediate values should be regarded as inconclusive
- The AET cutoff and number of reflux episodes are variable since it depends on the software used for analysis and the presence or absence of PPI use.
- The other impedance parameters include bolus exposure, baseline impedance, and post-reflux swallow-induced peristaltic wave (PSPW) and may be useful in reflux assessment. However, supporting evidence is yet inadequate.

1.18.4 Reflux–Symptom Association

- Temporal association between the symptoms and reflux episodes is obtained using both pH monitoring and combined pH-impedance monitoring. The time window considered for reflux–symptom relationship analysis is 2 min, which may be applicable for crisp symptoms such as heartburn and regurgitation but may not be appropriate for others such as cough.
- A quantitative way to show this association is the **Symptom Index (SI) which** is the percentage of symptom events related to reflux episodes, and the optimal SI cut-off for heartburn is 50%. Nevertheless, SI is limited by the fact that the frequency of reflux episodes is not taken into consideration.
- The **Symptom Association Probability (SAP)** and the **Ghillebaert Probability Estimate (GPE)**, also known as the **Binomial Symptom Index (BSI)**, require statistical calculations to determine the probability of occurrence of symptoms of reflux episodes. Unlike SI, both SAP and GPE/BSI consider the total numbers of symptom events, reflux episodes, and reflux-related symptom events for calculating the probability. However, statistical significance may not always translate into correct causality.
- Therefore, instead of using either test in isolation, the combination of a positive SI and positive SAP offers a better estimate of the association.

1.18.5 Novel Metrics

1. **PSPW (post-reflux swallow-induced peristaltic wave**)
    - This metric may enhance the diagnostic credibility of impedance–pH testing. This is particularly useful in distinguishing between GERD and functional heartburn because acid reflux elicits PSPW while functional heartburn does not. It is highly sensitive (95–99%).
    - **The PSPW index**, however, is limited by tedious calculation, since, currently, an analytical software program is unavailable for the same.
2. **Baseline impedance**
    - Baseline impedance estimates the permeability of the esophageal mucosa with low values associated with modifications of the normal histological integrity and reflux symptoms.
    - It is optimally determined using the mean of pH–impedance tracings obtained every 10 min with an hour's interval between each during the sleep and reading obtained, therefore, known as **mean nocturnal baseline impedance (MNBI)**.
    - Low MNBI aids in differentiating between different types of esophagitis, associated cough and NERD (non-erosive reflux disease) and functional heartburn.
    - Low MNBI (<2292 Ohms) also can estimate probable response to reflux therapy. It also demonstrates a better association between heartburn and reflux compared to AET. The values improve with enhanced integrity of the esophageal mucosa.

1.19 Combined Multichannel Intraluminal Impedance and Esophageal Manometry

Standard manometry gives information about pressure measurement in the segment of the esophagus and does not reveal whether a contraction led to an actual bolus passage. Therefore, the combined use of multichannel intraluminal impedance with esophageal manometry (MII-EM) is useful since it describes **both pressures and bolus transit** within the esophagus.

1.19.1 Indications

- It can be useful in the assessment of patients with dysphagia, heartburn/regurgitation, and noncardiac chest pain.
- Assessment of esophagus before anti-reflux surgery or endoscopic anti-reflux intervention.

The utility of MII-EM in the diagnosis of extra-esophageal disorders is not yet proven.

1.19.2 Principle

The basis of impedance testing involves the determination of alterations in the resistance to the alternating current as the bolus moves with a pair of metallic rings mounted on a catheter. The ions that exist on the surface of the esophageal mucosa conduct the electrical current between the two rings.

- Boluses such as liquid-containing ones display greater conductivity due to a higher proportion of ions indicating low impedance. Whenever the bolus is in the segment, the impedance is the lowest and returns to baseline when the bolus is removed. During the passage of the bolus from the segment, the impedance elevates transiently due to the contraction owing to the drop in intraluminal space.
- Gas passing momentarily by the impedance-measuring segments will result in rapid elevation, generally >3000 Ohms, due to its poor electrical conductivity.

Determining the impedance at multiple points provides the direction of bolus transit. Bolus entries progressing from proximal to distal or vice versa indicate antegrade and retrograde bolus movement, respectively.

1.19.3 Patient Protocol

- This is an outpatient procedure with the patient needing to fast for 4–6 h.
- The combined MII-EM catheter is slipped trans-nasally via the esophagus into the stomach.
- After the catheter is positioned appropriately, 10 swallows each of liquid and viscous solution are administered at a 30-s interval.
- Since, based on the source, the impedance of water may fluctuate, resulting in residual errors, normal saline is favored due to its standardized formulation offering reproducible outputs with no confounders.
- Using a viscous solution provides better outputs due to improved sensitivity and, therefore, enhances the identification of esophageal motility.

1.19.4 Technique

- The LES is tracked using the stationary pull-through technique, and the farthest sensor is positioned in the high-pressure zone of the LES.

Table 1.4 Classification of swallow by multichannel intraluminal impedance (MII)

Classification	Bolus entry site	Bolus exit site
Complete bolus transit	Proximal site (20 cm above the LES)	Recorded in all three distal impedance-measuring sites (i.e., 15, 10, and 5 cm above the LES)
Incomplete bolus transit	Proximal site (20 cm above the LES)	Not identified at any one of the three distal impedance-measuring sites
Transit abnormality	>30% of liquid or >40% of the viscous solution has incomplete transit	

Table 1.5 Classification of transit abnormalities on esophageal impedance manometry

Abnormal pressure and transit	Abnormal pressure only
Achalasia	Nutcracker esophagus
Scleroderma	Hypertensive LES
Ineffective esophageal motility	Hypotensive LES
Distal esophageal spasm	Poorly relaxing LES

- The configuration of the catheter regulates the location of the pressure and impedance-measuring segments.
- The traditional manometry nine-channel esophageal function testing catheter has five pressure sensors, each 5 cm apart starting at LES, and the four proximal pressure sensors mounted across the four impedance-measuring segments.
- The newer high-resolution impedance manometry that uses a software program applies multiple pressures and impedance points segregated by a 1–2-cm gap to enable a closer view of pressure and bolus transit from the pharynx to the stomach.

1.19.5 Impedance Results Parameters

Swallows are classified by MII as shown in Table 1.4.

1.19.6 Clinical Application

- The output data provided by MII-EM aid in the classification of transit abnormalities, as shown in Table 1.5.

1.20 Third-Space Endoscopy and Its Clinical Utility

What is the third space in endoscopy?

First space—luminal endoscopy.
Second space—endoscopy of ducts attached to the lumen (cholangioscopy).
Third space—endoscopy within the layers of the gut walls (submucosal endoscopy).

1.20.1 Third-Space Endoscopy Procedures

Various therapeutic procedures are accomplished through third-space endoscopy, a few of which are listed in Table 1.6.

1.20.2 Principles of Third-Space Endoscopy

The fundamental principle of third-space endoscopy consists of manipulating the space between the muscularis mucosae and the muscularis propria where the submucosa lies. Technically, it aims at creating a tunnel in the submucosa along the length of the site of the procedure by elevating the sub-mucosal cushion ahead of the site, incising across the cushion. This permits access to the muscular section farther to the site of entry, followed by the conduct of the required procedure and closure of the incision via endoscopy with clips or sutures. This third-space procedure averts the risk of mucosal perforation due to the separation of the mucosal incision and the point of muscle division.

Treatment outcomes after POEM for achalasia cardia

Table 1.6 Types of third-space endoscopy procedures and its utility

Disease/disorder	Acronym	Procedure
Achalasia	POEM/E-POEM	Per-oral endoscopic myotomy
Refractory gastroparesis	G-POEM/POP/POEP	Gastric per-oral endoscopic myotomy
Zenker diverticulum	Z-POEM/STESD	Zenker per-oral endoscopic myotomy
Hirschsprung disease/megacolon	PREM	Per-rectal endoscopic myotomy
Esophageal diverticulum	D-POEM	Diverticular per-oral endoscopic myotomy
Submucosal tumor	STER/POET	Submucosal tunneling endoscopic resection/per-oral endoscopic tunneling
Esophageal stricture	POETRE	Per-oral endoscopic tunneling for restoration of the esophagus

- POEM is a commonly used third-space endoscopic procedure to manage Achalasia Cardia. One-year studies from Asian countries have demonstrated a clinical success rate of 90–94%, while evidence from longer-term studies (4–5 years) has shown clinical success rates between 83% and 87.1%. This procedure has been associated with modest post-procedure GERD in 12% of patients.

1.21 Esophageal Stents

1.21.1 Indications

1. Esophageal carcinoma (palliative care and providing nutrition while the patient is on adjuvant therapy).
2. Benign esophageal strictures (peptic, radiation-induced, anastomotic, and caustic).
3. Tracheoesophageal fistulas.
4. Iatrogenic perforations.
5. External compression from extraesophageal tumors.
6. Postoperative leaks.

1.21.2 Types

1. Fully covered self-expandable metallic stents (SEMSs) reduce the risk of tumor ingrowth and overgrowth; however, they carry a greater risk of migration versus the partially covered SEMSs.
2. Partially covered SEMSs—overall efficacy is the same as fully covered SEMS.
3. A fully covered self-expanding plastic stent (SEPS) is used for benign refractory esophageal strictures, and this indication is approved, while for SEMS, this indication is not yet approved, hence, used as off-label.
4. Biodegradable stents—made up of polydioxanone and hydrolysis within 6 months.

1.21.3 Technique

- The procedure is generally performed under anesthesia.
- The patient is in the supine position. The stricture length is delineated with endoscopy using contrast. The proximal and the distal end of the stricture are marked with either external markers or internal injection of contrast agents sub-

mucosally under fluoroscopy. An ultra-thin endoscope may be required to transverse stricture.
- Guidewire passed across stricture under endoscopic and fluoroscopic guidance. Dilatation may be done if required to pass the endoscope across the stricture.
- Appropriate type and length of stent selected. Ideal to have a stent length ~4 cm longer than the stricture length (to have 2 cm margin on either side of the stricture).
- Under fluoroscopic guidance, the stent is inserted and placed over the guidewire. The gastroscope may be passed adjacent to the stent deployment system to monitor position and deployment.
- During stent deployment, foreshortening may occur.
- It is optimal to keep a distance of 2 cm between the proximal end of the stent and the upper esophageal sphincter to avoid globus sensation when treating stricture in the cervical esophagus. A proximal release system may be helpful.

1.21.4 Adverse Events

Early

- Perforation
- Bleeding
- Tracheal compression
- GERD with esophageal stents which cross the OG junction (stents with anti-reflux valves may be useful in situations where the distal end of the stent is likely to be placed across the OG junction)

Late

- Stent obstruction due to tumor in-growth or tissue hyperplasia at the proximal or distal end
- Stent migration (more common with fully covered stents)

Chapter 2
Stomach

Gourdas Choudhuri

2.1 Embryology

- The stomach develops from the caudal part of the foregut in the fourth week.
- It appears as a fusiform dilatation of the foregut.
- During the development, the stomach undergoes two rotations:
  - After *longitudinal stomach rotation* by 90°, the left side turns anteriorly and the right side posteriorly.
  - With *anteroposterior rotation*, the distal.

The portion moves to the right and up, while the cranial portion moves to the left and down.

- The dorsal border grows more rapidly than the ventral border, forming greater curvature and lesser curvature.
- Longitudinal rotation also pulls the dorsal mesogastrium to the left, forming a lesser sac.

2.2 Anatomy and Physiology

- The stomach is the most dilated part of the digestive tube.
- The position of the stomach changes subject to the contents within the stomach being anatomically divided into cardia, fundus, body, and pylorus.

G. Choudhuri (✉)
Department of Gastroenterology and Hepatobiliary Sciences, Fortis Memorial Research Institute, Gurugram, India

G. Choudhuri et al. (eds.), *Gastroenterology and Hepatology*,
https://doi.org/10.1007/978-981-99-9287-4_2

Gastric wall: It consists of four layers:

- *Mucosa* consists of an epithelial layer and lamina propria
- Submucosa
- Muscularis propria has three layers again:
  - Inner oblique layer
  - Middle circular layer
  - Outer longitudinal layer
- Serosa

2.2.1 Gastric Mucosal Anatomy

The gastric mucosa is lined by columnar epithelial cells. These cells secrete gastric acid, pepsin. They also secrete mucin and bicarbonate, which protects the epithelial cells from injury by gastric acid. The mucosal surface of the stomach is invaginated by foveolae. The foveolae lead to gastric glands.

There are three types of gastric glands:

1. Oxyntic or parietal glands
2. Pyloric glands
3. Cardiac glands

Oxyntic or Parietal Glands

Oxyntic glands are the most numerous. The glands contain parietal cells, mucous cells (mucous neck cells and mucous surface cells), chief cells, and endocrine cells.

Parietal Cells (Oxyntic Cells)

- These cells are part of epithelial fundic glands in gastric mucosa. The cells are large and round and have an eosinophilic cytoplasm with a centrally located nucleus.
- Parietal cells have large mitochondria, microvilli lacking in glycocalyx, and a cytoplasmic canaliculi system in contact with the lumen. These features distinguish them from other cells.
- Parietal cells produce gastric acid and intrinsic factors.

Chief Cells

- Chief cells or zymogen cells are located deep within oxyntic glands. The cells are basophilic and have a large amount of rough endoplasmic reticulum in their cytoplasm.

- Chief cells release pepsinogen I and II, gastric lipase, and renin. Pepsinogen is converted to pepsin within the lumen. Pepsinogen release is regulated by acetylcholine, cholecystokinin (CCK), and gastrin.

Endocrine Cells

D cells, enterochromaffin (EC), and enterochromaffin-like (ECL) cells are the different types of endocrine cells in the stomach.

D Cells

- Secrete somatostatin.
- Are present in the oxyntic glands.
- They are flask-shaped cells with long cytoplasmic extensions which end in processes similar to nerve terminals.

Somatostatin

Also known as growth hormone inhibiting hormone. This peptide hormone is produced by D cells in the stomach, small intestine, and pancreas.

- Secreted as a preprohormone, it has two active forms: somatostatin 14 and somatostatin 28.
- The receptor for somatostatin is a G-protein-coupled receptor. There are five known subtypes of somatostatin receptors [subtypes 1 and 5 found in the gastrointestinal (GI) tract].

  Meal ingestion and gastric acid secretion increase somatostatin release.
- Catecholamines inhibit somatostatin release, while cholinergic mediators increase release.
- Somatostatin inhibits GI motility, gallbladder contraction, and secretion of most GI hormones.
- *Helicobacter pylori* infection can decrease somatostatin production.
- Somatostatin released from the cells acts either locally to exert a paracrine effect or systematically to exert an endocrine effect.

Enterochromaffin Cells

- Enterochromaffin cells (EC) are enteroendocrine cells with agranular cytoplasm. Embryologically, they are derived from the neural crest.
- They produce and contain about 90% of the body's store of serotonin [5-hydroxytryptamine (5-HT)].
- These cells control intestinal motility, secretion, and absorption.

Serotonin

- Serotonin, or 5-HT, is a monoamine neurotransmitter.
- It is derived from tryptophan.

Ninety-five percent of serotonin in the body is located in EC cells of the gut. Platelets take up serotonin from EC cells and store it.

- Serotonin exerts its action by binding to 5-HT receptors distributed throughout the human body. In the GI tract, serotonin plays an important role in initiating peristalsis. Serotonin derived from EC cells also contributes to intestinal inflammation.
- By inhibiting osteoblast proliferation, EC cell-derived serotonin lowers bone mass.

Enterochromaffin-Like Cells or ECL Cells

- ECL cells are a form of enteroendocrine cells.
- ECL cells are found in oxyntic glands close to the parietal cell.
- The cells have granules that stain with silver in the presence of reducing substances.
- These cells contain histamine, which is released in a paracrine manner.

Pyloric Glands

These coiled glands predominate in the antrum of the stomach. The glands are lined by mucous-secreting epithelial cells, pepsinogen II-secreting oxyntic cells, and gastrin-secreting G-cells.

G-Cells

- Are located in deep portions of pyloric glands.
- Contain a large number of secretory granules containing gastrin.
- The apical surface of the gland contains microvilli with receptors regulating the release of gastrin.

Gastrin

- Gastrin is a peptide hormone produced by G-cells in the antrum.
- It is secreted as a large molecule which is further cleaved to yield two forms (G34 and G17). G34 is the circulating form of fasting, and G17 is in a fed state.
- Gastrin is released from G-cells in response to amino acids, especially aromatic amino acids, namely, phenylalanine and tyrosine. Its release is regulated by gastrin-releasing peptide (GRP) and somatostatin.

- Gastrin-releasing peptide stimulates, whereas somatostatin inhibits gastrin release.
- Gastric pH influences the release of gastrin. Low pH inhibits gastrin release.
- Gastrin acts via CCK2 receptors on ECL cells to release histamine, which in turn acts on the H_2 receptor on the parietal cell to release acid.
- Gastrin also acts directly on parietal cells to release acid. It has trophic action on ECL cells to induce hyperplasia.
- Chronic hypergastrinemia in rats increases the risk of carcinoid tumors in rats.

Physiology of Gastric Acid Secretion

Gastric acid secretion occurs in three phases:

(a) Cephalic phase: This phase is activated by sight, smell, taste, and even thinking about food.
(b) Gastric phase: The entry of food into the stomach and subsequent dilatation of the stomach initiates this phase.
(c) Intestinal phase: The entry of chyme into the intestine stimulates this phase. This phase can account for only a small proportion of gastric acid secretion.

Regulators of gastric acid secretion are as follows:

(a) Endocrine: Gastrin exerts its effect via the CCK2 receptor of parietal cells and ECL cells. It is a major stimulator of gastric acid secretions.
(b) Paracrine: The local somatostatin and histamine release also regulates gastric acid secretion. Histamine released from ECL cells acts on H_2 receptors in parietal cells to stimulate gastric acid secretion. Somatostatin released from D cells is a major inhibitor of gastric acid secretion.
(c) Neural: Acetylcholine stimulates gastric acid secretion by increasing gastrin and histamine release and inhibiting somatostatin release.
(d) Autocrine: Prostaglandins (PGs) inhibit gastric acid secretion. Prostaglandins exert their action by inhibiting histamine release. Transforming growth factor alpha also inhibits gastric acid secretion.

Mechanism of Acid Production

- Activation of parietal cells leads to intracellular cyclic AMP (cAMP) generation.

  Cyclic AMP activates cAMP-dependent protein kinases with resultant activation of secretory vesicles. These vesicles coalesce when activated to form canaliculi which drain into the apical lumen.

  The membrane of the canaliculi has an ATP-dependent proton pump, the hydrogen potassium ATPase pump (H^+–K^+-ATPase). The dissociation of water molecules within the parietal cell leads to the formation of hydrogen (H^+) and hydroxyl (OH^-) ions.

- Thus, the OH-ions combine with carbon dioxide (CO_2) in the presence of carbonic anhydrase to produce bicarbonate ion (HCO_3^-).
- Conductance channels transport chloride (Cl^-) K^+ into the canalicular lumen.
- H^+ released from water is exchanged for K^+ into the canalicular lumen by the H^+–K^+-ATPase.

  An osmotic gradient is created across the canalicular membrane by the accumulated H^+, which leads to an outward diffusion of water.
- The resulting gastric juice contains HCl acid.

Acid rebound: In some patients, a rebound increase in gastric acid secretion occurs after cessation of H_2 receptor antagonist (H_2RA) and proton pump inhibitors (PPIs). This is seen more often in individuals not infected with *H. pylori*. This is probably related to the increase in parietal and ECL cell mass secondary to hypergastrinemia.

Tolerance: Tolerance to H_2RA develops in some individuals within a few days of starting H_2RA. Tolerance is not seen with PPI use. The exact mechanism is not known. Tolerance probably contributes to the poor clinical efficacy of H_2RA noted in some individuals.

2.3 Gastric Mucosal Barrier

The gastric mucosa is exposed almost continually to acid and pepsin; despite this, the gastric mucosa can withstand damage. The gastric epithelial lining is only one cell thick but is protected from being digested by its juices (Fig. 2.1).

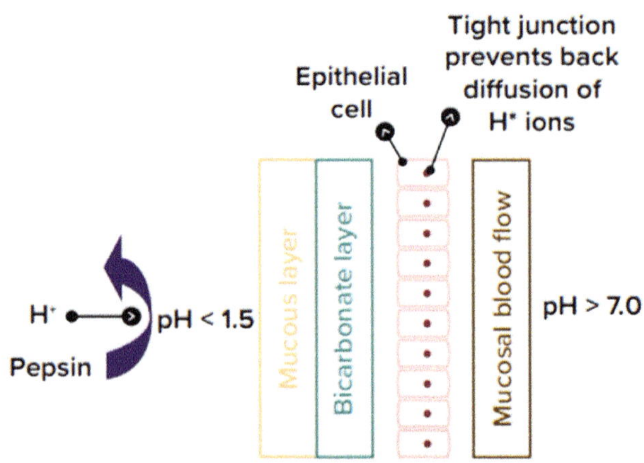

Fig. 2.1 Components of gastric mucosal barrier

2.3.1 Components of Gastric Barrier

(a) **Mucous layer**: This thick viscous layer lines the gastric epithelium. The mucus layer comprises two distinct forms of mucin from two different sources. MUC5AC is secreted by surface epithelial cells, while MUC6 is secreted from mucous neck cells. The mucus layer constitutes the first line of defense. A pH gradient exists within the mucus layer, with the pH decreasing from the mucosal surface to the lumen. The structure of the mucus layer is such that it prevents the diffusion of pepsin and slows the progression of H^1 ions (Fig. 2.1).
(b) **Bicarbonate**: Bicarbonate ions are produced in parietal cells during acid production. The amount of bicarbonate produced is equivalent to the acid produced by the parietal cells. These bicarbonate ions move into the interstitium through the basolateral surface. Fenestrated capillaries carry the bicarbonate ions to the epithelial surface from the interstitial space. These bicarbonate ions then diffuse into the mucus layer to increase the pH. Of the bicarbonate produced by parietal cells, only a small portion reaches the epithelial surface, and the majority passes out in urine as an alkaline tide.
(c) **Epithelium**: The apical membranes of gastric epithelial cells are highly resistant to damage by acid. Moreover, epithelial cells in the stomach have a high turnover rate. The cells are renewed every 2–3 days. Thus, damaged cells are continually replaced.
(d) **Mucosal blood flow**: The blood flow is an important component of the mucosal barrier. It is regulated by vasoconstrictors namely endothelin 1 (ET1) and vasodilators namely nitric oxide (NO). Endothelin 1 and NO between themselves regulate the vascular tone. The presence of gastric acid in the lumen is detected by sensory neurons in the epithelium; the neurons release calcitonin gene-related peptide (CGRP), a vasodilator peptide. Calcitonin gene-related peptide via NO mediates relaxation of arterioles, increasing blood flow. The acid is thus buffered by the bicarbonate carried by the mucosal vasculature. When there is a damage to the epithelial lining, mucus from damaged epithelial cells and plasma from mucosal vasculature coalesce to produce a protective barrier (mucoid cap) for the basement membrane. The mucosal blood flow helps keep pH in the vicinity of the cap high.
(e) **Restitution**: Injury to the epithelium results in the migration of healthy epithelial cells from gastric pits near ulcer margins to cover the denuded surface restoring epithelial continuity. This is mediated by PGs.

2.3.2 Prostaglandins and Their Role in Gastric Mucosal Defense

Prostaglandins play a major role in regulating the gastric mucosal defense systems mentioned above. Gastric mucosa produces PGE2 and PGI2. Prostaglandins have the following function in gastric mucosa:

1. Decrease gastric acid production
2. Increase mucous and bicarbonate production
3. Increase resistance to epithelial cells to acid-mediated damage
4. Anti-inflammatory effect:
   (a) Decreases production of proinflammatory cytokine via TNFα, IL-1, IL-8, and leukotriene
   (b) Inhibits leukocyte adhesion
5. Healing of ulcer by restitution and increasing mucosal blood flow

Prostaglandin production is regulated by two forms of cyclooxygenase (COX1 and COX2). COX1 is the predominant form of cyclooxygenase expressed in gastric mucosa. COX2 is expressed in inflamed tissue. Experimental studies have shown that in the presence of mucosal injury, COX2 production is upregulated in gastric mucosa.

2.4 Gastric Motility

Gastric motility has three functions:

- To allow entry of food into the stomach—receptive relaxation
- Decreasing size of ingested food—trituration
- Regulate flow of ingested food out of stomach—antral waves coordinated with antro-pyloroduodenal coordination

Gastric motility is regulated by the central nervous, parasympathetic, sympathetic, and enteric systems.

The pacemaker region of the stomach is located on the greater curvature between the fundus and the proximal body. Gastric smooth muscle contractions are regulated by pacemaker cells [interstitial cells of Cajal (ICC)]. The ICCs generate smooth muscle contractions called slow waves, propagating abnormally toward the pylorus. Slow waves occur at three cycles/min.

2.4.1 ICC

- It was first described by Cajal in 1911.
- ICCs develop from Kit-positive mesenchymal precursors and express receptor tyrosine kinase-kit (CD117).
- They are classified based on the location within muscle layers as
  - ICC-MY: Between circular and smooth muscle layers
  - ICC-IM and ICC-DMP: Within circular and longitudinal muscles
  - ICC-SM: Submucosal surface of circular layer of the colon
- In the stomach, they are more dense in the body and antrum. In antrum, ICC-MY and ICC-IM predominate, while ICC-IM is only found in the fundus.

- ICC-MY initiates the slow waves, which are then integrated and spread by the ICC-IM.
- ICC-IMs in the fundus act as stretch receptors—in conjunction with vagal efferents, participate in accommodation reflex.
- ICCs are abnormal in patients suffering from gastroparesis.

2.4.2 *Gastric Motility in Response to Solid Meal*

Receptive Relaxation

- Once solid food enters the stomach, the fundus relaxes to accommodate the food.
- Receptive relaxation permits a large amount of food to be accommodated in the proximal stomach.
- The vagus nerve mediates receptive relaxation.
- Inhibitory neural transmitters, namely, nitric oxide and vasoactive intestinal polypeptide, mediate the accommodation reflex.

Trituration

- During this phase, the food is triturated to a particle size of ~1 mm by antral contractions.
- Peristaltic contractions mix food with acid and pepsin.
- Peristaltic waves occur at three cycles/min.

The above two phases are called the lag phase. The duration of the lag phase depends on the following: the content of ingested food, that is, fat, protein, or carbohydrate and the particle size of food after chewing.

Linear Phase of Gastric Emptying

- Chyme moves from the stomach to the duodenum with each peristaltic contraction.
- The antral contractions mediate this movement.
- The amount of calories which the duodenum receives is ~3–4 kcal/min.

2.4.3 *Gastric Motility in Response to Liquid Meal*

Liquid emptying from the stomach is mediated by:

- A pressure gradient between stomach and duodenum
- Antral peristalsis

- The rate of gastric emptying of liquids is dependent on
  - the volume of liquid
  - osmolarity
  - calorie content (calorie-dense liquid is retained for a longer duration in antrum)
  - viscosity

2.5 Gastrin

- Gastrin is a peptide hormone produced by G-cells in the antrum. It is secreted as a large molecule (preprogastrin), which is further cleaved to yield progastrin from which amidated forms are derived (G34 and G17).
- Progastrin and glycine (Gly) extended gastrins have biological actions. Progastrin has been found to have antiapoptotic properties and has been shown to increase the proliferation of colonic epithelial cells in rats. Similarly, Gly gastrin has been shown to have proliferative actions on pancreatic cell lines and colonic mucosa.
- G34 is the circulating form of fasting, and G17 is in a fed state. Gastrin is released from G-cells in response to amino acids, especially aromatic amino acids, namely phenylalanine and tyrosin. Its release is regulated by GRP and somatostatin. GRP stimulates, whereas somatostatin inhibits gastrin release. Gastric pH influences the release of gastrin. Low pH inhibits gastrin release. Gastrin acts primarily via CCK2 receptors on ECL cells to release histamine, which acts on the H_2 receptor on parietal cells to release acid. Gastrin also acts directly on parietal cells to release acid.
- Gastrin has trophic action on ECL cells to induce hyperplasia. Chronic hypergastrinemia may lead to the formation of ECL cell tumors (carcinoids). Such tumors are usually benign. Management of such tumors is usually conservative; CCK2 receptor antagonists have been used to manage such tumors. Netazepide, a highly selective CCK2 receptor antagonist, has been used to manage such tumors.
- In addition to regulating acid secretion, gastrin exerts an effect on extracellular matrix remodeling by regulating the expression of metalloproteinases.

2.6 Ghrelin

Ghrelin is a 28-amino acid orexigenic peptide. It is a member of the motilin family of peptides and is a natural ligand for the growth hormone secretagogue receptor (GHS-R). The stomach and duodenum have 90% of the body's ghrelin. The P/D1 cells located within the oxyntic glands of the fundus produce ghrelin. The gene for ghrelin is located on chromosome 3p. The hormone is secreted as a prepropeptide which undergoes post-translational modification. Ghrelin exerts its action by binding to the GHS-R receptor. This G protein-coupled receptor exists in two forms:

GHS-R1a and GHS-R1b. GHS-R is expressed both in central and peripheral tissues. Ghrelin production increases during fasting. This increase is the highest in the immediate preprandial period. Secretion decreases following a meal. This post-prandial suppression of ghrelin production is the highest following a carbohydrate meal. Ghrelin is an orexigenic peptide which stimulates appetite. More than the amount of food, the frequency of food increases with ghrelin. It also increases gastric motility. Ghrelin levels are elevated in weight loss, cachexia, and anorexia.

2.6.1 Clinical Relevance

Prader–Willi Syndrome

It is syndromic obesity characterized by hyperphagia, obesity, growth hormone deficiency, hypogonadism, and behavioral problems. This is caused by the failure to express paternally active genes on chromosome 15q. In this, despite obesity, patients have high circulating levels of ghrelin. This may be responsible for these patients' diminished secretion of growth hormones.

Bariatric Surgery

Following gastric bypass and sleeve gastrectomy, patients have low serum ghrelin levels. The premeal rise of ghrelin is also decreased. This may be one of the factors responsible for the efficacy of bariatric surgery.

Capromorelin—a GHS-R agonist undergoing clinical trials—has been shown to increase body weight without increasing fat. GHS-R receptor antagonist is undergoing clinical trials as a treatment for obesity.

2.7 Gastric Microbiota

The environment within the stomach is not conducive to the growth of microorganisms. Studies have shown that the concentration of luminal bacteria increases several times from the stomach to the colon. Recent studies have identified a number of bacteria which colonize the stomach. Of them, *H. pylori* is probably one of the better-known microbes. Studies using bacterial 16SrRNA have shown that there exists within the stomach a wide variety of microbiota. The composition of microbes varies widely among individuals, and this depends on pH, diet, use of PPIs, antibiotics, and stomach diseases.

An estimated 10^{2-4} cfu/mL of bacteria is found in the stomach. The bacteria colonizing the stomach usually colonize the upper respiratory tract and pharynx. Within the stomach, there exist 7–13 phyla. The major phyla include *Firmicutes*,

Actinobacteria, *Fusobacteria*, and *Proteobacteria*. *Firmicutes*, *Actinobacteria*, and *Proteobacteria* predominate during the first week of life. However, after the first week, *Proteobacteria* tend to outgrow other bacteria becoming the predominant bacterial species by the fourth week.

There is conflicting data on the influence of *H. pylori* infection on the composition of gastric microbiota. Some studies have shown that *H. pylori* affects the composition of gastric microbiota. *Proteobacteria* is the predominant microbiome in *H. pylori*-positive patients, while *Actinobacteria* and *Firmicutes* predominate in *H. pylori*-negative individuals.

The growth of certain bacterial forms may influence the development of gastric adenocarcinoma in patients with achlorhydria. With achlorhydria, the pH within the stomach rises, resulting in the growth of organisms within the stomach. Certain bacterial forms can produce N-nitroso compounds which are potent carcinogens.

2.7.1 *Non-*H. pylori *Helicobacter Species*

Since the description of *H. pylori*, a number of non-*H. pylori* Helicobacter spp. (NHPH) have been identified. They have been classified into two groups: gastric NHPH and enterohepatic NHPH.

(a) Gastric NHPH includes *H. heilmannii*, *H. suis*, and *H. felis*.
(b) Enterohepatic NHPH includes *H. cinaedi*, *H. bilis*, and *H. canis*.

Although initially reported in animals, a number of studies in humans have demonstrated the existence of NHPH in humans.

2.7.2 *Clinical Relevance*

Gastric Non-*H. Pylori* Helicobacter Species

- In a large study from Poland involving children between 4 and 18 years, *H. heilmannii* infection was found in 0.2% of gastric mucosal biopsies. Of the children, 54.5% reported contact with dogs and/or cats. Most children with *H. heilmannii* developed nodular gastritis.
- *Helicobacter suis* has also been reported to produce chronic gastritis in humans.

Enterohepatic Non-*H. Pylori* Helicobacter Species

- Bacteremia secondary to enterohepatic NHPH has been reported in patients with either primary or secondary immunodeficiency states.

- A few studies have suggested that the role of Enterohepatic NHPH has been shown to colonize biliary trees in patients with biliary obstruction.
- Enterohepatic NHPH has also been associated with cardiac disease. *Helicobacter cinaedi* is reported to be a risk for atrial arrhythmia.

Comparison between *H. heilmannii* genome with *H. pylori* showed some similarities. However, they do not have VacA and Cag pathogenicity islands. Gastric NHPH can be successfully eradicated using triple-drug regimens, including combinations of PPIs, amoxicillin, clarithromycin, and metronidazole.

2.8 Acid Pocket

Despite high intragastric pH, most symptoms of reflux disease occur in the post-prandial period. In 2001, Fletcher et al. identified an area below the esophagogastric junction, which had very low pH in the post-prandial period. This was called an "acid pocket."

2.8.1 Mechanism

After ingestion of food, the proximal stomach relaxes to accommodate the food. This accommodation reflex prevents adequate mixing of ingested food with freshly secreted gastric juice. The chyme is thus layered with acidic gastric juice. This accounts for low pH in the proximal stomach—the acid pocket. However, peristaltic contractions mix chyme adequately with gastric juice in the body and antrum, so the pH is high (Fig. 2.2).

Several authors have found a good correlation between the pH of the refluxate and the pH in the acid pocket. The median time for the development of acid pocket

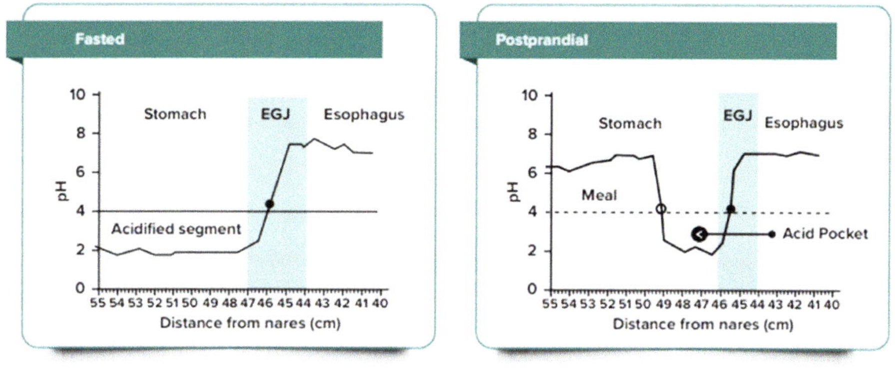

Fig. 2.2 pH study of acid pocket

is 14–17 min after a meal. The acid pocket usually lasts for 90 min after meals. Acid pocket is seen in healthy individuals and those suffering from gastroesophageal reflux disease (GERD). However, the extent of acid pockets varies among individuals. In patients with GERD, the proximal border of the acid pocket has been noted to extend beyond the esophagogastric junction (EGJ).

2.8.2 Gastroesophageal Reflux Disease and Acid Pocket

In post-prandial states, transient lower esophageal sphincter relaxation (tLESR) happens in healthy individuals and those suffering from reflux disease. Refluxate in the presence of tLESRs is more acidic in individuals with GERD than in healthy individuals. The acid pocket probably contributes to the low pH. Beaumont et al., in a study, demonstrated that hiatus hernia and proximal migration of acid pocket beyond the EGJ were the major risk factors for acid reflux during LESRs.

2.8.3 Management

Prokinetic agents, namely erythromycin and azithromycin, have been shown to decrease the acid pocket. Baclofen, a GABA agonist, decreases the extent of the acid pocket. Rohof et al. demonstrated that PPIs increase the pH and decrease the size of the acid pocket.

2.9 *H. pylori* and Pathogenesis of Peptic Ulcer

This was first described by Warren and Marshall in 1982. It is a persistent colonizer of the stomach. It remains in an extracellular location and is known to cause chronic gastritis, peptic ulcer, gastric adenocarcinoma, and lymphoma. It is a persistent colonizer of the stomach. It remains in an extracellular location and is known to cause chronic gastritis, peptic ulcer, gastric adenocarcinoma, and lymphoma. The prevalence of *H. pylori* infection in developing countries is high. The prevalence of *H. pylori* increases with age.

2.9.1 H. pylori

Transmission

- Fecal–oral transmission is the major route for acquiring *H. pylori* infection.
- Iatrogenic transmission through infected gastroscopes has also been reported.

Bacteriology

- *H. pylori* is a Gram-negative, micro-aerophilic, spiral-shaped flagellated bacterium.
- It is a fastidious, slow-growing organism which can be cultured in Skirrow's media or blood agar.

Adaptation of *H. pylori* to Stomach

H. pylori has evolved to survive the hostile environment in the stomach. It primarily uses three mechanisms:

(a) Urease

Urea is hydrolyzed by urease to produce ammonia, which helps to neutralize gastric acid around the bacteria. Urease accounts for approximately 5% of the bacterial protein. The activity of urease is triggered by exposure to gastric acid. Urel is an inner membrane protein which functions as an H^1-gated urea channel. When the pH in the environment falls, urel allows the entry of urea. The gene coding for urel exists in the urease coding region of *H. pylori.*

(b) Flagella

The flagella helps it to move through the mucin and reaches the gastric epithelium. *H. pylori* also produces mucolytic enzymes.

(c) Adhesion molecules

Pathogenic *H. pylori* harbor a number of adhesion molecules, namely BabA, SabA, AlpA/B, OipA, and HopZ. Adhesion molecules help *H. pylori* to attach to the epithelial surface. BabA mediates attachment to fucosylated Lewis B blood group antigens expressed on the epithelial cell; SabA mediates binding to glycoconjugates containing sialic acid. *H. pylori* also binds to MHC Class II antigens on host cells.

2.9.2 Pathogenesis of **H. pylori**-*Induced Mucosal Injury*

Factors mediating *H. pylori*-induced cellular damages are as follows:

(a) Enzymes

- Urease: Its action produces ammonium chloride and monochloramine, which damage cells. Urease can damage epithelial cells by acting on host immune cells.
- Phospholipases: Phospholipase A2 converts lecithin to lysolecithin. Alters phospholipid content of gastric mucus, thus increasing permeability.
- Catalase: Protects bacteria from the harmful effects of reactive oxygen species.

(b) Vacuolating cytotoxin

It is an 87-kDa protein. Gene coding VacA is present in all strains of *H. pylori.* VacA is, however, expressed by strains which also encode the cytotoxin-associated gene pathogenicity island (CagPAI). It induces vacuolation in eukaryotic cells and increases the permeability of gastric epithelium to urea by acting as a passive urea transporter.

(c) Cytotoxin-associated antigen (CagA)

Encoded by the cytotoxin-associated gene A. It is a 128–140 kDa protein. The cytotoxin-associated antigen is antigenic but not cytotoxic. With the help of a syringe-like pillus, the type IV secretory apparatus, *H. pylori* can inject CagA into host cells. Inflammation is more pronounced in individuals infected with strains producing both VacA and CagA. Duodenal ulcer is seen in almost 90% of individuals infected with CagA-positive strains. The risk of malignancy is also high.

(d) Outer-membrane inflammatory protein (OipA)

- OipA is a 34-kDa protein.
- Expressed in conjunction with CagA produces intense inflammation.
- OipA status of *H. pylori* is significantly associated with duodenal ulceration and gastric cancer.

(e) Heat shock proteins (HSP)

H. pylori expresses two heat shock proteins (Hsp A and B).

2.9.3 Immune-Mediated Injury

- *H. pylori* induces a significant inflammatory response.
- A *number* of antigenic substances are produced by *H. pylori* which can stimulate the immune system. Response to antigen results in the production of a large number of pro-inflammatory cytokines, notable among them being IL8. *H. pylori* activates NFK beta, which in turn increases IL8 production. IL8 is a potent pro-inflammatory cytokine which activates acute inflammatory response in the mucosa.
- *H. pylori* also increases the production of other pro-inflammatory cytokines, namely, IL1, IL6, and TNF.

2.9.4 Host Genetics and H. Pylori *Pathogenicity*

- *H. pylori* infection in patients with the HLA DQ5 genotype has been associated with atrophic gastritis and intestinal metaplasia.

- IL1B polymorphism (-511T/-31C) increases the risk for atrophic gastritis, intestinal metaplasia, and severe inflammation.
- IL1RN2 carriers are at increased risk of gastric cancer.

2.9.5 *Pathogenesis of Duodenal Ulcer*

H. pylori is a major etiologic factor for duodenal ulcers. But, only 10–15% of individuals infected with *H. pylori* develop duodenal ulcers. Individuals with the TNFA-308G/G genotype have a higher risk of developing duodenal ulcers.

(a) Increased gastric acid secretion:

   - Chronic *H. pylori* infection increases basal and maximal acid output.
   - This is probably related to an increase in gastrin release.
   - *H. pylori* infection also decreases somatostatin release from D cells.

(b) Gastric metaplasia:

   - Gastric epithelium develops in the duodenum.
   - Probably secondary to increased gastric acid.
   - This provides epithelium for *H. pylori* colonization.

(c) Immune response: Increased production of pro-inflammatory cytokines as described above.
(d) Down-regulation of mucosa defense: Proteases produced by *H. pylori* degrades protective mucous.

2.10 NSAID and Gastric Mucosal Injury

1. Nonsteroidal anti-inflammatory drugs (NSAIDs) induced stomach injury ranging from damage to the mucosal barrier to gastric ulcer.
2. The mucosal injury is mediated by

   (a) Topical effects:

      - Nonsteroidal anti-inflammatory drugs being weak acids are not ionized in gastric pH and are absorbed intact across the mucosa.
      - In the presence of neutral pH of the mucosa, the drug ionizes and is trapped within epithelial cells, thus damaging the cells.

   (b) Systemic effect:

      - Prostaglandins protect the gastroduodenal mucosa by stimulating mucin, bicarbonate, and phospholipid secretion by epithelial cells. Prostaglandins also improve mucosal blood flow, restitution, and epithelial cell proliferation.

- Cyclooxygenase (COX) is the rate-limiting enzyme for PG synthesis. There are two forms of COX: COX1 and COX2. COX1 is used to produce PGs the gastroduodenal mucosa. Nonsteroidal anti-inflammatory drugs inhibit COX activity leading to the injury of the gastroduodenal mucosa.

2.11 Pathophysiology of Functional Dyspepsia

The Rome IV criteria define dyspepsia as any combination of four symptoms: postprandial fullness, early satiety, epigastric pain, and epigastric burning that are severe enough to interfere with the usual activities and occur at least 3 days/week over the last 3 months with an onset of at least 6 months in advance.

The umbrella term "FD" comprises patients from the following categories:

1. Post-prandial distress syndrome (PDS) is characterized by meal-induced dyspeptic symptoms suggestive of a motility disturbance.
2. Epigastric pain syndrome (EPS), which refers to epigastric pain or epigastric burning that does not necessarily occur after meal ingestion, may occur during fasting and can be even improved by meal ingestion (rule out PUD and GERD).
3. Overlapping PDS and EPS, characterized by meal-induced dyspeptic symptoms and epigastric pain or burning.

2.11.1 Pathophysiology of FD

1. Gastroduodenal motor dysfunction
2. Sensory dysfunction
3. Impaired mucosal integrity

1. **Motor abnormalities**
   (a) Delayed gastric emptying—seen in about 30–50% of patients with FD.
   (b) Impaired accommodation—fundic accommodation is a vagally mediated process. Impaired accommodation is seen in ~40% of patients, leading to impaired distribution of gastric contents in the inhibited gastric fundus, leading to abnormally distended antrum (correlate partially with symptoms' quality and severity). Also, raised intragastric pressure due to poor fundic accommodation causes the activation of mechanoreceptors in the gastric wall, thus inducing symptoms.
2. **Sensory dysfunction**
   Visceral hypersensitivity—defined as abnormally enhanced perception of visceral stimuli.

(a) Hypersensitivity to gastric distention—patients with FD are hypersensitive to isobaric gastric distention due to the involvement of tension-sensitive mechanoreceptors and alteration at the visceral afferent nerve level.
(b) Altered duodenal sensitivity to lipids or acid—duodenal exposure to HCl induces nausea in patients with FD but not in healthy subjects. The excess exposure is attributed to impaired clearance of acid from the duodenum. Similarly, excess exposure to lipids enhances the perception of gastric distention due to the release of cholecystokinin.

3. **Increased permeability of the mucosal barrier** exposes the submucosal immune system to luminal noxious agents. *H. pylori* infection, stress, duodenal acid exposure, smoking, and food allergy have all been implicated in the pathogenesis of duodenal mucosal inflammation-related permeability changes.

Psychosocial disorders commonly associated include anxiety, depression, neuroticism, physical and emotional abuse, and difficulty coping with life events.

Genetic predisposition—The frequency of dyspepsia in first-degree relatives of affected patients is increased compared with the frequency in their spouses. Polymorphisms of the G-protein beta polypeptide 3 (GNB3) gene have been associated with the risk of functional dyspepsia.

2.12 Gastric Cancer Stem Cell

Gastric cancer is one of the leading causes of death worldwide. It is the second most common cancer in males and the third most common in females. There exists a wide geographic variation of this cancer. In India, the highest incidence of gastric cancer has been reported in northeastern Mizoram. Overall the incidence of the disease is higher in Northeast and South India. Although early detection and treatment have improved the immediate outcomes of this cancer, the 5-year survival of this cancer is less than 40%. The reason for poor survival rates in gastric cancer probably lies in the gastric cancer stem cell (CSC).

Stem cells are pluripotent cells capable of renewal and differentiation into various cell types. CSCs are similar to other stem cells in that they can regenerate and have the potential for multidirectional differentiation, drug resistance, and metastasis. Gastric CSCs are associated with the development of gastric cancer. The source of gastric CSC is not known.

One hypothesis is that gastric CSC probably develops from gastric stem cells. Gastric epithelial cells are renewed every 2–4 days. The older cells undergo apoptosis and are replaced by new cells derived from stem cells in the epithelium. In intestinal metaplasia, a gastric stem cell is found to differentiate into gastric and intestinal cells. It is thought that the malignant transformation of gastric CSC may be responsible for developing gastric cancer.

The other hypothesis is that gastric CSC probably originates from bone marrow-derived mesenchymal stem cells (BM-MSC). Bone marrow-derived mesenchymal stem cells are mesenchymal stem cells that can migrate, regenerate, and differentiate. In experimental models, it has been found that BM-MSC participates in the repair of injured gastric mucosa. During repair, the abnormal transformation of BM-MSC can give rise to gastric cancer.

CD 44^{+} is an adhesion molecule expressed on gastric CSC and is a marker for gastric CSC. Gastric CSCs also express CD133. Expression of CD133 correlates to malignancy.

Gastric CSCs play a role in cancer onset, migration, and drug resistance. Gastric CSCs express matrix metalloproteinase 1, matrix metalloproteinase 2, and cyclooxygenase 2, all of which play a role in tumor migration. Gastric CSC expresses aldehyde dehydrogenase, which functions to protect cancer cells. Aldehyde dehydrogenase confers resistance of cancer cells to 5-flurouracil and cisplatin.

Gastric CSC is an ideal candidate for the treatment of gastric cancer. Notch1 is expressed in CD 44^{+}-positive cells. β-Elemene, an inhibitor of Notch 1 expression, interferes with CD 44^{+} cell proliferation. Huang et al., in a study, showed that genistein reduces the chemoresistance of gastric CSC.

2.13 Pathogenesis and Molecular Classification of Gastric Cancer

Gastric cancer is a heterogeneous disease and the endpoint of a multistep process (Figs. 2.3 and 2.4). Gastric cancer stem cells have emerged as fundamental players in cancer development and contributors to gastric cancer heterogeneity.

2.13.1 Histologic Subtypes

Gastric cancer can be subdivided using the *Lauren classification* into two distinct histologic subtypes with different epidemiologic and prognostic features.

The *formation of gland-like tubular structures with features reminiscent of intestinal glands characterizes the intestinal type of cancer*. This type of gastric cancer is more closely linked to environmental and dietary risk factors, tends to be the predominant form in regions with a high incidence of gastric cancer, and is now declining worldwide.

The ***diffuse type*** of cancer lacks glandular structure and consists of poorly cohesive cells that infiltrate the wall of the stomach. It is found at the same frequency

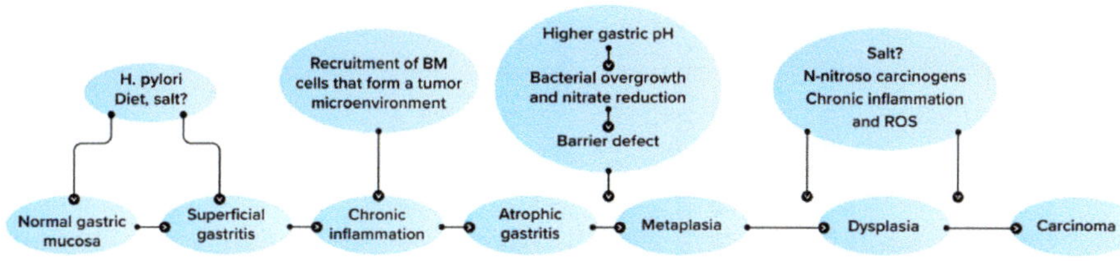

Fig. 2.3 Proposed Correa pathway of pathologic events in gastric adenocarcinoma

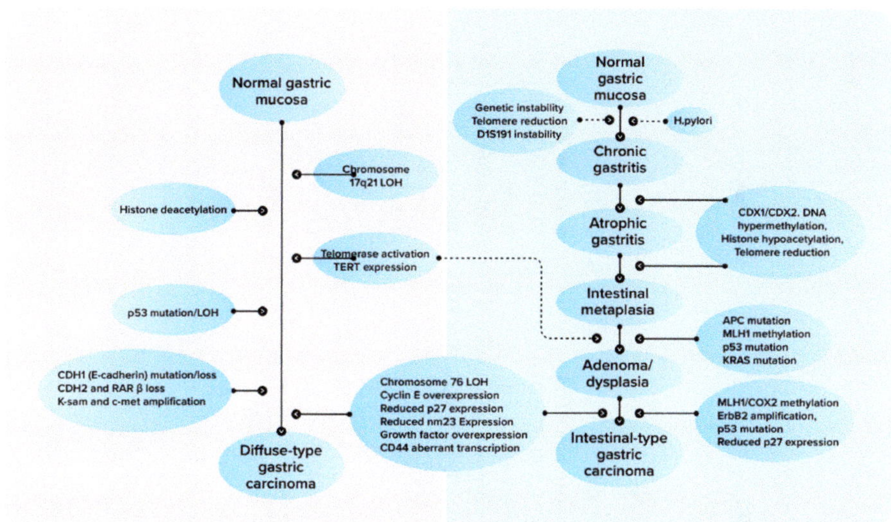

Fig. 2.4 Pathogenesis of intestinal-type and diffuse-type gastric cancer

worldwide, occurs at a younger age, and is associated with a worse prognosis than the intestinal form.

2.13.2 Newer Molecular Classification of Gastric Cancer

Recent insights into the genetic alterations noted in gastric cancer have led to new molecular classifications. The subgroups in each of these new classification systems have many unique identifying features and outcomes. The cancer genome atlas (TCGA) and Asian cancer research group (ACRG) subtypes have been published in the last decade (Tables 2.1 and 2.2).

Table 2.1 TCGA subtypes of gastric cancer

Subtypes	EBV positive	MSI	GS	CIN
Frequency	8.8%	21.7%	19.7%	49.8%
Demographic	Male patients (81%)	Old age (median 72 years)	Young age (median 59 years)	No special
Histology			Diffuse histology	Intestinal histology
Main location	Fundus or body (62%)			Gastroesophageal junction/ cardia (65%)
Molecular alterations	• EBV-CpG island methylator phenotype (CIMP) • PD-L1/2, JAK2 overexpression • Mutation in PIK3CA, ARIDIA, BCOR • CDKN2A silencing immune cell signaling • Rare TP53 mutations	• Gastric-CIMP • Hypermutation in TP53, PIK3CA, ERBB3, ARIDIA • MLHI silencing • Mitotic pathways activation commune changes in the genes of CMHI	• CDHI, RHOA mutation • CLDN18-ARHGAP fusion • Cell adhesion, angiogenesis pathways enriched • Rare TP53 mutations	• TP53 mutation • RTK-RAS activation • Mutations of SMAD4 and APC
Potential targets	PIK3CA, JAK2, PD-L1/ PD-L2	PIK3CA, ERBB2/3, EGFR, PD-L1, MLH1 silencing	RHOA, CLDN18	RTKs, EGFR, VEGFA, CCNEI, CCNDI, CDK6
Treatment reaction		No response to adjuvant chemotherapy		

ARID1A AT-rich interactive domain-containing protein 1A, *BCOR* B-cell lymphoma 6 corepressor, *CIN* chromosomal instability, *GS* genomically stable, *GC* gastric cancer, *CIMP* CpG island methylator phenotype, *DNMT3b* DNA methyltransferase 3b, *EBV* Epstein–Barr virus, *EGFR* epithelial growth factor receptor, *ERBB2* Erb-B2 receptor tyrosine kinase 2, *JAK2* Janus-associated kinase 2, *LMP2A* latent membrane protein 2A, *LELC* lymphoepithelioma-like carcinoma, *MSI* microsatellite instability, *PI3K* phosphatidylinositol-3-kinase, *RHOA* Ras homolog family member A, *TCGA* The Cancer Genome Atlas, *PD-L1/2* programmed death ligand-1/2, *CDK6* cell division protein kinase 6

Ref.: Wang Q et al. Molecular classification of gastric adenocarcinoma. Gastroenterol Res. 2019;12(6):275–282

Table 2.2 ACRG subtypes of gastric cancer

Subtypes	MSI GC	MSS/EMT GC	MSS/TP53-GC	MSS/TP53 + GC
Frequency	22.7%	15.3%	35.7%	26.3%
Demographic		Diagnosed at a significantly younger age	Male	Male
Histology	Intestinal histology (>60%)	Diffuse histology (>80%)	Intestinal histology	Intestinal histology
Molecular alterations	Silencing of the MLH1 gene Mutations in ARID1A, MTOR, KRAS, PIK3CA, ALK, PTEN, overexpression of PD-L1 T cell infiltrate	Loss of CDH1 Loss of cellular adhesion, angiogenesis, motility	The highest prevalence of TP53 and RHOA mutations APC, ARIDA1A, KRAS, PIK3CA, SMAD4, enriched	Frequent EBV infection Frequent mutations in ARID1A, PIK3CA, SMAD4, APC
Prognosis	Best overall prognosis, lowest recurrence rate (22%)	Worst prognosis, highest recurrence rate	Intermediate	Intermediate

ACRG Asian Cancer Research Group, *ARID1A* AT-rich interactive domain-containing protein 1A, *CDH1* E-cadherin, *GC* gastric cancer, *MSS/EMT* microsatellite stable/epithelial-mesenchymal transition, *MSS/TP53+* microsatellite stable/epithelial/*TP53* intact, *MSS/TP53* microsatellite stable/epithelial/*TP53* loss, *MSI* microsatellite instability, *PI3K* phosphatidylinositol-3-kinase, *RHOA* Ras homolog family member A, *PD-L1* programmed death ligand-1
Ref.: Wang Q et al. Molecular classification of gastric adenocarcinoma. Gastroenterol Res. 2019;12(6):275–282

2.14 *H. pylori* Treatment Guidelines: The Maastricht V/Florence Consensus Report

2.14.1 Indications for Testing for H. pylori

- Duodenal and gastric ulcer.
- *H. pylori* eradication is the treatment for localized stage gastric MALToma.
- Post-endoscopic resection for early gastric cancer—*H. pylori* eradication is recommended.
- First-degree relatives of patients with gastric cancer—test and treat for *H. pylori.*
- *H. pylori* gastritis has been associated with dyspeptic symptoms. *H. pylori* eradication produces long-term relief of dyspepsia in about 10% of patients. *H. pylori* gastritis has to be excluded before a diagnosis of functional dyspepsia can be made reliably.

- A test-and-treat strategy is appropriate for uninvestigated dyspepsia. This approach is subject to regional *H. pylori* prevalence and cost–benefit. Considerations. It does not apply to patients with alarm symptoms or older patients.
- Testing for *H. pylori* should be performed in aspirin and NSAID users with a history of peptic ulcer.
- Unexplained iron-deficiency anemia (IDA), idiopathic thrombocytopenic purpura (ITP), and vitamin B12 deficiency: *H. pylori* should be sought and eradicated.

2.14.2 Diagnosis

Tests for Identification of *H. pylori*

1. **Urea breath test (UBT)** is the most investigated and best recommended non-invasive test in the context of a "test-and-treat strategy." PPI should be discontinued for 2 weeks (minimum: 1 week) before testing for *H. pylori* infection. H_2 blockers and antacids do not interfere with *H. pylori* testing. Antibiotics and bismuth compounds should be discontinued at least 4 weeks before the test.
2. An **endoscopy-based strategy** should be considered in patients with dyspeptic symptoms, particularly in low-prevalence *H. pylori* populations and those with alarm symptoms or elderly patients.

   – In clinical practice, when there is an indication for endoscopy, the **rapid urease test (RUT)** is recommended as a first-line diagnostic test. One biopsy should be taken from the corpus and one from the antrum. The sensitivity of biopsy urease tests is approximately 90%. Recent GI bleeding or use of PPIs, antibiotics, bismuth-containing compounds, excessive atrophy, and intestinal metaplasia may result in false-negative reports. If RUT is to be performed, patients should be off antibiotics or bismuth for 4 weeks and off PPI therapy for 2 weeks. H_2 blockers and antacids do not interfere with RUT results. A negative result should not be used to exclude *H. pylori*.
   – Most cases of *H. pylori* infection can be diagnosed from gastric biopsies using **histochemical staining** alone. Immunohistochemical testing for *H. pylori* can be considered in chronic (active) gastritis cases where *H. pylori* is not detected by routine histochemistry. For the assessment of *H. pylori* gastritis, a minimum of five biopsies—two from the antrum (greater and lesser curvature 3 cm proximal to the pyloric region) and two from the middle of the body and an additional biopsy from the incisura has been recommended.
   – It is **recommended to perform clarithromycin susceptibility testing when standard clarithromycin-based treatment is considered the first-line therapy**, except in populations or regions with well-documented low clar-

ithromycin resistance (<15%). This test can be performed either by a standard method (antibiogram) after culture or by a molecular test directly on the gastric biopsy specimen.

3. **Monoclonal stool antigen test** can be used. Patients should be off antibiotics or bismuth for 4 weeks and off PPI therapy for 2 weeks. H_2 blockers and antacids do not interfere with RUT results.
4. Serological tests can be used only after validation. Rapid ("office") serology tests using whole blood should be avoided. Serological tests (ELISA-based) presenting high accuracy and locally validated can be used for non-invasive *H. pylori* diagnosis.

2.14.3 Treatment

H. pylori resistance rates to antibiotics are increasing in most parts of the world.

- Clarithromycin resistance undermines the efficacy of triple and sequential therapy; metronidazole resistance undermines the efficacy of sequential therapy.
- Dual clarithromycin and metronidazole resistance undermine the efficacy of sequential, hybrid, and concomitant therapy.

First-Line Treatment (Table 2.3)

- **In areas of low clarithromycin resistance**
  - Triple therapy is recommended as a first-line empirical treatment. Bismuth-containing quadruple therapy is an alternative.
  - The use of high-dose PPI twice daily increases the efficacy of triple therapy. Esomeprazole and rabeprazole may be preferred in Europe and North America, where extensive PPI metabolizers are prevalent.
  - The treatment duration of PPI-clarithromycin-based triple therapy should be 14 days.

Table 2.3 *H. pylori* first-line treatment regimens

Triple therapy (7–14 days)	Sequential therapy (10 days)	LOAD therapy (7–10 days)	Quadruple therapy (7–14 days)
PPI + amoxicillin + clarithromycin	First 5 days: PPI + amoxicillin Second 5 days: PPI + clarithromycin + metronidazole	Levofloxacin + omeprazole + nitazoxanide + doxycycline	Bismuth based: PPI + bismuth + metronidazole + tetracycline Non-bismuth (concomitant): same as triple therapy + metronidazole or tinidazole

- **In areas of high (>15%) clarithromycin resistance**
  - Bismuth quadruple
  - Non-bismuth quadruple, concomitant (PPI, amoxicillin, clarithromycin, and a nitroimidazole) therapies are recommended
- **In areas of high dual clarithromycin and metronidazole resistance**
  - Bismuth quadruple therapy (BQT) is the recommended first-line treatment. The treatment duration of bismuth quadruple therapy should be extended to 14 days.
  - Currently, concomitant therapy (PPI, amoxicillin, clarithromycin, and a nitroimidazole administered concurrently) for 14 days should be the preferred non-bismuth quadruple therapy, as it has shown to be the most effective to overcome antibiotic resistance.

First-Line Treatment Failure: Management

- **After the failure of bismuth-containing quadruple therapy**, a fluoroquinolone-containing triple or quadruple therapy may be recommended. In cases of high quinolone resistance, the combination of bismuth with other antibiotics, or rifabutin, may be an option.
- **After a non-bismuth quadruple therapy fails**, either a bismuth quadruple therapy or a fluoroquinolone-containing triple or quadruple therapy is recommended.
- **After the failure of PPI-clarithromycin-amoxicillin triple therapy**, a bismuth-containing quadruple therapy or a fluoroquinolone-containing triple or quadruple therapy is recommended as a second-line treatment.

Second-Line Treatment Failure: Management

- **After the failure of second-line treatment**, culture with susceptibility testing or molecular determination of genotype resistance is recommended to guide treatment.
- **After the failure of the first-line treatment (clarithromycin-based) and second-line treatment (with bismuth-containing quadruple regimen)**, it is recommended to use the fluoroquinolone-containing regimen. In regions with a known high fluoroquinolones resistance, a combination of bismuth with different antibiotics or a rifabutin-containing rescue therapy should be considered.
- After the failure of the first-line treatment (triple or non-bismuth quadruple) and the second-line treatment (fluoroquinolone-containing therapy), it is recommended to use the bismuth-based quadruple therapy.
- After the failure of the first-line treatment with bismuth quadruple, the second-line treatment (fluoroquinolone-containing therapy) is recommended to use a clarithromycin-based triple or quadruple therapy. A combination of bismuth with different antibiotics may be another option.

Treatment in Patients with Penicillin Allergy

- In patients with penicillin allergy, in areas of low clarithromycin resistance, for the first-line treatment, a PPI-clarithromycin-metronidazole combination may be prescribed, and in areas of high clarithromycin resistance, bismuth quadruple therapy should be preferred.
- **Rescue regimen in patients with penicillin allergy**: A fluoroquinolone-containing regimen may represent an empirical second-line rescue option in the presence of penicillin allergy.

Tests for Confirmation of *H. pylori* Eradication

1. **A UBT is the best option to confirm *H. pylori* eradication, and a monoclonal stool antigen test is an alternative**. It should be performed at least 4 weeks after completion of therapy.
2. A RUT is not recommended for confirmation of *H. pylori* eradication.

2.15 Morphology of Superficial Neoplastic Lesions: Paris Classification

In the esophagus, stomach, and colon, neoplastic lesions of the digestive tract are called "superficial" at endoscopy when the endoscopic appearance suggests either a small cancer or a non-invasive neoplastic lesion (dysplasia/adenoma). If invasive, "superficial" tumors correspond to the T1 stage of the TNM classification, in which invasion is limited to the mucosa and submucosa. "Superficial" tumors are nonobstructive, usually asymptomatic, and often are detected as an incidental finding or by screening.

2.15.1 Historical Perspective

In Japan, neoplastic stomach lesions with a "superficial" endoscopic appearance are classified as subtypes of "type 0." The term "type 0" was chosen to distinguish the classification of "superficial" lesions from the Borrmann classification, proposed in 1926 for "advanced" gastric tumors, which included types 1–4.

Within type 0, there are polypoid and non-polypoid subtypes. The non-polypoid subtypes include lesions with a small variation of the surface (slightly elevated, flat, and slightly depressed) and excavated lesions (Fig. 2.5).

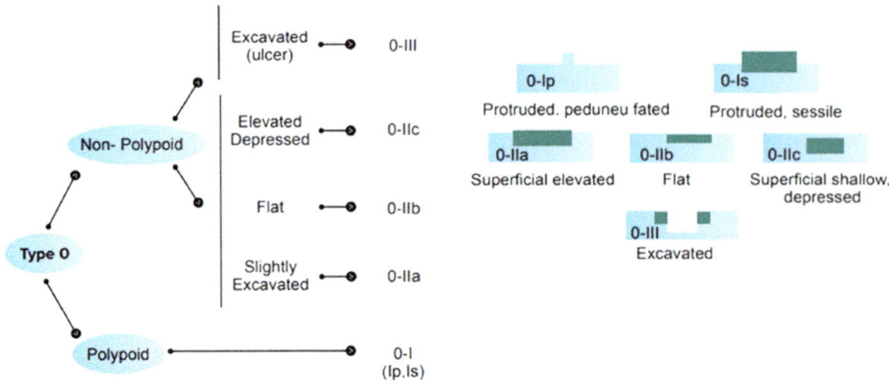

Fig. 2.5 Paris classification of superficial neoplastic lesions based on morphology

The classification of gastric "superficial" neoplasia was promptly applied to esophageal tumors and, later, when the incidence of colorectal cancer increased in Japan, to large bowel tumors.

Schematic representation of the major variants of type 0 neoplastic lesions of the digestive tract: polypoid (Ip and Is), non-polypoid (IIa, IIb, and IIc), non-polypoid, and excavated (III).

2.15.2 *Practical Considerations*

- With small and elevated neoplastic lesions, the respective classification in the polypoid sessile type 0-Is or the non-polypoid elevated type 0-IIa is made easier by placing biopsy forceps next to the lesion as a calibrating gauge. This standard applies to the height of the lesion and not to its diameter. **Lesions protruding above the level of the closed jaws of the biopsy forceps (approximately 2.5 mm) are classified as 0-Is**; lesions protruding below this level are classified as 0-IIa.
- In the esophagus, neoplasia in the stratified squamous epithelium is classified into identical major subtypes but with a distinct quantitative scale. The standard of comparison is a single cup of the opened biopsy forceps. **Lesions protruding above the level of the cup (approximately 1.2 mm) are classified as 0-Is**. The depth of depressed lesions is compared with half the level of a single cup (approximately 0.6 mm).
- Special attention is attached to depressed type 0-IIc lesions. The distinction between a depressed (0-IIc) and ulcerated lesion (0-III) during endoscopy is based on the depth of the depression and the analysis of the epithelial surface in the depressed area.

Table 2.4 Distribution of morphological types in different locations in the GI tract

Paris type	Squamous epithelium of esophagus (%)	Barrett's mucosa (%)	Gastric mucosa (%)	Colonic mucosa (%)
0-Ip + s	16	21	3	57
0-IIa, b	34	61	17	39
0-IIc	45	13	78	4
0-III	5	5	2	0

  - Superficial erosions in a depressed lesion involve only the most superficial layers.
  - In the ulcerated lesion, there is loss of the mucosa and often of the submucosa.

- Mixed types associate two distinct types of morphology. The pattern consisting of an elevation (IIa) and a depression (IIc) is easily diagnosed at endoscopy.
  - A depressed lesion with elevated borders or a central elevation is classified as type 0-IIc + IIa.
  - An elevated lesion with a central depression at its top is classified as type 0-IIa + IIc. This mixed type includes relatively depressed lesions in which depressed areas do not reach below the level of the normal mucosa. As a rule, type IIa + IIc lesions have a poorer prognosis, with a risk of large invasion in the submucosa, than all other types of lesions, including the IIa pattern.
- 0-IIa, b is more common in superficial lesions arising in Barrett's mucosa. 0-IIc is the most common subtype seen in superficial gastric neoplastic lesions (Table 2.4).

2.15.3 *Endoscopic Staging Based on Morphology*

The morphology of a type 0 neoplastic lesion has a predictive value for depth of invasion into the digestive tract wall, providing an "endoscopic staging." Thus, endoscopic descriptive morphology can assist treatment decisions involving endoscopic resection or surgery. The primary role of endoscopic staging is to predict the risk of submucosal invasion and the associated risk of nodal metastases.

- For a type 0-I lesion, the diameter is a reliable predictive criterion. The risk of submucosal invasion increases with the diameter.
- On the other hand, with type 0-II lesions, the morphologic subtypes have greater importance. Invasion of the submucosa is more frequent in depressed lesions (IIc).

The less-than-perfect reliability of endoscopic staging can be improved by EUS, particularly with high-frequency probes (20 MHz). Both endoscopic morphologic staging and EUS have their limits. Endoscopy tends to understage superficial lesions, and EUS tends to overstage them. When the two methods agree, the predictive value is high.

2.15.4 Superficial Neoplastic Lesions in Columnar Epithelium Barrett's Esophagus

Available data on the morphology of superficial neoplastic lesions in Barrett's esophagus are scarce; type 0-IIa,b lesions are the most frequent (70%), but depressed type lesions (0-IIc) are uncommon.

2.15.5 Superficial Neoplastic Lesions in Sub-cardiac Stomach

Most superficial neoplastic lesions in the sub-cardiac stomach are type 0-II, and most (70–80%) are depressed (0-IIc). Type 0-I (polypoid) lesions and type 0-III (ulcerated) are rare. Polypoid adenomas are rare precursors of invasive cancer.

- The endoscopic morphology of a type 0-II lesion has a predictive value for the risk of submucosal invasion. The risk is higher for type 0-I and combined type 0-IIa + IIc, and lower for type 0-IIb lesions (Table 2.5).
- Concerning clinical relevance, non-depressed (type 0-IIa or 0-IIb) lesions with a diameter less than 2.0 cm can be treated safely by endoscopy; the safety limit is lowered to 1.0 cm for all types of lesions with a depressed (0-IIc) morphology.

2.15.6 Superficial Neoplastic Lesions in the Colon

Although depressed (0-IIc) lesions are uncommon (1–6% of non-polypoid lesions), their risk of submucosal invasion is the highest: the overall risk is reported to be 27–35.9% compared with 0.7–2.4% in flat (0-IIa) lesions. More than 40% of small (6–10 mm) depressed (0-IIc) lesions contain submucosal invasive cancer; virtually all large (>20 mm) depressed (0-IIc) lesions have a submucosal invasion. Some elevated lesions may reach a large (>10 mm) lateral diameter without increasing their height or protrusion above the mucosa. In the colon, these are called "lateral spreading type" (Table 2.6).

Table 2.5 Morphology and risk of submucosal invasion in superficial lesions according to location

Paris type	Squamous epithelium of esophagus (%)	Gastric mucosa (%)
0-Ip + s	79%	57%
0-IIa, b	15–48%	25%
0-IIc	27%	37–65% 0-IIa + c > IIc + a > IIc
0-III	84%	40%

Table 2.6 Lateral spreading tumor

Laterally spreading tumor (LST)	Laterally growing superficial neoplasm (instead of upward or downward growth) ≥10 mm in size
LST-granular-homogenous (LST-G-H)	LST polypoid type that corresponds to Paris subtype 0-IIa
LST-granular-nodular mixed (LST-G-NM)	LST type that corresponds to a combination of Paris subtypes 0-IIa and 0-Is
LST-non-granular-flat elevated (LST-NG-FE)	LST non-polypoid type corresponds to Paris subtype 0-II
LST-non-granular-pseudo depressed (LST-NG-PD)	LST non-polypoid type corresponds to a combination of Paris subtypes IIa and 0-IIc

LST-G-H has the lowest risk (0.5%; 95% confidence interval [CI], 0.1%–1.0%), whereas LST-NG-PD has the highest risk of submucosal invasion (31.6%; 95% CI, 19.8–43.4%)
The Paris classification of superficial neoplastic lesions of esophagus, stomach and colon-Gastrointest Endosc 2003;58(6);S1–43

2.16 Proton Pump Inhibitors: Indications and Side Effects

There are six FDA-approved proton pump inhibitors:

1. Omeprazole
2. Pantoprazole
3. Rabeprazole
4. Lansoprazole
5. Esomeprazole
6. Dexlansoprazole

2.16.1 Indications

Indications of proton pump inhibitor use are the following:

1. Peptic ulcer disease
2. Zollinger–Ellison syndrome
3. Stress ulcer prophylaxis
4. Prevention of NSAID-induced or ASA-induced gastroduodenopathy
5. *H. pylori* eradication
6. Upper gastrointestinal bleeding
7. GERD (symptomatic)
8. Esophagitis, Barrett esophagus, and peptic stricture
9. Uninvestigated dyspepsia and functional dyspepsia
10. Esophageal eosinophilia and suspected eosinophilic esophagitis
11. Exocrine pancreatic insufficiency—along with pancreatic enzyme supplements

Table 2.7 Side effects of proton pump inhibitors

1	Structural or functional abnormalities	Increased risk of fundic gland polyps
		Increase in parietal cell mass
		Hypergastrinemia, hyperchromograninemia
		Enterochromaffin-like cell hyperplasia
2	Effects on *H. pylori* gastritis	Shift from antrum-predominant to corpus-predominant gastritis, exaggerated increase in gastrin
3	Gastrointestinal infection	Increased risk of *Clostridium difficile* infection (OR 1.5–1.8)
		For other pathogenic bacteria, >2–4-fold increased risk
4	Pneumonia	Conflicting data, all from retrospective studies
		In large retrospective RCT and cohort studies, increased risk not confirmed
5	AIN	Considerable (fivefold increased) risk of AIN, re-exposure to be avoided
6	CKD	Low risk of development of CKD
7	Hypomagnesemia	Consider PPIs as a cause of hypomagnesaemia in patients with CKD on diuretic therapy
8	Hypocalcemia	No population-based or controlled studies
9	Fractures and osteoporosis	PPI dose and duration dependency in older patients (aged >65 years) Low risk in patients without any other risk factors
10	Myopathies	Possibly idiosyncratic reaction The underlying mechanisms leading to rhabdomyolysis with PPI use are unclear but are possibly related to interactions with concomitantly administered drugs such as HMG-CoA reductase inhibitors (statins)
11	Reduced clopidogrel activation with increased cardiovascular risks	No increase of risk in randomized clinical trials
		Slightly increased risk in retrospective cohort studies
12	Dementia	Increased risk (OR 1.6) with high bias

2.16.2 Side Effects

There have been a lot of concerns about the potential side effects of proton pump inhibitors, including infections, risk of renal injury, and effects on bone health and drug interactions (Table 2.7).

2.17 Prokinetics: Mechanism and Side Effects (Table 2.8)

Table 2.8 Prokinetics—mechanism of action and side effects

S. no.	Class of drugs	Drugs	Comments	Side effects
1	Dopamine antagonists	Metoclopramide	Crosses blood–brain barrier	Extrapyramidal effects, including tardive dyskinesia Lactorrhea
		Domperidone	Does not cross the blood–brain barrier	
		Itopride	Associated anti-cholinesterase activity	
		Levosulpiride	Associated 5HT4 agonist	Extrapyramidal effects, including tardive dyskinesia Lactorrhea
2	Serotonin (5 HT4) agonists	Cisapride	Withdrawn from market	QTc prolongation leading to arrhythmia—withdrawn from the market
		Mosapride		
		Prucalopride	Predominantly used in IBS-C	Abdominal cramps and diarrhea; no significant cardiac side effects as it has low affinity for hERG K+ channels
		Tegaserod	Initially withdrawn; Reintroduced for IBS-C IN 2019	Withdrawn from the market due to cardiac side effects like MI and stroke
3	5HT3 antagonists	Ondansetron	Used in chemotherapy-associated nausea and vomiting	Constipation; rarely causes serotonin syndrome
		Granisetron		
4	Cholinesterase inhibitors	Acotiamide	Associated M1, M2 autoreceptor on cholinergic nerve ending	Occasional constipation
5	Motilin agonist	Erythromycin	For acute gastroparesis-post op.	Tolerance; antibiotic use
6	Neurokinin antagonist	Aprepitant	Mainly for chemotherapy-induced emesis, it improves gastric accommodation without modifying gastric emptying time	
		Tradipitant	For gastroparesis	
7	5HT1A antagonist	Buspirone	Additional weak D2 antagonist; improves gastric accommodation	Tardive dyskinesia

(continued)

Table 2.8 (continued)

S. no.	Class of drugs	Drugs	Comments	Side effects
8	Antidepressants	Amitriptyline	No improvement in gastroparesis	
		Nortriptyline	No improvement in gastroparesis	
		Escitalopram	No improvement in gastroparesis	
		Mirtazapine	Additional central adrenergic and serotonergic activity; promising results in gastroparesis	Promising results in gastroparesis

2.18 Newer Antiplatelets and Anticoagulants: What the GI Physician Needs to Know

Newer oral antiplatelets are P2Y12 inhibitors, namely ticagrelor and prasugrel.

Newer oral anticoagulants, also called direct-acting oral anticoagulants, are anti-thrombin inhibitors (dabigatran) and oral anti-Xa inhibitors apixaban, rivaroxaban, and edoxaban). The advantage of newer oral anticoagulants is the absence of the need for monitoring prothrombin time and fewer drug interactions as compared to warfarin.

2.18.1 Risk of GI Bleeding with Newer Antiplatelets

Systematic review with meta-analysis of the risk of gastrointestinal bleeding in patients taking third-generation P2Y12 inhibitors compared with clopidogrel showed the following:

1. Third-generation P2Y12 inhibitors are associated with a higher risk of GI bleeding (RR: 1.28 95% CI 1.13–1.46). Findings were consistent for upper (RR 1.32, 95% CI 1.05–1.67) and unspecified GIB (RR 1.25, 95% CI 1.01–1.53), but not lower GIB (RR 1.25, 95% CI 0.95–1.65).
2. Subgroup analysis showed higher GIB risk in prasugrel studies (RR 1.40, 95% CI 1.10–1.77) than in ticagrelor studies (RR 1.15, 95% CI 0.94–1.39).
3. Apixaban was noted to have a lower incidence of GI bleed compared to other newer anticoagulants and warfarin. The risk of anticoagulant-associated GI bleed was lower when co-administered with proton pump inhibitors.

Table 2.9 Risk of GI bleeding with various endoscopic procedures

Low risk of GI bleed	High risk of GI bleed	Ultra-high risk of GI bleed
Diagnostic endoscopy with biopsy	Polypectomy	ESD
EUS without FNA	ERCP with sphincterotomy	EMR of polyps >2 cm
ERCP with biliary and pancreatic stenting	Dilatation of strictures	
Diagnostic enteroscopy/capsule endoscopy	Banding or injection of varices	
Esophageal, enteric, and colonic stent	PEG/PEG-J	
APC	EUS FNAC	

Table 2.10 Recommended modification of antiplatelet and anticoagulants before the planned procedure

Procedure-related risk of bleeding	High		Moderate-low
Risk of thrombosis on stopping antiplatelets/anticoagulants	Low	High	Low
Aspirin	Continue	To discuss in multi disciplinary meeting	Continue
P2Y12 inhibitors	Stop 5 days prior	To discuss in multi disciplinary meeting	Stop 5 days prior
Warfarin	Stop 5 days prior	Heparin bridging	Stop 5 days prior
DOAC	Stop 2 days prior	No heparin bridging	Stop 2 days prior

2.18.2 Procedures Associated with the Risk of GI Bleed (Table 2.9)

2.18.3 Recommended Modification of Antiplatelet or Anticoagulant Use Before the Planned Procedure

The modifications to be adopted before the planned endoscopic procedure in patients who are on antiplatelet or anticoagulant medications are summarized in Table 2.10.

2.18.4 *Strategies for Reversal/Antidote to Newer Anticoagulants During a Life-Threatening Bleed*

- FFP or prothrombin complex concentrate is used if INR >2.5, and there is a life-threatening bleed.
- Activated charcoal could be used in case of overdose if direct oral anticoagulant intake was taken less than 3 h earlier.
- Idarucizumab is a specific antidote for dabigatran.
- Dialysis can be used for dabigatran overdose.
- Andexanet alfa is a modified recombinant inactive form of human factor Xa used to reverse factor Xa inhibitors.

Chapter 3
Small Intestine

Piramanayagam Paramasivan

3.1 Development and Malrotation

Endoderm, mesoderm, and ectoderm contribute to the development of the gastrointestinal system. The epithelial layer of the gut is derived from endoderm. Lamina propria, muscular layer, and mesentery are derived from mesoderm. The neural crest cells of the ectoderm form the origin of the enteric nervous system. Gastrointestinal (GI) tract is derived from the foregut, midgut, and hindgut.

- The foregut consists of the complete oesophagus and stomach, along with the first two components of the small intestine, which includes the duodenum.
- The midgut consists of the distal parts of the small intestine and the initial part of the large intestine, namely, distal duodenum, jejunum, ileum, cecum, ascending colon, and more than 60% of the transverse colon.
- The hindgut includes the last of the GIT, the remaining part of the transverse colon, the complete descending colon, the sigmoid colon, and the rectum terminating in the anal canal.

3.1.1 Timeline of Development (Table 3.1)

3.1.2 Intestinal Malrotation

Rotation may get halted in various stages of development resulting in malrotation (Table 3.2).

P. Paramasivan (✉)
Department of Medical Gastroenterology, Apollo Hospitals, Chennai, India

G. Choudhuri et al. (eds.), *Gastroenterology and Hepatology*,
https://doi.org/10.1007/978-981-99-9287-4_3

Table 3.1 Timeline of small intestine development

Week	Milestones	Ancillary development
4	The division into fore, mid, and hindgut	Neural crest cells enter the foregut
4–5 weeks	The rapid growth of the midgut and herniation of the midgut into the umbilical cord forms a loop with its mesenteric vessel The superior limb of the loop forms a small intestine The inferior limb of the loop forms the colon	
5–10 weeks	90° counter-clockwise rotation of herniated midgut loop—superior limb turns to right; inferior limb turns to left	Neural crest cells reach the hindgut by week 7; cloaca formation begins by week 9; villi formation begins
10 weeks	As the midgut retracts into the abdominal cavity, further 180 ° counterclockwise rotation. The cecum is positioned in the upper right quadrant Further enlargement of the colon pushes the cecum into the right iliac fossa	
11 weeks		A mature smooth muscle layer is formed in the small intestine

Table 3.2 Mechanism of intestinal malrotation

Anomaly	Reason
1. Intestinal nonrotation	Initial counterclockwise rotation does not occur. Midgut growth without this rotation causes mesentery to grow from a narrow base-predisposes to midgut volvulus
2. Intestinal malrotation	Incomplete rotation during the retraction of midgut loops into the abdominal cavity. Bands from the cecum to the duodenum may obstruct the duodenum. Small bowel may protrude through caecal or colonic mesentery resulting in mesocolic (para duodenal) hernia
3. Incomplete fixation	If mesentery of the ascending, descending colon or duodenum is not fixed to the retroperitoneum, potential internal hernial pouches arise Non-fixation of the cecum can lead to torsion and volvulus of the midgut

3.2 Meckel's Diverticulum

When the duct communication between the intestine and the yolk sac continues past the embryonic phase, anomalies of the omphalomesenteric or vitelline duct may occur, such as the following:

1. Central cystic dilatation or an omphalomesenteric or vitelline cyst: in this, the central space of the duct is closed at the ends.
2. Umbilical-intestinal fistula occurs when the duct remains patent through its entire length.
3. Blind omphalomesenteric duct, or Meckel's diverticulum.
4. Omphalomesenteric band: complete expunging of the duct leading to the formation of a fibrous cord that elongates from the ileum to the umbilicus.

3.2.1 Surgical Anatomy of Meckel's Diverticulum

- Meckel's diverticulum originates from the antimesenteric surface of the ileum, approximately 40 cm from the ileocecal valve in infants, and nearly 50 cm in adults. In about 4% of cases, this measurement may be <15 cm or even as high as 167 cm.
- The diverticulum, in certain cases, may be very short (1 cm) or too long (26 cm): in three-fourths of the cases, it will measure between 1 and 5 cm; among the remaining, it could be longer. Despite these wide variations, there are limits to the possible size and position of vitelline duct remnants.

3.2.2 Types

Histology

- Gastric mucosa is present in 80%. Ectopic gastric mucosa with parietal cells leads to ileal ulceration and is an important conduit of the pathology of the diverticulum to the adjacent ileum. Ileal, duodenal, or pancreatic tissue may also be present.
- The independent blood supply of Meckel's diverticulum originates from an intestinal arcade.
- Incidentally found that Meckel's diverticula are equally distributed between the sexes.

Clinical Presentation

Meckel's diverticular disease often has myriad presentations:

- Ulceration accounts for about 40% of Meckel's diverticulum disease. It is produced by secretions from ectopic gastric mucosa in the diverticulum. Ulceration acts on the exposed ileal mucosa in the diverticulum or in the proximal ileum.
- Bright red or brick-red rectal bleeding secondary to ulceration is painless and episodic.

In 90% of cases of bleeding associated with a Meckel's diverticulum, Tc-99m pertechnetate (nuclear scan with technetium pertechnetate) confirms the clinical diagnosis by demonstrating the ectopic gastric mucosa in the diverticulum.

- Obstruction and intussusception together account for about 32% of Meckel's diverticulum disease. Obstruction can be caused by volvulus.
- Appendicitis-like inflammation, often from the presence of a foreign body and narrow base, accounts for 17% of Meckel's diverticulum disease.

- Neoplasms are found in about 6% of Meckel's diverticulum disease. In order of frequency, these neoplasms are leiomyomas, leiomyosarcomas (malignant gastrointestinal stromal tumours), carcinoid tumours, and adenocarcinomas. Carcinoid tumours may be the most common neoplastic lesions of the diverticulum.
- Umbilical symptoms (umbilical leakage from an omphaloileal fistula, weeping from an umbilical polyp, or infection of an umbilical sinus) account for the remaining 5% of Meckel's diverticulum disease.

Treatment

- Surgical resection is necessary for managing all types of vitelline duct abnormalities.
- Asymptomatic Meckel's diverticulum should preferably be excised if found during an exploratory laparotomy.

3.3 Enteric Nervous System

- The neural system of the GIT, also known as the *enteric nervous system,* lies in its wall. It originates at the oesophagus and traverses its entire length up to the anus (Fig. 3.1). The neuronal population (~1 million) in the enteric nervous system is the same as in the complete spinal cord.
- It consists primarily of two plexuses: the exterior and the interior.
- The enteric nervous system is composed mainly of two plexuses:
  - The outer plexus, known as *myenteric plexus* or *Auerbach's plexus,* lies between the longitudinal and circular muscle layers of muscularis externa.
  - The inner plexus, called the *submucosal plexus* or *Meissner's plexus*, lies between the circular muscles and the mucosa.
- The myenteric plexus controls the GI movements, while the submucosal plexus regulates the GI secretion and local blood flow.
- Both the plexus are regulated by the autonomic nervous system (sympathetic and parasympathetic nerve fibres). Nevertheless, the enteric nervous system can function autonomously.
- The sensory nerve endings arise in the epithelium of the GIT. The afferent fibres link these nerve endings to the enteric plexuses, the prevertebral ganglia of the sympathetic nervous system, and the spinal cord via the Vagus nerve up to the brain stem.
- These sensory nerves can draw local reflexes within the gut wall itself and other reflexes that are passed on to the gut from either basal regions of the brain or the prevertebral ganglia.

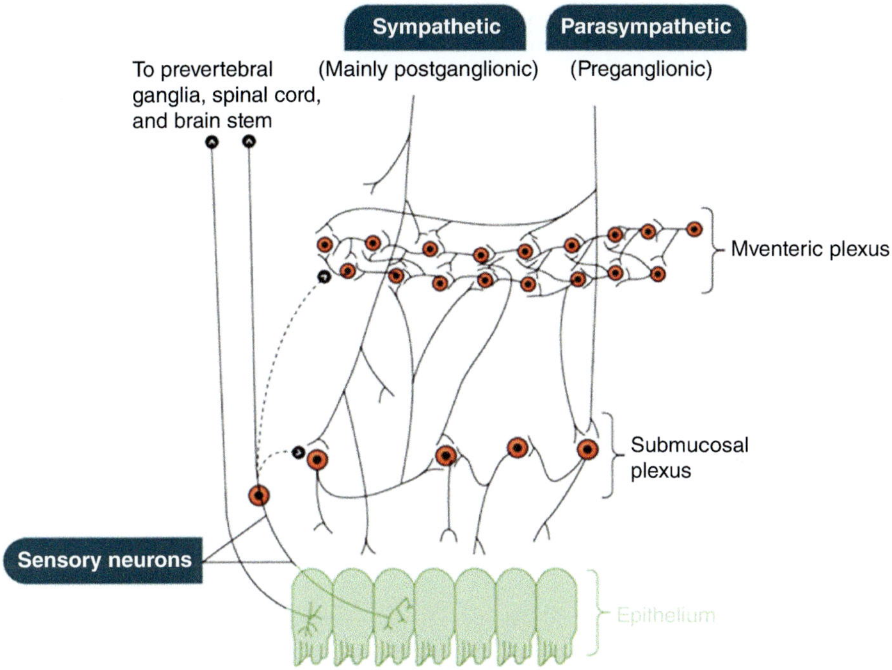

Fig. 3.1 Enteric nervous system

3.3.1 Functions of Myenteric Plexus

The *myenteric plexus* regulates muscle activity across the entire length of the GI tract.

Its principal effects are as follows:

- Helps to maintain muscle tone throughout the intestinal wall.
- Along the gut wall, it elevates the intensity of the periodic contractions with a slight effect on its rate, increases the velocity of the excitatory waves, increases rapid movement of the gut peristaltic waves, and increases digestive secretions.

The neurons of the myenteric plexus are excitatory and include inhibitory ones. An inhibitory transmitter is secreted by their fibre, possibly a *vasoactive intestinal polypeptide*. These inhibitory signals are responsible for regulating the activity of the intestinal sphincter muscles (pyloric sphincter and ileocecal valve) that allow passage of food along consecutive segments of the GI tract.

3.3.2 Functions of Submucosal Plexus

- Unlike the myenteric plexus, the *submucosal plexus* predominantly regulates the function of the small intricate segments of the intestine.

- The GI epithelium transmits several sensory signals via the submucosal plexus, which regulates *i*ntestinal secretion and absorption along with the contraction of the submucosal muscle.

3.4 Villi

- Minute projections (0.5–1.6 mm) originating from the intestinal mucosa are called villi, with microscopic extensions obtruding from its epithelial cells called microvilli.
- Due to their structural nature, these projections increase the surface and, consequently, the intestinal wall absorptive area.
- The villi of the jejunum are the most distinctly finger-like or long and tongue-shaped projections.
- The villi of the ileum are shorter and blunter in shape, disappearing gradually as the ileocecal valve is approached.

3.4.1 Function

- The villi and microvilli, due to the cumulative effective surface area, increase the absorption capacity of the small intestine 30-fold and 600-fold, respectively.
- Along with microvilli, enzymes are also present on the intestinal surface to enhance digestion. (Table 3.3).
- The capillaries lining the villus capillaries absorb the simple sugars and amino acids into the systemic circulation.
- The absorbed chylomicrons are accumulated by villus lacteals (lymph capillary), which are lipoproteins composed of triglycerides, cholesterol, and amphipathic proteins that the lymph fluid passes to the rest of the body.

Table 3.3 Brush border enzymes and substrates

Enzyme	Substrate	Products
Lactase	Lactose	Glucose + galactose
Maltase (glucoamylase)	α-1,4-Linked oligosaccharides	Glucose
Sucrase	Sucrose	Glucose + fructose
Amino-oligopeptidases (at least two types)	Cleave amino acids from the C terminal end of 3–8 amino acid peptides	Amino acids and dipeptides
Aminopeptidase A	Cleaves dipeptides with acidic amino acids at the N-terminal end	Amino acids
Dipeptidase I	Cleaves dipeptides containing methionine	Amino acids

Table 3.3 (continued)

Enzyme	Substrate	Products
Dipeptidyl aminopeptidase IV	Cleaves proline-containing peptides with free a-amino groups	Peptides and amino acids
Carboxypeptidase P	Cleaves proline-containing peptides with free C-terminal end	Peptides and amino acids
Gamma-glutamyl transpeptidase	Cleaves gamma-glutamyl bonds and transfers glutamine to amino acid or peptide acceptors	Gamma-glutamyl amino acid or peptide
Dipeptidases (several types)	Cleave most dipeptides	Amino acids
Aminotripeptidase	Cleaves tripeptides	Amino acids
Proline dipeptidase	Cleaves proline-containing dipeptides	Proline and amino acids

3.5 Microvilli

Microvilli are microscopic projections that form the brush border, enhancing the surface area of the intestinal cells.

The functions of microvilli are as follows:

- absorption
- secretion
- cellular adhesion
- mechano transduction

3.5.1 Location

- Microvilli are found on the apical surface of some epithelial cells associated with absorptive function, namely, proximal tubules of nephrons and the enterocytes in the small intestine.
- Apart from the intestinal microvilli, these are also found in the sensory cells of the inner ear (as stereocilia), olfactory receptors, and taste buds.
- These are also found on the surface of the ovum to aid in the attachment of spermatozoa after penetration of its outermost membrane. The elongated microtubules assemble to enable the fusion of egg and sperm by bringing and holding them closer.
- Migration of white blood cells is facilitated by the microvilli facilitate present on their surface.

3.5.2 Structure

- Plasma membrane (enclosing cytoplasm and microfilaments cover microvilli).
- Microvilli have little or no cellular organelles.

- The structural core of the microvilli consists of a dense bundle of actin filaments cross-linked by the fimbrin and villin proteins.
- This structural core is pinned to the plasma membrane across its length by fibres composed of myosin 1a and Ca^{2+} binding protein calmodulin.
- Myosin 1a acts via two binding sites, filamentous actin and a lipid-binding domain on either end.
- The microvilli have plus and minus ends. The plus end is covered by the capZ proteins, while the spectrin and myosin-containing sections hold the minus ends.

3.5.3 Functions

- Due to the greater surface area, the microvilli are a major site of nutrient absorption. The GI tract has microvilli as the primary surface for the absorption of nutrients.
- The membrane of the microvilli (packed with enzymes) helps in the easy absorption of food through the breakdown of complex nutrients. The microvilli enterocyte surface has a high number of glycosidases and peptidases.
- In the kidney proximal tubule, microvilli can act as mechano-sensors. There are biochemical responses that convert the fluid flow (sensed in the tubule lumen) into reabsorption.
- Microvilli help in the migration of white blood cells, showing the importance of being on the surface of white blood cells.

3.5.4 Glycocalyx

Glycocalyx, which contains peripheral glycoproteins, covers microvilli. These glycoproteins can attach to a membrane. This cover may help in the binding of substances to protect against harmful substances or to adhere to nutrients.

3.5.5 Destruction of Microvilli

- Osmotic diarrhoea and nutrient malabsorption, along with febrile illness, occur when the repositioning of the cytoskeleton in host cells leads to the elimination of the microvilli in certain diseases like celiac disease, infections caused by enteropathogenic *Escherichia coli* (EPEC) subgroup *Escherichia coli*, and microvillus inclusion disease.
- Microvillous atrophy occurs due to a congenital lack of microvilli. This is a very rare fatal condition in newborn babies.

3.6 Intra-Epithelial Lymphocytes

Intra-epithelial lymphocytes (IELs) reside closely in the lumen of the intestine and constitute a large population of lymphocytes. In eosin and haematoxylin-stained sections, IELs display a basophilic nuclear chromatin pattern. These also have a perinuclear halo and an irregular nuclear outline.

Phenotypically, IELs are constituted by >70% CD4–/CD8+ T-lymphocytes, and <20% are CD4+ T-lymphocytes. The majority of these are T-cell receptor (TCR) αβ positive. B-cells are absent. These lie between underlying follicles and M-cells. The IELs form the initial line of the host-immune defence system.

3.6.1 The Upper Limit of the Normal Range of IELs

The lymphocytes that line the basement membrane are regarded as IELs, and these are always expressed as mean lymphocyte count per 100 enterocytes. In the small intestine, normally 40 IELs per 100 epithelial cells are found in the small intestine. However, recent evidence indicates it to be between 20 and 25 IELs per 100 enterocytes, and immunohistochemistry with CD3 antibody increases it slightly. The haematoxylin and eosin staining shows a lower count of IEL than when followed by immunohistochemistry with CD3 antibody. It is labelled as borderline intra-epithelial lymphocytosis when the count is 25 and 29 IELs per 100 enterocytes and pathologic intra-epithelial lymphocytosis when it exceeds 30 IELs per 100 enterocytes in the duodenum. Architectural distortion is avoided by obtaining biopsy specimens from later parts of the distal duodenum. For a healthy microvillus, more than 300 enterocytes should sequentially align along the length of the epithelium.

3.6.2 Associated Diseases with Increased IELs

Increased IELs and Celiac Disease

Higher IEL are characteristic manifestations of active celiac disease. When only IEL occurs in the absence of microvilli abnormality, it is categorised as latent or potential celiac disease. Dietary modifications are recommended in patients suspected of having latent celiac disease. The usual pattern of descending number of IEL along each microvillus is lost since the IEL line is largely at the tip across the length of the microvilli. This is a unique characteristic of latent or potential celiac disease.

Increased IELs and Refractory Sprue

It is characterised by a population with abnormal IEL containing intracytoplasmic CD3 chain (CD3c). However, it does not exhibit CD4/CD8 or a T-cell receptor. This condition is known as low-grade or cryptic T-cell lymphoma derived from IELs. While a normal IEL phenotype (CD3+ CD8+) is found in patients with celiac disease and healthy subjects with small bowel biopsy specimens but in patients with refractory sprue, the IEL phenotype is abnormal, mostly CD3+ CD8–, and rarely CD3+ CD8+. In patients with abnormal CD3+ CD8–, confirmation of refractory sprue diagnosis can be obtained only with TCR gene rearrangement studies. Besides, a rigorous follow-up is necessary due to a weak prognosis.

Increased IELs and Non-celiac Disease-Associated Disorders

- **Intolerance to protein in the food protein**: Intolerance to the protein in food is noted in young infants just after the introduction of cow milk. Such intolerance leads to hypertrophy of the villus along with concomitant crypt hyperplasia and inflammation not only in the epithelial cells but also in the lamina propria, with an increase in eosinophils in several cases.
- **Auto-immune enteropathy**: Clinically, infants with this autoimmune condition suffer from incessant watery diarrhoea and, in half of these cases, anti-enterocyte antibodies are present. Adults with autoimmune conditions such as those associated with type 1 diabetes, variable immunodeficiency, systemic lupus erythematosus, rheumatoid arthritis, haemolytic anaemia, and Hashimoto's thyroiditis may also experience refractory diarrhoea. This autoimmune condition may manifest mucosal lesions with damaged crypt cells.
- **Tropical sprue**: The histological results of small bowel biopsies in patients with tropical sprue may be similar to those of celiac disease. However, based on the positive response to antibiotic and folate therapy and negative serological tests for celiac antibodies are useful for differential diagnosis.
- **Enteropathy-type intestinal T-cell lymphoma (EITCL) or enteropathy-associated T-cell lymphoma (EATL)** may have derived from IELs, and an intermediate or 'cryptic' stage of this lymphoma in patients with refractory sprue has been reported. Histologically, anomalous clonal intra-epithelial T-cell population with markedly raised IELs characterises mucosal lesions in these patients. Patients do not respond to diet therapy (gluten-free) or relapse after the initial response.
- ***Helicobacter pylori*-related gastritis**: Patients with *H. pylori* chronic active gastritis show a high number of IELs in the proximal duodenum.
- **Others**: IELs could also be elevated in some patients with Crohn's disease (CD) peptic duodenitis, giardiasis, bacterial overgrowth, graft-versus-host disease, common variable immunodeficiency, and eosinophilic gastroenteritis. Infrequently, this elevation could also be drug-induced, for example. Nonsteroidal anti-inflammatory drugs (NSAID) induced gastro enteropathy.

3.7 Microfold Cells

- Microfold cells (or M-cells) are present in the follicular epithelium of Peyer's patch.
- M-cells are vital in stimulating mucosal immunity as they engulf the antigen from the lumen and conduct particles and organisms (from the GIT lumen) via transcytosis to immune cells, also called dendritic cells or antigen-presenting cells and lymphocytes, across the epithelial barrier.

3.7.1 Structure and Function

- M-cells are less frequent than enterocytes in the small bowel.
- Instead of microvilli, the M-cells have larger microfolds, as the name also suggests.
- M-cells do not secrete digestive enzymes or mucus. The antigens in the gut lumen easily approach the M-cells to reach Peyer's patch. The filamentous brush border glycocalyx is either absent or much thinner on M-cells.

3.7.2 Pathology

- M-cells act as a way for several pathogens, which include *Salmonella typhimurium*, *Shigella flexneri*, and *Yersinia pseudotuberculosis*, to infiltrate the intestinal epithelium.
- Enteropathogenic *Escherichia coli* consisting of plasmids with *E. coli* adherence factor (EAF) genes can attach to M-cells.
- Viruses like polio and reovirus also exploit them for propagation.
- M-cells only conduct CXCR4 tropic but not CCR5 tropic HIV across the epithelium.

3.7.3 Development

M-cells are formed in response to cues from immune cells of the developing Peyer's patch.

3.8 Paneth Cells

- Paneth cells in the intestinal tract contain zinc and lysozyme and large eosinophilic refractile granules within their apical cytoplasm. Although its exact function is unclear, due to lysozyme presence, it probably participates in and also helps in maintaining the mucosal integrity against infectious pathogens.

Fig. 3.2 Paneth cells

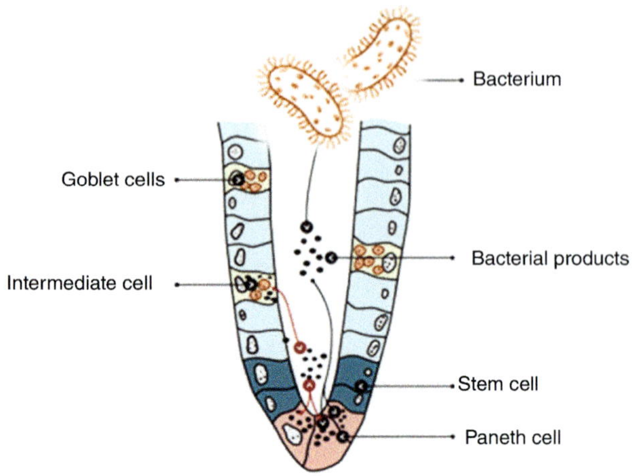

3.8.1 Relationship with Stem Cells

- Small intestinal crypts consist of stem cells that replace the worn-out and dead epithelial cells. Therefore, protecting the stem cells is necessary to ensure long-term maintenance of the intestinal epithelium.
- The strategic position of the Paneth cells in proximity to the stem cells indicates their participation in protecting the process of replacement of epithelial cells (Fig. 3.2).

3.8.2 Sensing Microbiota

Paneth cells contain the MyD88-dependent toll-like receptor (TLR), which enables recognition of bacterial pathogens, thus activating the antimicrobial mechanism.

3.8.3 Secretions

Defensins

- Alfa-defensins, also known as crypts, are the key defence components secreted by the Paneth cell on exposure to bacteria (Gram-positive and -negative) or its components [muramyl dipeptide (MDP) and lipid A and lipopolysaccharide].
- These peptides are positively charged and connect with the negatively charged phospholipids in the cell membranes inserting themselves, leading to the formation of and lysis of the cell via disruption of the cell membrane.
- These peptides selectively destroy only the bacterial cell membranes owing to their higher concentration of phospholipids than the cell membranes of host cells.

3.8.4 *Other Secretions*

- Paneth cells also secrete lysozyme and phospholipase A2, which also possess antimicrobial activity.
- Due to the presence of several defence components, Paneth cells are empowered to protect the intestinal cells from several pathogens other than bacteria, including fungi and enveloped viruses.

3.9 Zona Occludens

- The zonula occludens or tight junction (TJ) regulates the paracellular permeability. The TJ acts as a fence and forms a semipermeable, regulated barrier that separates protein (and partially lipid) components of the apical and basolateral plasma membrane domains. It regulates the paracellular movement of fluids and solutes.
- Under an electron microscope (EM), the TJ appears as a sequence of distinct links between the plasma membranes of adjoining cells and is also linked to the apical peri-junctional actomyosin ring. TJ is composed of a multiprotein complex. Since it possesses charge-selective properties, TJ regulates the overall permeability.
- TJ proteins identified so far include cytoplasmic plaque proteins ZO-1, ZO-2, ZO-3, transmembrane proteins occludin and claudin, cingulin, and 7H6.

3.9.1 *Physiological Regulation by Na^+/Nutrient Cotransport*

Experimental studies and those in human volunteers have shown that the small intestinal permeability (SIP) regulation is dependent on the Na^+-glucose cotransport.

3.9.2 *Influence of Cytokines*

Interferon

Studies in patients with IBD have found an upregulation of IFN in the mucosa of the colon, which may lead to enhanced paracellular permeability.

Hepatocyte Growth Factor

It is primarily produced by the mesenchyme-derived cells and modulates the epithelial permeability in a paracrine manner via ligation with its receptor, c-*met*. The decrease in the transepithelial electrical resistance (TER) induced by IFN corresponds to that induced by HGF.

Tumour Necrosis Factor

Tumour necrosis factor induces an increase in the intercellular gaps (associated with the dispersion of VE-cadherin).

Other Cytokines

Transforming growth factor (TGF)-1 may enhance barrier function in human enterocytes.

Epidermal growth factor (EGF) enhances the assembly of intercellular junctions of sub-confluent monolayers. It also redistributes apical actin and phosphorylates cytoplasmic plaque proteins ZO-1 and ZO-2 of TJ.

Among the other cytokines such as vascular endothelial growth factor (VEGF), interleukin (IL)-1, IL-4, IL-13, and insulin-like growth factor (IGF)-I and (IGF)-II may diminish the barrier characteristics of epithelial/endothelial cells.

3.9.3 Modulation by Leukocytes and Bacteria

- Inflammation is associated with an influx of leukocytes and secretion of several cytokines. The paracellular permeability markedly increases upon stimulation of polymorphonuclear leukocyte (PMN) to migrate across endothelial monolayers.
- Several GI-related conditions (e.g. lymphocytic colitis, chronic idiopathic conditions, and sprue-like disorders) are associated with chronic inflammation, which involves increased IELs and consequent loss of epithelial barrier integrity. Some interleukins like IL-10, IL-4, and interferons are mediators secreted by the intra-epithelial T-lymphocytes, which regulate the transepithelial migration of PMN.
- Mediators (secreted by bacteria) may also alter paracellular permeability, greatly affecting both absorptive and secretory processes. The activation of protein kinase C by different agonists or the *Vibrio cholera* zonula occludens toxin (ZOT) may result in alterations of the peri-junctional actin ring that constitutes the TJ apparatus.

3.10 Toll-like Receptors

TLRs, which constitute innate immunity, belong to the group of pattern recognition receptors (PRPs). They identify and interact with pathogen-associated molecular patterns (PAMPs), conserved motifs in various microbiological organisms. TLRs are transmembrane proteins. The ectodomain recognises PAMPs. The transmembrane component is in continuity with the cytosolic component (Toll IL1R) domain which initiates multiple intracellular pathways. There are two main pathways: the

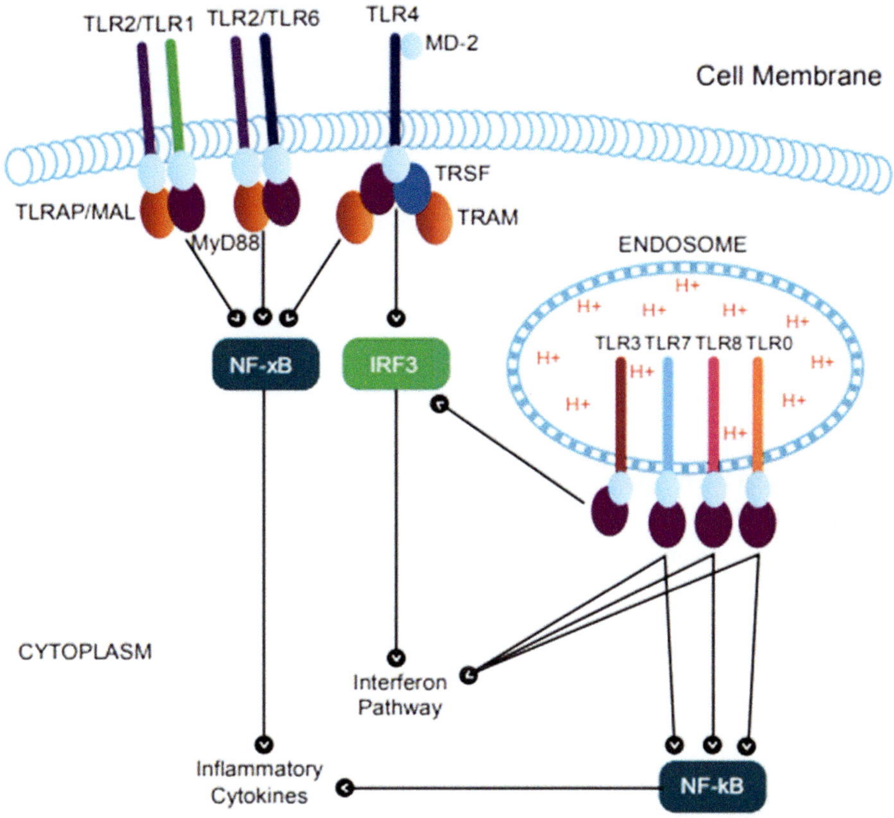

Fig. 3.3 Toll-like receptors. (Adapted from: https://www.biovision.com/products/diabetes-obesity-metabolic-syndrome/signaling-pathways-in-diabetes-and-obesity/toll-like-receptor-tlr-signaling.html)

Table 3.4 Toll-like receptors and corresponding ligands (PAMPs)

TLR	PAMPs
TLR3,7,8,9	ssRNA, dsRNA, DNA
TLR1	Bacteria, mycobacteria
TLR2/6	Gram-positive bacteria, fungi
TLR4	Gram-negative bacteria
TLR5	Bacteria

MyD88 pathway and the TRIF pathway. They result in the activation of NFkB and IFN pathways, producing inflammatory cytokines.

There are ten known functional TLRs in humans (Fig. 3.3). TLRs are expressed either on the cell surface or intracellular vesicles. TLR1–6 are located on the cell surface and recognise microbial membrane components; TLR3 and 7 and TLR8 and 9 are located in vesicles within the cells and recognise nucleic acids. TLRs bind to corresponding distinct pathogen-associated molecular patterns, as described in Table 3.4 and clinical relevance of TLRs and their association with diseases is listed in Table 3.5.

Table 3.5 Toll-like receptors and their association with diseases

TLR	Disease
Defects in proximal TLR signalling molecules (MyD88)	Primary immunodeficiency with increased risk of pyogenic infection
Mutation in TNF. TRIF, IFN pathway molecules	Higher risk of HSV-1, influenza pneumonia
Defects in the distal component of TLR signalling pathway—NFKB	Increased susceptibility to bacteria, mycobacteria, viruses, and opportunistic pathogens
Increased expression of TLR4 in the intestinal epithelium	Crohn's disease and ulcerative colitis
TLR4 expression upregulation	*H. pylori*-related gastric cancer, multiorgan dysfunction in acute pancreatitis, PBC, PSC, NASH, and alcoholic hepatitis

3.11 Migrating Myoelectric Complexes

- Unique manometric patterns of activity characterise the motor functions of the small intestine and stomach.
- The inter-digestive period or fasting is typified by a cyclic motor phenomenon which lasts 60–90 min.
- There are three phases of this complex:
  - A period of quiescence forms the first phase
  - Intermittent pressure activity constitutes the second phase
  - Contraction of the stomach (3/min) and small intestine (8–12/min) occurs at the highest frequency in this phase
- The giant migrating complex is another characteristic of inter-digestive motor pattern, which transports the residue bolus forms from the ileum into the large intestine.

3.12 Intestinal Pacemaker

An intestinal pacemaker is a group of cells in the intestine that creates the basal electrical rhythm leading to the contraction of smooth muscles (peristalsis) of the intestine. It is formed by specialised mesenchymal cells called interstitial cells of Cajal (ICC). The frequency of pacemaker activity of these ICCs depends upon their location:

- 3/min in the stomach and the large intestine
- 12/min in the duodenum
- 10/min in the ileum

3.12.1 Interstitial Cells of Cajal

Pleomorphic mesenchymal cells form an intricate matrix via elongated, narrow cytoplasmic projections. ICC form electrical gap junctions since these lie adjacent to both nerve axons and myocytes.

3.12.2 Functions

ICC regulate the motility of the small intestine in two ways:

1. First, ICC creates an electrical slow wave that acts like a pacemaker to retain the rhythmicity of intestinal contractions.
2. These cells conduct all nerve signals (excitatory and inhibitory) to the myocytes.

3.12.3 Types

Three separate functional groups of ICCs exist:

1. **ICCMY** (myenteric) is located between longitudinal and circular muscular layers. They generate slow waves in smooth muscle and act as pacemakers of the small intestine.
2. **ICCIM** (intramuscular) are predominantly innervated by the intrinsic enteric motor neurons.
3. **ICCDMP** are subtypes of ICCIM and lie mostly in the inner surface of the circular muscle layer in close proximity to the deep muscular plexus. These are also innervated by both inhibitory and excitatory enteric nerve terminals.

3.13 Gut Mucosal Barrier and Conditions with Increased Gut Permeability

The mucosal surface of the gut is protected by epithelial cells, which form a barrier between the external and the internal environment. Its selective permeability enables it to absorb nutrients and secrete waste products.

3.13.1 Anatomy of Gut Mucosal Barrier

The gut mucosal barrier is made up of extracellular and cellular components.

(a) **The extracellular component** of the barrier is made of mucins secreted by two cells, gastric foveolar in the stomach and goblet cells in the intestine.

- It develops a selectively permeable physical barrier against large particles, including bacteria permitting passage of only smaller molecules, thus guarding the epithelial cells.
- An unstirred layer is formed due to restricted bulk fluid flow across the mucosa, which permits only slow diffusion of smaller molecules and ions.

Over the mucosal cells of the stomach, a stretch of relative alkalinity is formed by the bicarbonate secreted by the epithelial cells at the mucosal surface. In the small intestine, mucins regulate nutrient absorption by controlling the rate at which nutrients reach the microvillus brush border.

(b) **Cellular components** of the mucosal barrier include the epithelial cell, inter-epithelial cell junctions, and intra-epithelial lymphocytes.

3.13.2 Physiology of Solute Transport Across the Gut Mucosal Barrier and Their Regulation

1. **Transcellular route**: The plasma membrane of the epithelial cell is impervious to the majority of hydrophilic solutes and requires transporters to permit passage of these. It also activates the signalling events between cells to control the TJ barrier and provides the driving force for paracellular transport.
2. **Paracellular route**: It consists of tight and subjacent adherens junctions, which together form the apical junctional complex.

   (a) **Tight junctions:**

   - Multi-protein complexes comprise proteins extending across the plasma and the peripheral membrane, namely claudin (CLDN), occludin transmembrane proteins in the plasma membrane, and zona occludens 1 and 2 in the plasma membrane. It also consists of kinases which are regulatory enzyme proteins.
   - Among these, CLDN is the most important of the transmembrane proteins that regulate the permeability of the TJs, which is tissue specific. It is expressed in an organ-specific manner (Table 3.6). Depending on the similarity of the sequence CLDNs are either classic or non-classic
     - Classic CLDN include CLDN 1–10, 14, 15, 17, and 19
     - Non-classic CLDN include 11, 12, 13, 16, 18, 29, and 24
   - Since TJs limit solute flux along the paracellular pathway, they regulate the rate-limiting step and permeability of the transepithelial membrane.

Table 3.6 Expression pattern of claudins in the human gastrointestinal tract

Part	Claudin expression pattern
Oesophagus	CLDN1, 4, 7 (CLDN12, 17)
Stomach	CLDN1, 2, 3, 4, 7, 8, 17, 18 (CLDN6, 9, 11, 23)
Small intestine	CLDN1, 2, 3, 4, 7, 8, 12, 15
Colon	CLDN1, 3, 4, 7, 8, 15, 23, (CLDN2, 9, 12, 14, 17, 20)

- In patients with IBD, those with neoplastic transformation, tumour growth, and metastasis, elevated CLDN-1 expression by intestinal epithelial cells occurs.
- CLDN expression is elevated in patients with colitis due to increased inflammatory interleukins IL-13 and IL-17.
- TNF-α increases para flux activity via the leak pathway resulting in failure of barrier integrity, which may be the underlying pathology in several conditions such as IBD, intestinal ischemia, and graft-versus-host disease.

(b) **Adherens junction**

- It is formed by a complex interaction of E-cadherin, catenin δ1, β-catenin, and α-catenin 1. It performs local actin set-up and peri-junctional actomyosin ring.
- Helps maintain cellular juxtaposition and facilitates intercellular communication.

Transport via TJ

(a) Leak pathway—it permits passage of large solutes, such as proteins and bacterial lipopolysaccharides, via paracellular transport. It does not display charge selectivity and relies on cytokine activation, IFNγ and TNF-α.
(b) Pore pathway—unlike the leak pathway, the pore pathway functions based on charge selectivity and is formed by TJ-associated CLDN proteins.

3.13.3 Diseases Associated with Increased Gut Permeability

1. **Gastroesophageal reflux disease (GERD):**
    - Non-erosive reflux disease (NERD), the most prevalent type, has shown dilated intercellular spaces towards the base of the stratified epithelial cell layers even in the absence of evident mucosal injury.
    - Basal impedance measurement, which measures oesophageal barrier function, is decreased in GERD.

2. **Barrett's oesophagus and oesophageal cancer:**
    - Barrett's oesophageal—increase of CLDN2 and CLDN3 and decrease of CLDN1.
    - Squamous cell carcinoma—CLDN1 immunohistochemical staining is enhanced.
    - Adenocarcinoma—CLDN3 and CLDN4 were increased.
    - **Eosinophilic esophagitis (EoE)**: Oesophageal intra-epithelial eosinophil infiltration forms a unique feature of EoE. A recent study demonstrated loss of mucosal integrity resulting in barrier dysfunction. Oesophageal mucosal impedance and dilated intercellular spaces (DIS) are indicators of active EoE. They are associated with CLDN1 down-regulation.
3. **Gastric cancer**: Changes in the phenotypes of CLDNs are noted during oncogenesis.
    - Two types of gastric CA: the expressions of CLDN1, 3, 4, 6, and 7 were increased in intestinal-type gastric cancer, and CLDN18 was generally down-regulated or sometimes absent.
    - CLDN1 was down-regulated in the metastasis of gastric cancer origin, showing increased cell–cell dissociation by a reduction in CLDN expression.
4. **Colon cancer**: Normal colonic mucosa has CLDN1, 3, 4, 7, 8, 15, and 23. TJ proteins are prognostic biomarkers of clinical significance.
    - CLDN1 gene is upregulated in colon cancer at all cell–cell boundaries aligned with the lateral membranes and in the cytoplasm of cancer cells, unlike the healthy ones in which these lie at apical borders of lateral membrane. Expression of the CLDN1 gene at the lateral membrane may mark colorectal tumorigenesis. It is a strong negative prognostic marker indicating a higher probability of recurrence and associated mortality.
    - CLDN18 (normally expressed in gastric mucosa)—loss of CLDN18 is associated with poor survival. It is also associated with serrated-pathway-associated adenocarcinoma of the colon.
5. **Irritable bowel syndrome (IBS)**: Abnormalities of intestinal permeability occur in some cases of FD (functional dyspepsia) and IBS.
    - Occludin and CLDN1 proteins were diminished in IBS, with diarrhoea as the predominant symptom (IBS-D).
    - Intestinal permeability is elevated in post-infectious IBS (PI-IBS) patients, which is attributed to the sustained loss of TJ integrity due to the acute infectious phase delaying structural recovery.
    - In IBS-D and IBS-A patients, colonic JAM-A expression was lessened, and it was indirectly proportional to the abdominal pain. IBS patients also demonstrate reduced expression of ZO-1 and E-cadherin.
6. **Inflammatory bowel disease (IBD)**: IBD is a chronic inflammatory condition of the intestine with an unidentified aetiology. Recent GWAS revealed that bar-

rier dysfunction of the intestinal membrane in ulcerative colitis (UC) may be associated with susceptible single-nucleotide polymorphisms in genes. These include HNF4, CDH1, and LAMB1.

- Expression of CLDN4 and CLDN7 was decreased, while CLDN2 was increased in patients with UC.
- JAM-A expression decreased in UC and Crohn's disease (CD).
- CLDN1 and CLDN2 may be elevated in UC and CD.

TNF-α increases CLDN2 expression; anti-TNF-α antibody alleviates inflammation, thereby reducing intestinal mucosal permeability. Low-dose sulfasalazine in experimental studies has demonstrated attenuation of TNF-α-induced TJ disruption.

3.14 Small Intestinal Permeability (SIP) and Its Assessment

Permeability is a membranal attribute that allows the conduct of a solute via diffusion without the need for transporters.

3.14.1 Factors That Determine Permeability

- Concentration gradient on either side of the membrane.
- Available surface area of the epithelium for diffusion.
- Available duration for permeation.
- The inherent properties of the barrier also affect permeability.

The critical rate-limiting factor in SIP is the structural and functional aspect of the TJs adjoining the small intestinal epithelial cells.

3.14.2 Measurement of Small Intestinal Permeability

Various chemicals can be used to measure SIP, called probes.

Properties of the ideal probe include the following:

- Small, water soluble
- Noncharged compound
- Not destroyed in the gut
- Nontoxic
- Cheap and easily available
- Does not undergo biotransformation or complex after absorption and is eliminated by the urinary system

- Identifiable in urine and selectively differentiated from useful components and removed

Examples include sucrose, mannitol, lactulose, and sucralose. They are absorbed from various parts of the gut and are used to test the permeability of these specific regions.

3.14.3 Principles

- Aqueous pores are distributed in the villi and crypts of the small intestine. Pores at the tips of villi allow only small molecules (e.g. mannitol) to pass through. Pores in crypts selectively allow larger molecules (e.g. lactulose) to diffuse, while those in the base of villi allow intermediate size molecules to diffuse.
- Since enterocytes mature and progress from the crypt to the tip of villi, measuring the permeation of small molecules that pass through pores at the tip of villi (e.g. mannitol) is an indirect measurement of mature intestinal absorptive surface area.
- SIP is expressed as a fraction of large-to-small molecule excretion (e.g. lactulose: mannitol). In diseases associated with villous atrophy like CD or tropical sprue, there is a considerable decrement in the fractional excretion of smaller molecules.

3.14.4 Methods of Measurement

- High-performance liquid chromatography (HPLC)
- Nuclear magnetic resonance spectroscopy

3.14.5 Clinical Relevance

Abnormal small intestinal permeability is found in many diseases:

- **Diabetes**
- **Crohn's disease**: in ileocolonic disease, higher lactulose–mannitol ratio is observed than in those with only colonic disease; 10–20% of otherwise healthy first-degree relatives of CD patients have elevated IP. Interestingly, elevated IP is associated with a higher relapse rate.
- **Celiac disease**: more than one-third of otherwise healthy first-degree relatives of patients with celiac disease have membrane permeability abnormalities.
- **Alcohol**: ethanol and acetaldehyde alter the signal-transduction pathways within cells by

  - Causing disorganisation of the epithelial TJs
  - Increasing paracellular transport of macromolecules

- **Cirrhosis**: the intestinal permeability index, which is the proportion of polyethene glycol (PEG) 3350 to PEG 400, is elevated in patients with both cirrhosis and concomitant ascites those with cirrhosis alone.
- **Irritable bowel disease**: differentiation between organic and functional diarrhoea.
- **NSAID use**

3.15 Mucosal Defence

The mucosal defence comprises innate and adaptive immunity, the components of which are shown in Table 3.7.

Table 3.7 Components of mucosal defence

Innate immunity		Adaptive immunity
Physicochemical • Mucous • Tight junction • Epithelial membranes • Brush border enzymes • Bile salts • pH ranges • Trefoil factors	Cellular • NK cells • Macrophages • Polymorphonuclear cells • Epithelial cells • Pattern recognition receptors (TLRs)	Adaptive Immunity • Intra-epithelial lymphocytes • Regulatory cytokines • IgA • Peyer's patches • Epithelial cells/antigen presentation

3.16 Immunoglobulin A

Immunoglobulin A (IgA), in its secretory form, is a crucial initial defence against pathogens in the GI tract. It is present in saliva, tears, colostrums, and GI secretions.

Immunoglobulin A exists in two isotypes:

1. IgA1: 90% found in serum and formed by bone marrow B-cells. IgA is produced by plasma cells in its polymeric form. The J chain links two monomers.
2. IgA2: 10% found in tears, colostrums, saliva; formed by mucosa-associated B-cells. Secretory IgA produced from mucosal B-cells contains an additional secretory component synthesised by enterocyte epithelial cells.

3.16.1 Structure (Fig. 3.4)

- Two heavy chains—alfa 1 and 2.
- Two additional chains—J chain and secretory component. J chain joins two molecules of IgA, and the secretory component is secreted by epithelial cells.

3.16.2 Functions

1. Performs immune exclusion by disrupting the interaction between the pathogen and mucosal epithelial cells.
2. Intracellular neutralisation and virus excretion.
3. In the serum, it acts as a defence against pathogens that have escaped the initial line of defence in the epithelial mucosa by eliminating them.

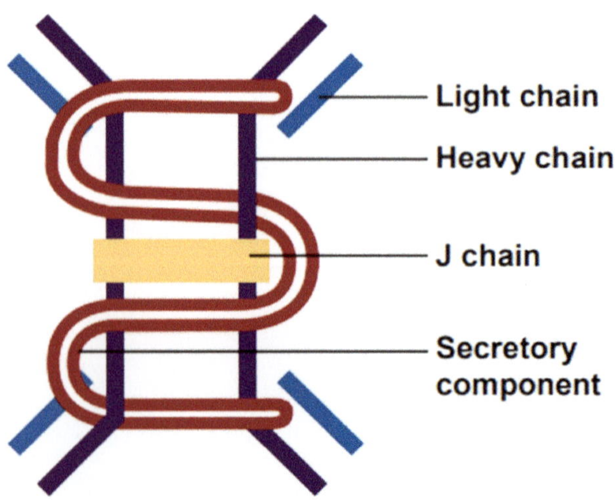

Fig. 3.4 Structure of IgA

3.16.3 Clinical Significance

Immunoglobulin A Deficiency

It is seen in 1 in 600 Europeans. Over the years, it can progress to common variable immunodeficiency and vice versa. Some parts of the population with IgA deficiency may seem healthy, while some may experience recurrent respiratory or GIT infections. These patients primarily suffer from atopic conditions like asthma or dermatitis. Symptoms guide the treatment. Since the use of exogenous IgA is not effective and, instead, may increase the formation of IgA antibodies, these are not used in clinical practice.

Immunoglobulin A and Celiac Disease

Of celiac disease patients, 2% have IgA deficiency and a 10–16-fold increased incidence of CD is seen in patients with IgA deficiency. In these patients, IgA-based serological tests may be falsely negative. If clinical suspicion is high, then IgG antigliadin antibodies should be done in such cases.

3.17 Defensins

- Defensins are small cysteine-rich cationic proteins of the innate immune system. These are secreted by epithelial cells and granulocytes with activity against all pathogens.
- Defensins bind to the membrane of pathogens and form pores, which cause the efflux of essential nutrients and ions, resulting in cell death.
- θ-Defensins, α-defensins, and β-defensins are identified as mammalian forms (Table 3.8). Their peculiar characteristics include small size with a high density of cationic charge and a six-cysteine-residue motif. Generally, they are encoded by two-exon genes.

3.17.1 Clinical Importance

Altered defensin production is an important feature of IBD. Patients with CD in whom the ileum is affected show reduction in α and β defensins. Reduced levels of defensin are probably due to the destruction of enterocytes secondary to the

Table 3.8 Expression of defensins

Defensins	Description
α-Defensins	Expresses in neutrophils, Paneth cells, and NK cells
β-Defensins	Leucocytes secrete the most common defensin, many epithelial cells, including the tongue, cornea, skin, respiratory tract, oesophagus, and kidney
θ-Defensins	Not seen in humans

NK natural killer

inflammatory process. Controversy exists regarding whether reduced defensin level in IBD is the cause or effect of IBD.

3.18 Enterokinase

Enterokinase belongs to the serine protease group of brush border enzymes. Crypts of Lieberkuhn and duodenal glands secrete it. An important function of enterokinase is the conversion of trypsinogen to active trypsin. Enterokinase is coded by the *PRSS-7* gene (Fig. 3.4).

3.18.1 Application

The high specificity of enterokinase enables the split of the fusion proteins at specific cleavage sites. This helps obtain the protein target.

3.18.2 Enterokinase Deficiency

Of infants (undergoing evaluation for suspected pancreatic insufficiency), 1–2% is observed with congenital enterokinase deficiency. The patients affected are seen with hypoproteinaemia, severe growth retardation, and malabsorption.

3.19 Cholecystokinin

Cholecystokinin (CCK) (pancreozymin) is a peptide hormone produced by I cells in the jejunum and duodenum. It plays an important role in the digestion of protein and fat.

Cholecystokinin is cleaved during post-transcriptional modification of the *CCK* gene product, pre-pro-cholecystokinin. Depending on the number of amino acids, cholecystokinin is identified as CCK8, CCK58, etc. CCK, similar to gastrin, bears the same five amino acids at its C-terminal.

3.19.1 Functions

Cholecystokinin receptors are divided into the CCKA receptor (CCKAR) and CCKB receptor (CCKBR).

The CCKA receptor mediates the action of CCK on:

1. secretion of pancreatic amylase

2. contraction of the gallbladder
3. inhibiting gastric emptying

Cholecystokinin B (CCKB) receptor activity is associated with

1. anxiety
2. increased neuronal firing
3. nociception

3.19.2 Clinical Significance

- In clinical practice, CCK and secretin are used for testing the pancreas since they stimulate its secretions.
- It is also used scintigraphically or radiographically to determine the contracting capacity of the gallbladder.
- Lower levels of CCK occur in patients with bulimia nervosa and CD, while these are elevated in a few of those with chronic pancreatitis due to interference in its negative feedback mechanism.

3.20 Secretin

When duodenal luminal pH falls below 4, the S-cells secrete a peptide hormone called secretin that has structural similarities to VIP, GIP, and glucagon.

3.20.1 Functions

1. Bile secretion from the liver is stimulated.
2. It increases watery bicarbonate solution from the pancreatic duct and duodenal Brunner's gland. It promotes the normal structure and growth of the pancreas.
3. Improves the effects of CCK, which results in the induction of bile secretion, amylase, and lipase.
4. Stabilises blood glucose concentration surges by stimulating insulin release.
5. Acid secretion is reduced in the stomach due to the inhibition of gastrin release.
6. Pepsin secretion gets stimulated by secretin. Glucagon release is stimulated along with somatostatin and pancreatic polypeptide.

3.20.2 Clinical Uses

1. **Pancreatic secretory functions**: Measurements of volume and bicarbonate secretion after intravenous secretin injection are used to measure pancreatic secretory function directly.

2. **Zollinger–Ellison syndrome**: Secretin stimulation test for secretin helps to recognise ZES when fasting serum gastrin levels are not severely elevated.
3. **Secretin-enhanced MRCP**: Secretin administration stimulates bicarbonate and fluid secretion by the exocrine pancreas. Evaluation of the pancreatic duct anatomy gets improved along with side-branch delineation.

3.21 Incretins

Incretins are hormones in the GI that increase the amount of insulin released from the pancreas after a meal. These are

1. Glucose-dependent insulinotropic peptide (GIP)
2. Glucagon-like peptide 1 (GLP1)

3.21.1 Sources

Released from neuroendocrine cells of the GI tract, e.g. GLP1 is released by L-cells of the small intestine. Food in the intestinal lumen stimulates the secretion of incretins. It has a short half-life. It is inactivated by the enzyme dipeptidyl peptidase.

3.21.2 Actions

- Augmentation of insulin secretion
- Increase beta cell mass
- Glucagon inhibition
- Delay gastric emptying

3.21.3 Clinical Use

Liraglutide, semaglutide and exenatide, and GLP1 agonists are used for the treatment of diabetes mellitus.

3.22 Vitamin B12 Absorption

3.22.1 Sources

Cobalamin exists in three forms: methylcobalamin, hydroxocobalamin, and adenosylcobalamin. Important sources are the liver, beef, kidney, fish, milk, and eggs; the typical daily requirement is 1–2 μg.

Vitamin B12 may be absorbed actively or passively.

3.22.2 Physiology

- B12 combines rapidly with haptocorrins which is a salivary glycoprotein. Pancreatic trypsin processes the haptocorrin in the intestine, and the released cobalamin is transferred to the intrinsic factor (IF).
- Intrinsic factors released from parietal cells bind to cobalamin. Glycosylation resists proteolysis by pancreatic enzymes and allows passage of the complex to the terminal ileum. IF–cobalamin complex binds to specific (cubilin–amnionless complex) ileal receptors. After binding, cobalamin enters the cell intact by translocation. Cobalamin binds to transcobalamin II within enterocytes and is transported to portal circulation.

Cobalamin deficiency causes the following (Table 3.9):

Clinical signs and symptoms of deficiency involve the GI tract, the blood, and the nervous system.

- Megaloblastic anaemia.
- Anorexia, glossitis, discolouration over knuckles, and paraesthesia.
- Demyelination, axonal degeneration and neuronal death occur successively. Sites of involvement include the spinal cord, cerebrum, and peripheral nerves. The Romberg and Babinski's signs, which reflect abnormalities in neurological functions regulated by the corticospinal tract, may be positive, and position and vibration senses decline.
- Psychological abnormalities range from slight irritability and forgetfulness to extreme psychosis or dementia. Such impairment may be noted in patients in the absence of abnormal haematocrit and RBC indices.

3.22.3 Diagnosis of Cobalamin Deficiency

Peripheral Blood Film

The peripheral blood film reveals pronounced anisocytosis, poikilocytosis, macroovalocytes, basophilic stippling, hypersegmented neutrophils, and an occasional nucleated RBC. MCV will be high (>100 fl). The reticulocyte index is poor; leukopenia and thrombocytopenia may occur in patients with severe anaemia.

Table 3.9 Causes of vitamin B12 deficiency

A. Malabsorption				
A. Defective release of cobalamin from food 1. Gastric achlorhydria 2. Partial gastrectomy 3. Drugs that block acid secretion	**B. Inadequate production of IF.** 1. Pernicious anaemia 2. Total gastrectomy 3. Congenital absence or functional abnormality of IF (rare)	**C. Disorders of the terminal ileum** 1. Tropical sprue 2. Nontropical sprue 3. Regional enteritis 4. Intestinal resection 5. Neoplasms and granulomatous disorders (rare) 6. Selective cobalamin malabsorption (Imerslund's syndrome) (rare)	**D. Competition for cabalamin** 1. Fish tapeworm (diphyllobothriumlarum) 2. Bacteria: 'Blind loop' syndrome	**E. Drugs: p-aminosalicylic acid, colchicine, neomycin**
B. Inadequate intake: vegans (rare)				

IF intrinsic factor

Serum B12 Levels

Normally the serum cobalamin lies between 300 and 900 pg/mL. A value of 200 pg/mL indicates a significant deficiency, which needs clinical attention. However, serum levels will be normal even when stores are depleted. In cobalamin deficiency, serum concentrations of methylmalonic acid and total homocysteine are elevated. Levels of these metabolites tend to normalize within 1–2 weeks after the institution of replacement therapy, and some authors have suggested that their measurement can be used to distinguish between cobalamin and folate-deficiency states.

3.22.4 Schilling Test

- It helps in differentiating causes of vitamin B12 deficiency between intrinsic factors and the malabsorption in the ileum.
- Results of the test are not influenced by vitamin B12 replacement therapy.

Principle

To saturate vitamin B12 stores with intramuscular dose; subsequent, small oral radiolabelled vitamin B12 dose is given, and its excretion in urine is measured. If less than 7–10% of the administered dose is excreted in urine, it suggests vitamin B12 malabsorption.

- If vitamin B12 absorption can be normalised by oral IF administration, this suggests gastric cause (pernicious anaemia).
- If the absorption normalised after administration of pancreatic enzymes, it suggests pancreatic exocrine insufficiency.
- If vitamin B12 absorption values improve to normal after a 5-day course of tetracycline, it suggests blind loop syndrome.
- If vitamin B12 malabsorption persists to be abnormal despite the administration of IF and pancreatic enzymes, it suggests an ileal cause.

3.22.5 Treatment

- Correct underlying cause.
- Restoration of body stores should occur with six 1000-μg i.m. injections of hydroxocobalamin given at 3 days to a week's gap.
- For maintenance therapy, 1000-μg hydroxocobalamin i.m. once every 12 weeks is adequate.

3.23 Iron Absorption and Metabolism

- In humans, iron in haemoglobin acts as the oxygen carrier and is an important constituent of several enzymes in the mitochondrial cytochrome system.
- People consuming meat have approximately 20–30 mg of iron intake per day, predominantly in organic form as myoglobin or haemoglobin. However, iron in wheat and vegetables is inorganic and less readily available for absorption. Iron is present in a combined form as lactoferrin is in breast milk.
- More than 0.5 mg/day is required for growing children and adolescents to develop reserve iron stores to adult values. Females need 1 mg/day during reproductive life to compensate for menstrual losses.

3.23.1 Iron Absorption and Transport

- Nearly all dietary iron is absorbed in the duodenum, predominantly in the ferric form.
- Fe^{3+} is highly insoluble under physiologic conditions; the uptake requires conversion to Fe^{2+} via the enzyme ferrireductase (Fig. 3.5).

3.23.2 Iron Transport Across the Apical Membrane of Enterocytes

- Three significant iron pathways transport across the apical membrane are via divalent metal transporter 1 (DMT1), iron-responsive element (DMT1 IRE) mRNA, and non-IRE designated DMT1 (non-IRE) mRNA. The DMT1 mRNA

Fig. 3.5 Iron transport across the enterocyte

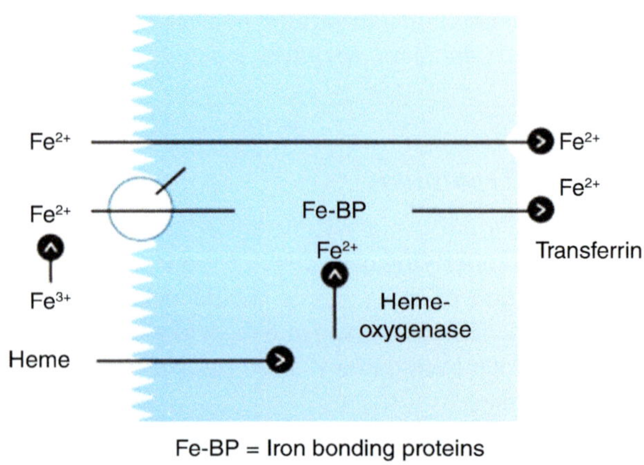

and the protein are predominantly found in the proximal duodenum, with a progressive decrease towards the distal axis. Hence, primarily, significant absorption occurs at the proximal segment.
- Iron in haeme form is readily absorbed across the brush border of the enterocytes, which is subsequently lysed by haeme oxygenase and the iron is released.
- Minute quantities of iron also diffuse via the passive paracellular diffusion route into the cells.

3.23.3 Iron Transport Across the Basolateral Membrane of Enterocytes

- Transporters or enzyme proteins are used for translocating iron across the basolateral plasma membrane of villus enterocytes, namely: ferroportin 1 and an enzyme ferroxidase, also known as hephaestin.

3.23.4 Regulation of Iron Stores

When there is iron deficiency, a low level of hepcidin in enterocytes stimulates ferroportin activity and increases transferring receptor production.

The causes of iron deficiency are provided in Table 3.10. The causes of iron overload are provided in Table 3.11.

Table 3.10 Causes of iron-deficiency anaemia

Chronic blood loss	Decreased iron intake	Decreased iron absorption	Increased demand
• Gastrointestinal—oesophagitis, peptic ulcer, malignancy, IBD, hookworm infestation • Genitourinary—heavy menstrual bleeding, postpartum haemorrhage, intravascular haemolysis • Systemic—hereditary haemorrhagic telangiectasia, eating disorder, chronic schistosomiasis	• Improperly balanced vegan or vegetarian diet • Iron-poor diet • Eating disorder	• Coeliac disease • Atrophic gastritis • Gastric and/or intestinal bypass • IBD • *Helicobacter pylori* colonisation • Concomitant drug use, e.g. proton pump inhibitors • Increased demand	• Pregnancy • Infancy and adolescence • Menstrual loss • Blood donation • Endurance sports

Adapted from: https://doi.org/10.1111/tog.12368

Table 3.11 Causes of iron overload

Chronic liver disease
Iron-overloading anaemia—thalassemia major, sideroblastic anaemia, haemolytic anaemia
HFE-related hereditary hemochromatosis Hepcidin mutations
Parental iron overload
Transferrin receptor mutation

3.24 Copper Metabolism

- Copper is an integral component of key enzymes like ferroxidase, superoxide dismutase, dopamine hydroxylase, and transport proteins like ferroprotein. The daily requirement is 1 mg/day.
- Liver, shellfish, nuts, bran, and legumes are rich sources of copper.

3.24.1 Absorption of Copper

- The precise mechanisms of copper absorption are not clear. Absorption probably takes place by active transport.
- There is a competition between zinc and copper for intestinal absorption.

3.24.2 Copper Transport

- Copper is absorbed into enterocytes by endocytosis mediated by human ctrl 1–190 amino acid protein receptor.
- Zinc superoxide dismutase, Atox1, CCS (copper chaperone for Cu^{2+}), Cox 17, and membrane-associated copper-transporting ATPases (ATP7A and ATP7B) are various chaperones involved in copper metabolism.
- Copper, after crossing the basolateral membrane, is bound to albumin and transferred to the liver.
- Ceruloplasmin, a major copper-containing protein, is synthesised in the liver. It can bind six copper molecules.
- Incorporation into ceruloplasmin and excretion into bile is mediated by the protein encoded by *ATP 7B* gene (13q chromosome).

3.24.3 Role of Copper

- Important for iron metabolism
- Synthesis of melanin
- Production of energy

- Synthesis of neurotransmitter
- Synthesis and cross-linking of elastin and collagen fibres
- Eliminating superoxide radicals

3.24.4 Copper Deficiency: Causes and Diagnosis

- Copper-deficiency anaemia occurs in patients with diseases causing malabsorption and nephrotic syndrome. It also is reported in patients treated with high oral zinc doses for Wilson's disease for a prolonged duration since zinc interferes with copper absorption.
- Low serum levels of copper (<65 g/dL) and low ceruloplasmin levels (<20 mg/dL) indicate copper deficiency. Serum levels of copper may be physiologically increased during pregnancy. Stress also increases serum copper levels since ceruloplasmin which binds 90% of the copper, is an acute phase reactant.

3.24.5 Copper Toxicity

Accidental ingestion of excessive copper leads to toxicity, characterised by acute liver failure, renal failure, and coma.

3.24.6 Disease Associated with Copper Metabolism

1. Menkes kinky hair syndrome
    - X-linked metabolic disturbance of copper metabolism
    - Caused by mutations in the copper-transporting *ATP7A* gene characterised by mental retardation, hypocupraemia
    - Infants usually die within 5 years of age because of cardiovascular complications
2. Aceruloplasminemia
    - Rare autosomal recessive disease
    - The gene for ceruloplasmin is localised at chromosome 3
    - Iron overload, mental decline, and microcytic anaemia are important manifestations
    - Poor serum iron and copper concentrations
3. Wilson's disease
    - Autosomal recessive disorder with defect at *ATP 7B* gene located at 13q14
    - Characterised by a failure of copper to incorporate into ceruloplasmin and failure in biliary excretion

- Excessive copper deposition in the liver, lens, and cornea of the eyes and brain
- Present with hepatocellular dysfunction, neurological symptoms, or haemolysis

3.25 Vitamin D Metabolism

Vitamin D can be synthesised de novo from the skin by converting 7-dehydrocholesterol by U–V light on sun exposure. Vitamin D ingested from food sources is absorbed in the intestine and undergoes 25-α-hydroxylation in the liver. 25 (OH) cholecalciferol is the most important reserve form of vitamin D. Subsequent 1-α hydroxylation is done in the kidneys. 1,25-Dihydroxy cholecalciferol is the active form of vitamin D. Vitamin D is inactivated by 24-hydroxylation, which is abundant in most tissues. Polar metabolites of 1,25-dihydroxy cholecalciferol are released in bile and subsequently undergo reabsorption in the ileum, thus undergoing enterohepatic circulation.

3.25.1 *Role of Vitamin D*

- Vitamin D and its effect on bone metabolism: 1,25-dihydroxy cholecalciferol improves the absorption of calcium and phosphorus from the intestine. This helps in the mineralisation of the bone matrix. 1,25 $(OH)_2$ vitamin D downregulates parathormone synthesis. Rickets is a disease of childhood caused by vitamin D deficiency characterised by soft deformed bones due to poor mineralisation of bones. In adults, osteomalacia is seen due to poor mineralisation of the bone matrix.
- The physiological effects of vitamin D are mediated through its receptors (VDR) which are distributed across almost all tissues. Vitamin D receptors control the genetic expression of many molecules. Vitamin D is known to express its effect on immune cells and activates innate immunity. It is known to regulate the renin–angiotensin system. It has antiproliferative action on cancer cells.

3.25.2 *Diagnosis of Vitamin D Deficiency*

- Low levels of vitamin D stored: serum 25(OH) vitamin D level below 15 ng/mL.
- Low calcium and phosphorous levels: low serum calcium levels activate parathyroid hormone production. Raised PTH levels increase the reabsorption of calcium from renal tubules in exchange for renal excretion of phosphorus.
- Increased serum levels of 1α hydroxylase activity: raised PTH hormone increases 1α hydroxylase activity. So, 1,25 $(OH)_2$ vitamin D levels may paradoxically be normal in a vitamin D-deficient state.

- Radiological manifestation of vitamin D deficiency: widened, expanded growth plates in the costochondral junction (rachitic rosary); decreased cortical density of long bones; pseudo-fractures due to poor mineralisation (Looser's zone); delayed fusion of skull calvaria bones.

3.25.3 Treatment

- If vitamin D deficiency, weekly 60,000 IU of cholecalciferol for 8 weeks and subsequently every 4 weeks.
- Daily recommended intake: 400–600 IU/day.

3.25.4 Sources of Vitamin D

- Fifteen minutes of exposure to sunlight (10 am–3 pm): over face, arms, shoulder and upper back. Food sources: cod liver oil, egg yolk, liver, fish (sardines, salmons), milk, and cereal.

3.26 Cholesterol Metabolism

Cholesterol is an important component of the cell membrane due to its combined polar and non-polar constituents, namely lipoproteins. Both the cholesterol obtained from dietary sources and synthesized de novo are circulated in combination with lipoproteins.

3.26.1 Sources

Our daily cholesterol consumption from dietary sources such as eggs or meat lies between 200 and 250 mg.

3.26.2 Cholesterol Synthesis

- Less than 50% of total cholesterol pool is synthesized in the body and the remaining are obtained from diet or recirculation.
- Cholesterol synthesis occurs in the microsomes (ER) and cytoplasm from the two-carbon acetate group of acetyl-CoA.

  In otherwise healthy adults, 1 g/day of cholesterol is synthesised.
  The process of cholesterol synthesis is depicted in Fig. 3.6.

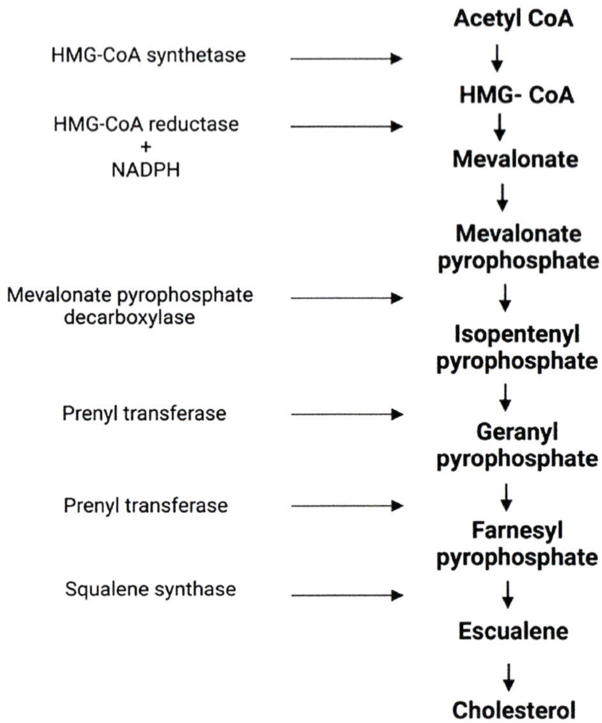

Fig. 3.6 Synthesis of cholesterol

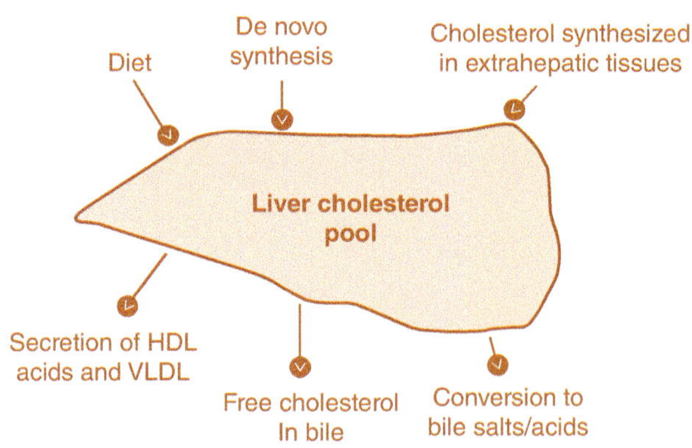

Fig. 3.7 Cholesterol metabolism. *HDL* high-density lipoprotein, *VLDL* very low-density lipoprotein

3.26.3 Regulation of Cholesterol Synthesis

- Conversion of acetyl Co A to 3-hydroxy-3-methylglutaryl Co-A reductase is the rate-limiting step.
- The excess intracellular free cholesterol is converted either to cholesteryl esters by the Acyl-CoA: cholesterol acyltransferases (ACAT) or regulated through HDL-mediated reverse transport in the periphery and LDL receptor-mediated uptake (Fig. 3.7).

3.26.4 *Role of Cholesterol*

- It is an antecedent moiety for several other steroidal components synthesised physiologically, e.g. bile acids, certain hormones, steroids, and vitamin D.
- It is an important plasma membrane component due to its amphipathic nature.

3.27 Van de Kamer Test

The Ven de Kramer test, first described in 1949, estimates the fat malabsorption in the faeces expressed as grams of faecal fat per 24 h. Although tedious, it is still the gold standard for quantitative fat estimation in the faeces since it estimates all the faecal fat content.

3.27.1 *Principle*

The test relies on the saponification of wet faeces using alcoholic sodium or potassium hydroxide, which releases fatty acids that are titrated against a base like sodium hydroxide.

3.27.2 *Technique*

The patient is given a 75–100 g fat-rich diet for 3 days before the start of the test, followed by 72-h stool collection while on a 100-g fat-rich diet. The 72-h collection is performed to ensure day-to-day variations do not affect the analysis results.

3.27.3 *Precautions*

- No barium study before the procedure
- No castor oil intake
- Low-fat diet can give a false negative result

3.27.4 *Result*

- After a diet containing 100 g/day of fat, faecal fat excretion of <7 g/day is normal

- In case of secondary fat malabsorption, this may rise to 14 g/day, for instance, if the patient suffers from diarrhoea
- In CD, it remains less than 9.5% while increases beyond it in chronic pancreatic insufficiency

3.27.5 Reasons for the Limited Clinical Use

1. Cannot differentiate pancreatic from enteric causes of steatorrhea reliably
2. Normal faecal fat excretion does not rule out malabsorption
3. Diet diary to check adherence to an 80–100 g diet is necessary to accurately judge compliance with a fatty diet
4. The test is routinely available at limited centres

3.27.6 Clinical Application

1. To establish malabsorption when conspicuous and conventional characteristics of intestinal or pancreatic disorders such as osteoporosis, osteomalacia, anaemia, and weight loss are not evident.
2. Monitoring treatment in an established malabsorptive disorder patient, such as short bowel syndrome (SBS) or exocrine pancreatic insufficiency.

3.28 Breath Tests for Malabsorption

3.28.1 Carbohydrates Malabsorption

Carbohydrates present in food sources are either in the form of polysaccharides or disaccharides. To be sufficiently absorbed, enzymatic digestion absorption is required. The unabsorbed part of the colon is fermented by bacteria. Tests for carbohydrate malabsorption are:

(a) Lactose breath test
(b) Fructose breath test

Lactose Breath Test

Principle

- Lactose, a disaccharide composed of galactose and glucose, is obtained from milk and its products.
- It is assimilated by the intestinal brush border enzyme.

- If lactase is deficient, lactose cannot be absorbed and fermented in the colon, which may cause a rise in fluid secretion and hydrogen gas.

Technique

- A lactose solution of concentration between 5% and 25% to lactose (25–50 g in 100–500 mL of water).
- Determine hydrogen gas concentration in the end-expiratory air at 15–30 min intervals for 4–5 h of test.
- An elevation of 10–20 ppm above the baseline at the end of expiratory air in more than two samples suggests lactose malabsorption. Based on the peak duration after ingestion, an early peak indicates rapid small bowel transit, small intestinal bacterial overgrowth (SIBO), or substrate fermentation by the oral flora, while a peak between 2 and 4 h suggests lactose malabsorption.

Fructose Breath Test

- Monosaccharide is naturally available in fruits and also as a synthetic sweetener.
- Fructose malabsorption has been frequently recognised as a cause of GI manifestations.

Technique

Although the exact dosage of fructose solution in the breath test is not defined, a recent study found a 25-g dose to be a reliable indicator of fructose malabsorption. It was validated based on the hydrogen and methane estimation and a symptom questionnaire. A rise of 20 ppm for hydrogen and 5 ppm for methane over the baseline in consecutive samples is highly suggestive of malabsorption.

3.28.2 Breath Test for Fat Malabsorption

^{14}C-Triolein Breath Test

- Triglycerides are absorbed from the proximal portion of the jejunum after they undergo hydrolysis by pancreatic enzymes and the formation of micelle.
- On administration of ^{14}C-labelled triolein (TG), in combination with 20–30 g of fat, its normal metabolism releases water and carbon dioxide which is detectable in the breath.

  A decrease in this ^{14}C-carbon dioxide is suggestive of fat malabsorption.

Drawbacks

1. Concomitant metabolic and pulmonary disease may confound the results.
2. Not commonly used in clinical practice due to false positives in patients with delayed gastric emptying and healthy individuals (20% in IBS).
3. Poor sensitivity hence mild cases may not be detected.
4. Requires use of radiation.
5. Technically expensive requirements (both the substrate and the equipment).

3.29 Breath Test for Small Intestinal Bacterial Overgrowth

The clinical manifestations of small intestinal bacterial overgrowth (SIBO) include diarrhoea, bloating, and infrequent signs of malabsorption. To classify as SIBO, at least 10^5 colony-forming units (CFU)/mL of bacteria should be found in intestinal aspirates. The bacterial flora in SIBO resembles that found normally. Commonly found bacteria include both aerobic and anaerobic ones: *Streptococcus*, *E. coli*, *Staphylococcus*, *Micrococcus*, *Klebsiella*, *Proteus*, *Lactobacillus*, *Bacteroides*, *Clostridium*, *Veillonella*, *Fusobacterium*, and *Peptostreptococcus*.

3.29.1 Breath Tests in Small Intestinal Bacterial Overgrowth

1. Glucose hydrogen breath test.
2. Lactulose breath test.
3. ^{14}C-xylose and ^{13}C-xylose breath tests.

Principles of Breath Test

- Biotransformation of carbohydrates by bacteria, being the sole source of hydrogen, is the basis of this test.
- In the presence of carbohydrate malabsorption, excess carbohydrates will be fermented by the bacteria, which leads to the generation and release of hydrogen in the expired breath that can be determined.
- In cases with SIBO, the bacterial population translocates to the proximal part of the small intestine, thus gaining access to the non-metabolised carbohydrates.

Patient Preparation and Procedure

- Not on antibiotics for 4 weeks
- Low roughage diet on day priors
- Stop eating 12 h before the test

Precautions

Avoid the following:

- Smoking (false positive)
- Poor oral hygiene—antibacterial mouthwash
- Exercise (false negative)
- High-fibre diet

Procedure

- Glucose: 50–75 g or lactulose 10 g or lactose 25–50 g dissolved in water.
- Determine H_2 in the end-expiratory breath samples every 15 min for 3 h.

Problems with Breath Tests

1. Of the population, 15% are methane producers who are colonized with *Methanobrevibacter smithii*; In this subset of the population, the hydrogen is consumed to form methane and, therefore, detectable hydrogen is decreased compared to the methane-non-producer population.
2. The intestinal transit time interferes with the accuracy of the results.
3. Consumption of non-absorbable forms of carbohydrates such as lactulose produces an acidic milieu in the intestine, attenuating the metabolism by the bacteria.
4. Diet, smoking, and exercise just before the tests affect the baseline hydrogen level in the breath of the patients.

3.29.2 Glucose Hydrogen Breath Test (GHBT)

- Glucose is a substrate with the most desirable attributes since it is easily absorbed in the proximal small bowel, seldom reaching the colon without severe transit abnormalities.
- Generally, 50–75 g of glucose is used, and breath samples are obtained for 2–3 h.
- An increment of more than 20 ppm of hydrogen or methane over the baseline 12–15 ppm breath after glucose ingestion indicates SIBO.
- False-positive tests occur only if the intestinal time is extremely short to allow glucose absorption prior to reaching the colon.
- Sensitivity and specificity of GHBT in SIBO are 45% and 80%, respectively.

3.29.3 Lactulose Hydrogen Breath Test (LHBT)

- Lactulose is typically metabolised by bacteria in the proximal colon, producing a delayed peak in exhaled hydrogen.
- However, in patients with SIBO, the hydrogen peak occurs earlier due to metabolism in the small intestine.
- Generally, 10 g of lactulose is used with a breath sampling duration of 2- to 4-h.
- Display of two peaks, the first representing the lactulose metabolism in the small intestine and the second in the caecum, indicating SIBO. An increment of 20 ppm in the early peak suggests SIBO.
- The sensitivity and the specificity of LHBT are 68% and 44%, respectively.

3.29.4 ^{14}C-Xylose and ^{13}C-Xylose Breath Tests

- Breath test measures ^{14}C-carbon dioxide produced by bacteria's breakdown of labelled substrates. Normally D-xylose is wholly absorbed in the small intestine and rarely metabolised. Only Gram-negative bacteria catabolise it making it a relevant substrate for breath testing for SIBO.
- Isotopes used are radioactive (^{14}C) or stable (^{13}C), in children stable (^{13}C) is used.
- The dose of substrate D-xylose used in the breath test is 1 g.
- An excess of 4.5% over the administered radioactive substrate at 4 h indicates SIBO positivity.
- The sensitivity ranges from 42% to 100%, and specificity rates between 40% and 100%.

3.29.5 Bile Acid Breath Test for Small Intestinal Bacterial Overgrowth

Principles

- Glycocholic acid breath test involves using the bile acid ^{14}C glycocholic acid.
- Bacterial deconjugation results in ^{14}C glycine absorbed through the ileal mucosa, and $^{14}CO_2$ is expelled into the air.
- In bacterial overgrowth, there will be elevated levels of $^{14}CO_2$.

3.30 Intestinal Adaptation

- Intestine is intrinsically capable of adaptation to different environmental stimuli by undergoing structural and functional modifications.

- Intestinal adaptation may occur due to fasting, partial loss of bowel length (short bowel syndrome), sublethal doses of abdominal exposure to radiation, metabolic diseases like diabetes, advanced age, and malnutrition.
- The changes in intestinal adaptation correlate with the length of the resected intestine, location, and dietary lipid composition.
- It balances nutrient absorption by increasing the efficiency in available surface area, thus reducing malabsorption and enhancing the patient's survival.
- Nonetheless, such adaptation may have detrimental effects on glycaemic control in patients with metabolic disorders since it increases the uptake of nutrients.

3.30.1 Definition

An adaptive process involves enhanced fluid and nutrient absorption to compensate for the partial loss of the small intestine, which usually occurs as (1) mucosal hypertrophy increasing the surface area for absorption, and (2) an increase in specific transport processes in the presence of particular dietary components or requirements.

3.30.2 Physiologic Changes

- Mucosal hypertrophy results in more significant numbers of villus enterocytes and their height, resulting in a five-fold increase in absorptive potential.
- Jejunal adaptation following ileal resection appears to be less efficient and vice versa.
- Intestinal resection also influences the passive permeability characteristics of the brush border mucosa.

3.30.3 Mechanism

- Signals of adaptation may correspond to changes in hormone levels, factors needed for transcription, molecular energy levels, or alterations in the concentration of substances in the intestinal lumen.
- The manner of adaptive changes that occur vary with the particular section of the small intestine up to the cellular level, for example, between the villous crypts and enterocytes.

3.30.4 Changes in Dietary Intake Influencing Mucosal Digestive and Absorptive Processes

- The enzymatic activities of sucrase and maltase increase in proportion to carbohydrate consumption, but enzymes metabolising protein metabolism remain unaffected.
- Sucrase levels initially are elevated in the crypt cells approximately 24 h after refeeding sucrose succeeded by a starvation phase. The production of disaccharidases is triggered, and their lysis is also reduced.
- Lactase levels remain unaffected by any modification of lactose consumption through food.
- Depending on the dietary levels of vitamins and trace elements, the relevant absorptive mechanisms either increase or decrease, increasing when consumption is low and vice versa. Signals for intestinal adaptation are as follows:
  - Luminal nutrition
  - Pancreaticobiliary secretions
  - Gut hormones like entero-glucagon and GLP
  - Gastrin and CCK display trophic effects on the GI tract
  - The mucosal hypertrophy after intestinal resection occurs due to local polyamine synthesis and release
  - Other signals described are growth hormone (GH), EGF, EGF receptor, prostanoids, peroxisome proliferator-activated receptor-α (PPAR-α), IGF-1 receptors, and IGF-binding proteins

Signal Molecules in Intestinal Adaption

- Glutamine is a principal energy source for enterocytes and mediates cellular nucleic acid synthesis and proliferation.
- Growth hormone: It has been suggested to possess proadaptive properties.
- Insulin growth-like factor: The expression of IGF-1 is regulated by the GH, and it possibly has enterotrophic effects post-resection.
- Glucagon-like peptide: GLP-2 enhances the adaptive response to extensive intestinal resection in animals.

3.31 Ileal Brake

The ileal brake is a distal to proximal feedback system that regulates the proximal GI motility and secretion to increase the digestion and absorption of nutrients effectively.

Ileal brake stimulation results in retarding the gastric emptying by

- reduction in the frequency of antral and duodenal peristaltic pressure waves;
- reduction in antro-pyloroduodenal pressure waves;
- engenders a rise in pyloric sphincter pressure.

Associated changes are as follows:

- Reduction in the frequency of jejunal contractions, along with a reduction in the proportion and distance of each propagation.
- Ileal and duodenal motility is unaffected.
- Diminished meal-dependent secretion of pancreatic enzymes.
- Diminished dose-dependent bile acid output.
- Suppression of both basal and physiologically triggered secretion of gastric acid.

3.31.1 Physiological Role of Ileal Brake

The caloric value, composition of the consumed meal, and rate of gastric emptying determine the proportion of nutrients passaged to the ileum. The gastric emptying rate is inversely associated with the nutrients reaching the ileum and the colon.

3.31.2 Mediators of Ileal Brake

Peptide YY: It is a peptide composed of 36 amino acids. It is secreted by L-cells in the ileum and colon after food intake depending on its calorie content and composition, with a greater amount secreted for fat-containing than protein or carbohydrate-containing foods.

GLP 1: It is secreted by L-cells in the distal small intestine and colon, based on the passage of carbohydrates and fat-containing food. GLP-1 suppresses gastroduodenal motility, gastric acid, and exocrine pancreatic secretion; relaxes the proximal stomach; and induces phasic and tonic motility of the pylorus.

Oxyntomodulin: L-cells secrete oxyntomodulin along with PYY and GLP-1. Its intravenous administration reduces gastric acid secretion and suppresses gastric emptying and duodenal postprandial motility.

3.31.3 Effect of Dietary Component on Ileal Brake

Fat: Perfusion of fat into the ileum suppresses the small intestinal transit, gastric emptying, pancreaticobiliary and gastric acid secretion, and food consumption.

Protein: The presence of protein also stimulates the ileal brake.

Carbohydrates: The presence of carbohydrates like starch and maltose suppresses gastric emptying and decreases both stimulated and basal gastric acid secretion, small intestinal motility, and vagal-induced pancreatic exocrine secretion.

3.31.4 Clinical Implications

- **Weight loss after bariatric surgical procedures**
  Bariatric surgical procedures, after which the ileum is exposed to an increased amount of nutrients, are as follows:
  1. The jejunoileal bypass (JIB): proximal jejunum anastomosed to the distal ileum.
  2. The ileal transposition (IT) partial ileum transposed into the duodenum or jejunum. Increased exposure to nutrients enhances stimulation of the ileal brake and elevates PYY and GLP-1 levels, thus reducing food consumption.
- **Glycaemic control**
  It shows a beneficial effect of GLP-1 in improvement in glycaemic control.

3.32 Genetics in IBD

3.32.1 Evidence in Favour of Genetic Susceptibility of IBD

- Approximately 10–20% of patients diagnosed with IBD have a family history.
- The first-degree relatives of patients with IBD have a 15-fold higher risk of IBD than the general population.
- Family history is more frequently found in IBD than Crohn's disease.
- There is more excellent concordance among monozygotic than dizygotic twins.

3.32.2 NOD2/CARD15 Gene in IBD

NOD, nucleotide oligomerization domain; ***CARD***, caspase-activating recruitment domain.

- Nine genes have been identified so far to be involved in the pathogenesis of IBD: *IBD1* to *IBD9*.
- *IBD1* gene, also called as *NOD2/CARD15*, has been studied most extensively. This gene was previously known as *NOD2* and later renamed *CARD15*. It is a pattern-recognizing receptor. It comprises two CARD domains, a nucleotide-binding domain (NBD), and ten leucine-rich repeats (LRR).

- Three main single nucleotide polymorphisms (SNPs) of *NOD2/CARD15* accounts for >80% of mutations: SNP8, SNP12, and SNP13. Penetrance of these SNPs is <10%.
- *CARD15* is involved in identifying bacterial ligands and secretion of alfa defensins invasion and at the same time, it has an inhibitory effect on the TLR-signalling pathway. *NOD2/CARD15* mutations interfere with recognising MDP (a degradation product of gram-negative bacterial cell wall). TLR-signalling pathway inhibition is lost, resulting in overexpression of pro-inflammatory cytokines and unchecked inflammation.
- *NOD2/CARD15* mutations may account for up to 20% of Crohn's disease in the white and Jewish population. *CARD15* is associated with CD only and not with UC.
- *NOD2/CARD15* mutation is associated with (1) premature disease onset, (2) ileal disease, and (3) increased probability of stricture formation.

3.33 Role of Genetics in IBD: Indian Scenario

Indian data are very sparse. A few studies have examined the frequency of three NOD2/CARD15 gene SNPs in Indian IBD patients.

(a) Juyal et al.

- Case–control study.
- Patients with UC ($N = 298$), CD ($N = 25$), and otherwise healthy subjects ($N = 262$) were investigated.
- Among these, three NOD2 SNPs were either completely absent (SNP8 and SNP13) or rarely present (SNP12). A significant association was found between SNP5, which had a comparatively high frequency (12%).

(b) Pugazhendhi et al.

- Patients with CD ($N = 82$) and otherwise healthy subjects ($N = 149$) were investigated.
- None of the patients with CD demonstrated NOD2 mutations.

3.34 Genetics of Celiac Disease

3.34.1 Evidence for the Genetic Basis

- Ethnic differences in prevalence: high in Caucasians; very low prevalence in Japan.
- Familial aggregation: 5–15% of first-degree relatives affected.
- Twin studies concordance rate: 83–86% in monozygotic twins; 11% in dizygotic twins.

- The association was found with other genetic syndromes, e.g. Down's and Turner's syndromes.
- Identification of genetic loci by GWAS: chromosomes 2, 5, 6, 9, 15, and 19.

3.34.2 Genes Associated with Celiac Disease

- **CELIAC 1 locus**: Chromosome 6p 21—coding for MHC Class II molecule: HLA DQ-2 (90–95%) and DQ-8 (5–10%). In 30% of normal individuals, only 3% develop celiac disease. So, HLA DQ genes are necessary for the development of celiac disease but not sufficient.
- **CELIAC 2–13 locus** identified by GWAS technique: an association not consistent across the population.

3.34.3 Clinical Relevance

- HLA-DQ8 or HLA-DQ2 or typing is currently clinically pertinently. The other genetic loci could not be reproduced invariably.
- Human leukocyte antigen (HLA) typing has a high, nearly 100% negative predictive value to preclude the diagnosis of celiac disease.
- First-degree relatives of celiac disease patients should be tested for HLA. Serologic tests might be helpful in asymptomatic patients if celiac disease cannot be ruled out (i.e. HLA-DQ2 or HLA-DQ8 positive). It should be performed every 5–10 years.
- In asymptomatic HLA-DQ2/DQ8-positive first-degree relatives aged >20 years, screening should be performed at least once before age 50 since there is a high risk of complications after this age.
- A low threshold for biopsy is indicated when a first-degree relative manifests symptoms.

3.35 Lactase Gene

- Lactose is a disaccharide present in milk. Lactase, a brush border enzyme, hydrolyses dietary lactose into galactose and glucose for transport across the cell membrane.
- The lactase gene is located on the long arm of chromosome 2(2q21) and is 50 kb in size approximately. Lactase-phlorizin hydrolase gene (lactase gene) as an enzyme has two active sites: one hydrolysing phlorizin (an aryl α-glucoside) and another hydrolysing lactose and a range of dietary glycolipids.

- In the human intestine, detectable lactase activity at the mucosal surface occurs by week 8 of gestation, which peaks by week 34 and birth. Lactase expression is at its peak with the increase in activity by week 34 and by delivery. This helps neonates to digest the high lactose load of the mother's breast milk.
- There is a dramatic decrease in lactase after weaning in all land mammals. Chinese and Japanese lose 80–90% lactase activity within 3–4 years after weaning. The enzymatic disappearance of lactase is a mechanism under genetic control. This is at the post-translational level (in a few cases) or transcriptional level. Lactase levels at birth are lost by 90–95% by early childhood in humans. Lactase activity sustains beyond weaning and into adulthood in 30% of the population, primarily North Europeans.

  Two main theories explain lactase persistence:

1. ***Culture–historical hypothesis***: Reliance on mammalian milk, especially during poor harvest.
2. ***Reverse cause***: The theory proposed that those with antecedent lactase persistence embraced milk consumption and dairy farming. Currently, lactase non-persistence is considered an ancestral type. Lactase persistence is associated with two SNPs in the lactase gene C/T 13910 and G/A 22018 substitutions occurring 14 and 22 kb upstream of the 5′-end of the lactase gene. Intermediate lactase activity is seen in the individuals heterozygous for either SNP. These individuals are more vulnerable to lactose intolerance, especially during GI infection or stress. Individuals with homozygous lactase persistence may infrequently develop lactose intolerance. Therefore, it can be argued that this polymorphism may not completely justify these findings.

3.36 Gut Microbiota

- Human intestine harbours—300–500 species of bacteria
- Microbiome larger than the human genome

3.36.1 Distribution of Normal Microflora

- The stomach and upper gut have very few bacteria; stomach—no more than 10^3/mL; duodenum/upper jejunum—no more than 10^4/mL–10^5/mL (just after a meal)
- The bacterial population in the small intestine is dependent on
  - the rate of entry
  - the rate of removal by body defences
- Large intestine—10^{11}–10^{12} bacteria per g of luminal contents

3.36.2 Composition of Gut Flora

- Anaerobes outnumber aerobic bacteria
  - Anaerobes: bacteroides, bifidobacterium, eubacterium, clostridium, peptococcus, and peptostreptococcus.
  - Aerobes: *Escherichia*, *Enterobacter*, *Enterococcus*, *Klebsiella*, *Lactobacillus*, and *Proteus*.
- The individual's composition pattern is usually constant
- Uniqueness in composition likely reflects
  - Consistent physiologic and immunologic idiosyncrasies of people
  - The genetic constitution of the host

3.36.3 Proximal Small Intestine

1. **Gram-positive aerobes/facultative anaerobes:**
   - Oral *Streptococcus* spp., *Staphylococcus* spp., *Micrococcus*, *Neisseria* spp., *Lactobacilli*, *Veillonella* spp., *Corynebacterium* spp., *Actinomyces* spp., and *Fusobacterium* spp.
   - *Candida* spp.
2. **Gram-negative bacteria:**
   - Enterobacteriacae, *Pseudomonas* spp., and occasionally bacteroides.

3.36.4 Large Intestinal Bacterial Flora

- Bacteroides/prevotella group, *Bifidobacterium* genus, *Clostridium histolyticum* group, *Clostridium lituseburense* group, *Coriobacterium* subgroup, *Escherichia coli, Enterococcus faecalis, Eubacterium cylindroides* group, *Eubacterium rectale–Clostridium coccoides* group, *Fusobacterium prausnitzii* group, *Lactobacillus*, *Ruminococcus* subgroup, and *Veillonella.*

The faecal microflora usually corresponds to that of the distal colon. The caecal content has quite a few facultative anaerobes.

3.36.5 Evolution of Gut Flora

- Early colonisation by facultative anaerobes include
  - enterobacteria

  - coliforms
  - lactobacilli

- After the first year—increase in *Bacteroides*, *Veillonella*, and *Fusobacterium.*
- In adults, *Bacteroides are* 25% of the microbiota.

3.36.6 Factors Influencing Colonisation Pattern

1. Type of delivery
2. Type of diet
3. Developed versus developing countries

3.36.7 Methods of Identification

- Traditional methods of culture → unsuitable as a significant proportion are nonculturable.
- Newer methods of identification
  - Ribonucleic acid dot blots
  - Fluorescence in situ hybridisation and epifluorescence microscopy
  - Terminal restriction fragment length polymorphism (T-RFLP)
  - Polymerase chain reaction (PCR) combined with denaturing gradient gel electrophoresis (PCR/DGGE)
  - Polymerase chain reaction combined with temperature gradient gel electrophoresis (PCR/TGGE)
  - 16S rRNA gene library
  - Real-time quantitative PCR
  - High-density DNA microarrays

3.36.8 Main Functions of Microflora

- Metabolic
  - Fermentation of nondigestible dietary residue
  - Salvage of energy as short-chain fatty acid (SCFA)
  - Ion absorption
  - Vitamin K production
- Trophic
  - Epithelial cell differentiation and proliferation
  - Homeostasis and development of the immune system

- Protective
  - Bacteriocin production
  - Host responses to the normal flora
  - Importance of phagocyte killing for limiting the consequence of bacterial translocation from the gut. Levels of secretory IgA and numbers of mucosal IgA-producing plasma cells are both low in germ-free experimental animals
- Oral tolerance—immunoregulation at mucosal sites
  - Deficient in germ-free mice; reversed by reconstitution of flora
  - Gut mucosa—a state of controlled immune reactivity guided by the intestinal flora
  - Control by multiple mechanisms acting at different levels
  - The gut displays an intrinsic predisposition towards Th2 or Th3/Tr T-cell responses
  - Regulatory cytokines: IL-10 and TGFβ
  - Constant association between DCs and the gut epithelium

3.37 Opportunistic GI Pathogens in HIV

An opportunistic infection occurs in immunocompromised individuals such as those with HIV disease with a variation in the GI pathogens with progressive deterioration.

3.37.1 Syndromic Manifestations of Opportunistic Infections

- **Diarrhoea**. It is the most frequent GI symptom of opportunistic infections in HIV patients.
  - Protozoa: *Cryptosporidium*, *Microspora*, *Isospora*, and *Cyclospora* (refer to Figs. 3.8, 3.9, and 3.10 for the life cycle of intestinal protozoal parasites).
  - Nematode: *Strongyloides* (refer to Fig. 3.11).
  - Virus: CMV and herpes simplex.
  - Fungus: Coccidiomycosis and cryptococcus Histoplasmosis.
  - Bacteria: *Salmonella*, *Shigella*, *Campylobacter*, *Clostridium difficile*, and *Mycobacteria.*
- **Odynophagia and dysphagia**. *Candida albicans*, *Cytomegalovirus*, *Herpes simplex*, and *Histoplasma capsulatum.*

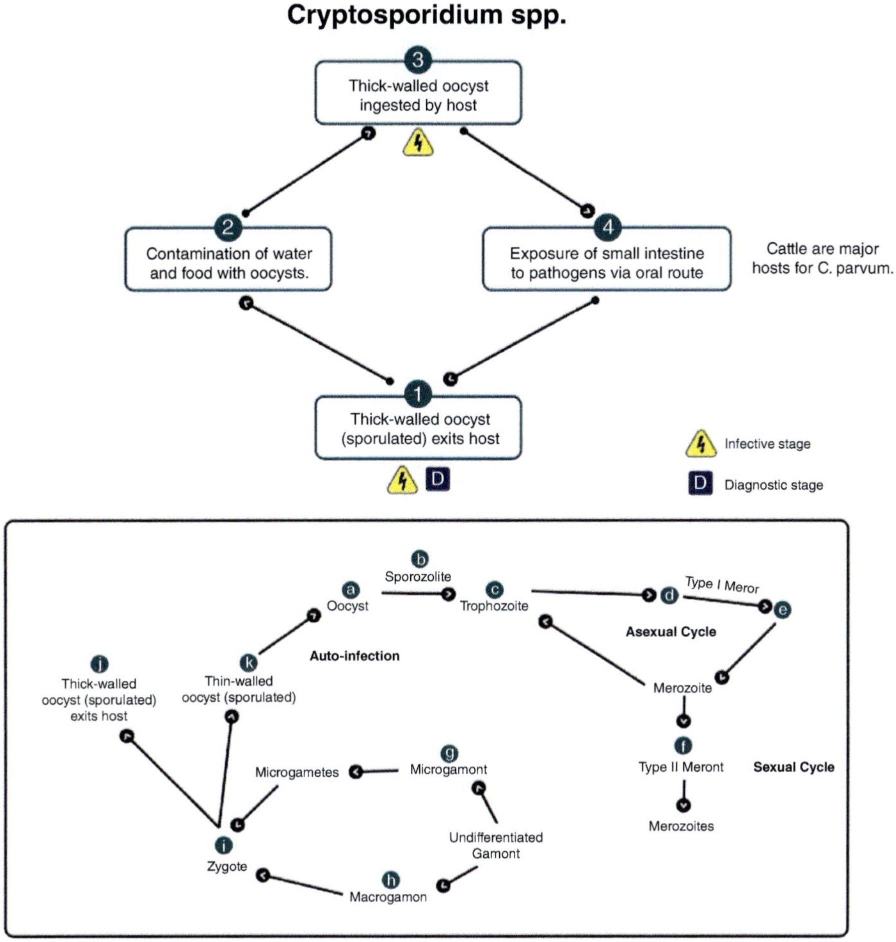

Fig. 3.8 Life cycle of cryptosporidium. (Adapted from CDC: https://www.cdc.gov/dpdx/cryptosporidiosis/index.html)

- **Anorectal disease**. More common among homosexuals
  - Bacteria: *Chlamydia trachomatis, Lymphogranuloma venereum, Neisseria gonorrhoeae, Shigella flexneri,* and *Mycobacterium tuberculosis.*
  - Protozoa: *Entamoeba* and *Leishmania.*
  - Virus: *Herpes simplex* and CMV.
  - Fungus: *Candida albicans* and *Histoplasma albicans.*
- **Other manifestations**. Gastrointestinal bleeding, abdominal pain, hepatomegaly, and liver function tests.

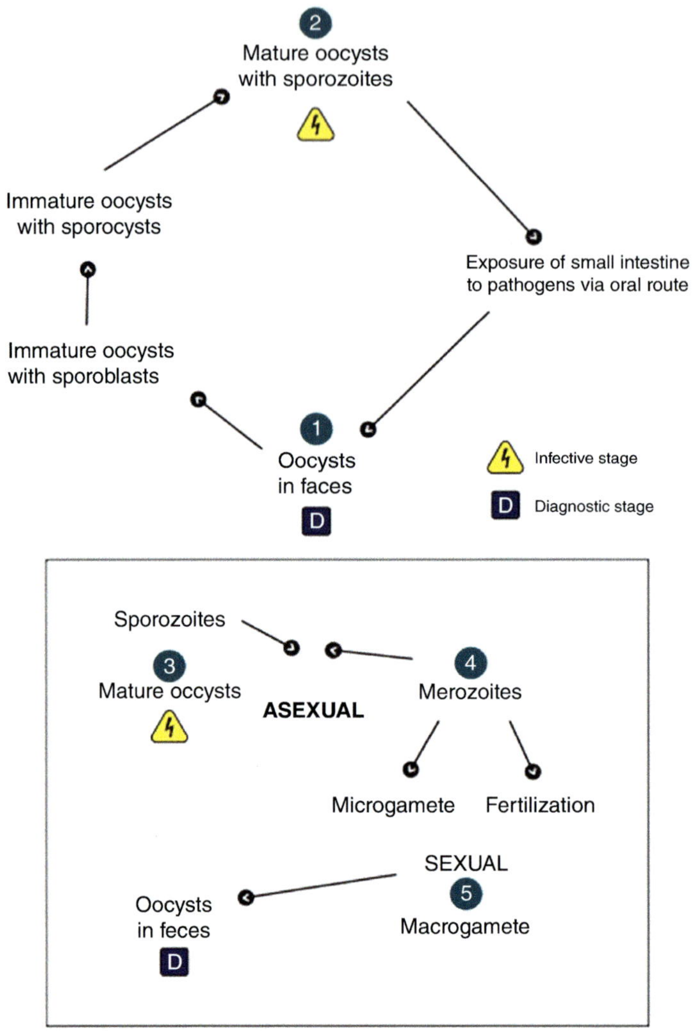

Fig. 3.9 Lifecycle of isospora. (Adapted from CDC: https://www.cdc.gov/dpdx/cystoisosporiasis/index.html)

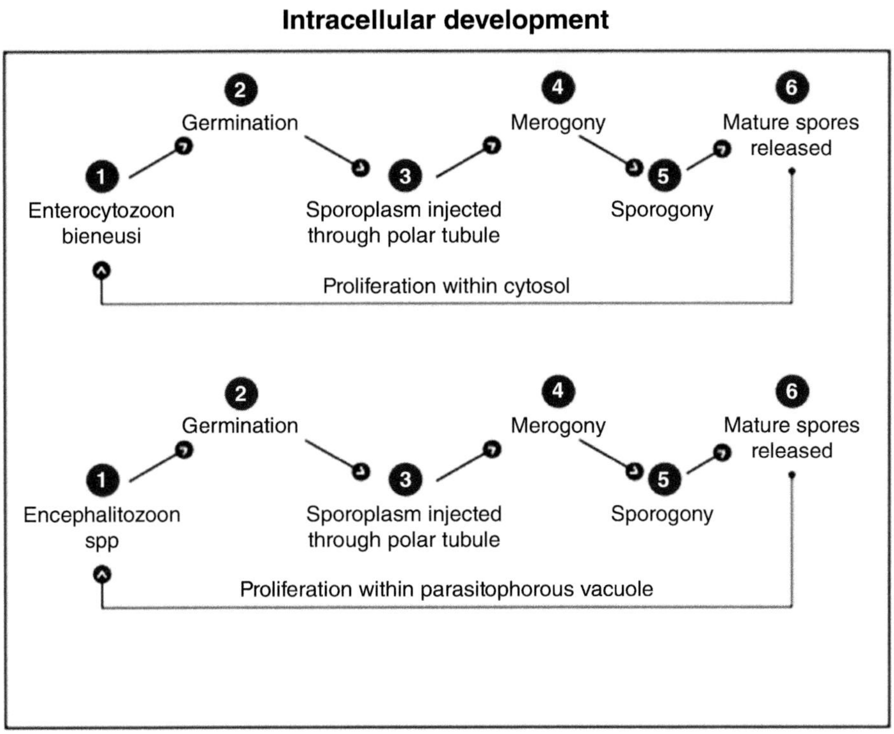

Fig. 3.10 Lifecycle of microsporidia. (Adapted from: https://www.cdc.gov/dpdx/microsporidiosis/index.html)

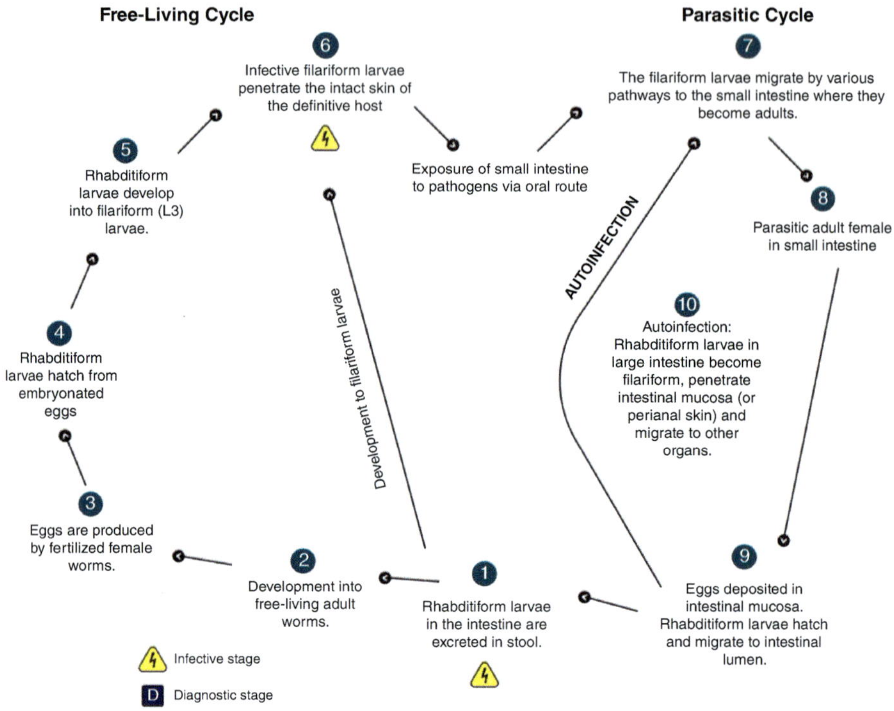

Fig. 3.11 Life cycle of Strongyloides. (Adapted from CDC: https://www.cdc.gov/parasites/strongyloides/biology.html)

3.38 Anti-saccharomyces Antibody and Anti-neutrophil Cytoplasmic Antibody: Role in IBD

Several auto-antibodies have been reported to be found in IBD, especially Crohn's disease. Still, the most studied antibodies are anti-neutrophil cytoplasmic (p-ANCA) and anti-saccharomyces cerevisiae (ASCA).

Following is a list of common auto-antibodies described in IBD.

3.38.1 *Anti-*Saccharomyces cerevisiae *Antibodies*

Anti-*Saccharomyces cerevisiae* antibodies target the mannose residue from the phosphor-peptido-mannan of the cell wall of yeast *S. cerevisiae*. Patients with CD and their otherwise healthy relatives have ASCA more commonly (20–25%) than patients and healthy relatives of those with UC (0–5%).

3.38.2 *Anti-neutrophil Cytoplasmic Antibodies*

Anti-neutrophil cytoplasmic antibodies target neutrophil granule constituents. The assay is done by standardised indirect immunofluorescence (IIF) microscopy on ethanol-fixed neutrophils. If the titre exceeds 1:20 dilution, the serum is considered positive in the IIF assay. IIF observes three main staining patterns: a speckled (sANCA) pattern, a perinuclear (pANCA or atypical ANCA), and a cytoplasmic granular (cANCA). Perinuclear ANCA is the most standard test used in IBD.

3.38.3 *Outer Membrane Porin C*

Outer membrane porin C (OmpC) is the principal outer membrane protein (isolated from *E. coli*), antibodies linked with CD.

3.38.4 *Possible Roles of Auto-antibodies in IBD*

- Distinguishing CD from intestinal tuberculosis and UC
- Tracking the probable disease activation and its prognosis
- Assess the disease severity
- Predicting the activation of an inactive disease
- Assess the risk of developing IBD in a serologically positive individual

3.38.5 *Anti-*Saccharomyces cerevisiae *Antibodies and Anti-neutrophil Cytoplasmic Antibodies*

Role in Diagnosis

- Anti-*Saccharomyces cerevisiae* and anti-neutrophil cytoplasmic antibodies are fairly precise for CD and UC, respectively. Both ASCA and pANCA testing aid in diagnosing IBD precisely due to high specificity (>90%) for Crohn's disease and UC, respectively. Besides, its positive output also helps differential CD diagnosis from intestinal tuberculosis (TB). However, given the low sensitivity, a negative output does not preclude a diagnosis of CD. Additionally, some evidence in the Indian population has found an association between ASCA positivity and TB, so its use in differentiating CD from TB is again questionable.

- The ASCA and pANCA testing sensitivity is slightly more than 50% in CD and UC, respectively. Considering the ASCA positivity demonstrated in other luminal conditions like celiac disease and Behcet's disease, it should be evaluated cautiously. Currently, ASCA and ANCA are not advised to be used for diagnosing IBD.
- In a fraction of IBD patients (12%), a diagnosis based on clinical manifestations is not possible, particularly in the early disease state. This underscores the need for serologic tests in differentiating between CD and UC.

Role in Predicting Disease Behaviour

High ASCA titres have been reported in patients with premature disease onset, internal penetrating, stricturing disease, and need for small bowel surgery. More recently, anti-I2, anti-CBir1, and anti-OmpC have also been reported to correlate with disease complications.

Role in Predicting Response to Therapy

High p-ANCA levels have predicted poor response to infliximab and a higher risk of chronic pouchitis.

Role in Disease Activity Monitoring

Serial measurement of ANCA titres in UC does not help monitor disease activity and predict relapses in contrast to systemic Crohn's disease since it is stable over time independent of disease activity, severity, and duration.

Role as Subclinical Disease Markers

The most critical risk factor for IBD could be a first-degree relative with IBD. Compared to general populations, the relative risk of developing IBD to a sibling is 15–35 times in CD and 7–12 times in UC. Hence, looking for a subclinical marker to determine the risk of IBD in clinically asymptomatic siblings of index cases seems reasonable.

Although the current status of ASCA and ANCA as subclinical markers is not defined, the prevalence of ASCA and ANCA positivity in patients' relatives is higher than in the general population.

3.39 Protein-Losing Enteropathy

Protein-losing enteropathy encompasses a diverse group of disorders characterised by the disproportionate elimination of serum proteins into the GI tract. Daily enteric loss accounts for 2% of total protein pool turnover.

3.39.1 *Mechanisms*

- Protein loss is due to increased permeability following mucosal ulcerations or cell damage regardless of protein molecular weight. Exudation of protein-rich fluid in lymphatic obstruction leads to protein loss. The rate of intravascular protein loss is constant irrespective of molecular weight.
- However, serum levels may vary depending on the half-life of the protein. Protein levels with rapid turnover (e.g. insulin and clotting factors) will not be reduced, while those with slow turnover (e.g. albumin and Ig except IgE) will be reduced.

3.39.2 *Pathophysiology*

Loss of epithelial matrix component (heparin sulphate from the basolateral surface of intestinal epithelial cells) is hypothesised to be a common causative factor in various disorders causing protein-losing enteropathy.

3.39.3 *Laboratory Studies*

- A decrease in serum globulin and albumin is the most prominent laboratory abnormality
- A GI disorder could also contribute to hypoalbuminemia in the absence of nephrotic syndrome, malnutrition, and chronic liver disease
- Stool alfa-1 antitrypsin is an essential marker for protein-losing enteropathy. Alfa-1-antitrypsin is not secreted or absorbed from the GI tract. So, its presence suggests increased mucosal permeability or breach of the intestinal mucosa.
- Plasma clearance (PC) of alfa-1-antitrypsin is shown by measuring stool alfa-1-antitrypsin concentration and stool volume. Monitoring response to therapy can be done by the PC of alfa-1-antitrypsin.

3.39.4 Imaging Studies

- Protein exudation can be estimated from stool collected after radionuclide-labelled serum albumin administered intravenously.
- Lymphatic obstruction may be detected in a CT scan. Chest radiography, radionuclide scanning, and echocardiography of the heart detect cardiac disease.
- Radiographic contrast studies, endoscopy, and biopsies can be used to diagnose ulcerative or erosive conditions.

3.40 Eosinophilic Gastroenteritis

- The manifestations of eosinophilic gastroenteritis, a rare benign disorder, include eosinophilic infiltration of the GI tract wall and infrequently peripheral eosinophilia.
- It affects children and adults with peaks between the age of 30 and 50 years.
- Diagnostic criteria, as suggested by Klein et al., include:
  - GI symptoms
  - Lack of signs of parasitic or extraintestinal disease
  - Eosinophilic infiltration on biopsy from the oesophagus to the colon, with peripheral eosinophilia on radiography
- Pathologic type is based on affected GIT layer—mucosal, muscularis, and subserosal predominant. The mucosal disease may present with anaemia secondary to GI blood loss, weight loss, and malabsorption secondary to protein-losing enteropathy. Muscularis predominant disease may present as intestinal obstruction. Subserosal disease may manifest mainly as eosinophilic ascites.

3.40.1 Pathogenesis

- Eosinophilic gastroenteritis may be primary (allergic) or secondary to the conditions provoking eosinophilic inflammation. Here parasitic infections are the leading cause.
- Normally, eosinophils are only found in the lamina propria of the GIT.
- Within GIT, most eosinophils are seen in the cecum and appendix. However, in non-inflammatory conditions, the oesophagus is devoid of eosinophils.
- Eosinophil-derived granule proteins (EDGPs) demonstrate antiviral and ribonuclease activity and can cause intestinal cellular damage. These EDGPs may also result in mast cells' degranulation and release of inflammatory mediators like cytokines, chemokines, leukotrienes, platelet-activating factors, substance P, and vasoactive active polypeptide. These include eosinophil-derived neurotoxin (EDN), eosinophil peroxidase (EPO), eosinophil cationic protein (ECP), and major basic protein (MBP).

3.40.2 *Role of Allergy*

- The response in the form of hypersensitivity may underlie the pathogenesis of gastroenteritis.
- History of food sensitivities, seasonal allergies, elevated serum IgE levels, asthma, eczema, and atopy are seen in 25–75% of eosinophilic gastritis patients.

3.40.3 *Genetic Factors*

Up to 10% of patients with eosinophilic gastroenteritis have a positive family history.

3.41 Pathophysiology of Celiac Disease

Gluten-sensitive disorders encompass a broad spectrum of conditions ranging from well-characterised celiac disease to wheat allergy and non-celiac gluten sensitivity (Fig. 3.12).

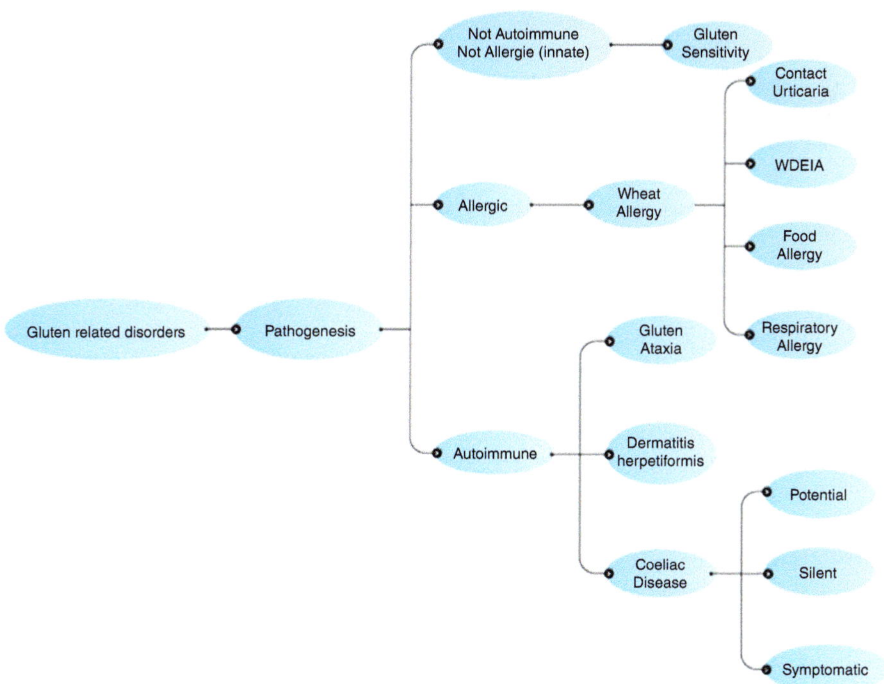

Fig. 3.12 Spectrum of gluten-related disorders

(a) **Environmental factors**

- Prolamins soluble in ethanol and glutenins soluble in dilute acidic or alkaline solutions are wheat proteins together classified as gluten.
- Prolamins from oats, barley, wheat, and rye have immunologic cross-reactivity. Oats contain comparatively lesser prolamins than wheat.
- Exposure to dietary gluten within 3 months of birth is linked to a higher risk for celiac disease compared to after 4 months of age.

(b) **Genetic factors**

- Among patients diagnosed with celiac disease, it is also found in 8–18% of their first-degree relatives, while the cooccurrence is very high in monozygotic twins.
- The HLA class II molecule DQ2 is reported in a more significant population with than without celiac disease (90% vs. 35%). *HLA-DQ8* is found in nearly all patients with celiac disease.
- Although DQ2 is found in a large proportion of the population, only a fraction of these develop celiac disease. This leads us to believe that genes other than those encoding for HLA-DQ may increase the susceptibility to celiac disease. There is some association between genes for celiac disease and type 1 diabetes.

(c) **Immune factors**

- Endomysium is the connective tissue structure surrounding smooth muscle. IgA antibodies to the antigen tTG-2 in this tissue are characteristic of celiac disease. tTG-2 deamidates gliadin to anionic glutamic acid residues for antigen presentation. These are presented by MHC Type II complex to CD4+ T cells.
- Abnormal permeability of the small intestine may result in anomalous immune responses to proteins (β-lactoglobulin, casein, and ovalbumin) in food.
- Activated T cells in celiac disease release potent proinflammatory mediators such as IFN-γ, TNF-α, IL-2, IL-6, and TGF-β.
- Activated T lymphocytes in celiac disease are majorly CD8+, while in healthy intestine, CD4+ dominate.
- There is an influx of primed memory T cells in the mucosa of untreated celiac disease patients. In healthy persons, more than 90% of IELs express the α/β TCR, whereas the expression of the γ/δ TCR by IELs in patients with untreated celiac disease is increased and is considered a hallmark of the disease.
- IL-15, an enterocyte- and macrophage-derived proinflammatory cytokine, is increased and bridges innate and adaptive immune responses in the pathogenesis of celiac disease. IL-15 activates an adaptive CD4+ T-cell response in the lamina propria as a response to gliadin peptides. It may induce IEL secretion of IFN-γ, which can result in direct epithelial cell injury.

In a host with an HLA predisposition, exposure to gluten results in enterocyte damage manifesting as celiac disease.

3.41.1 Non-celiac Gluten Sensitivity

Non-celiac gluten sensitivity (NCGS) or **gluten sensitivity** is defined as a condition in which symptoms arise on the consumption of gluten-containing food without celiac disease or wheat allergy, confirmed by the withdrawal of these foods and subsequent symptom resolution (Table 3.9). The pathophysiology of NCGS remains obscure (Fig. 3.13).

Gluten

Among patients presenting with functional GI disorders, ~14–30% have NCGS, as evidenced by worsening symptoms on gluten challenge.

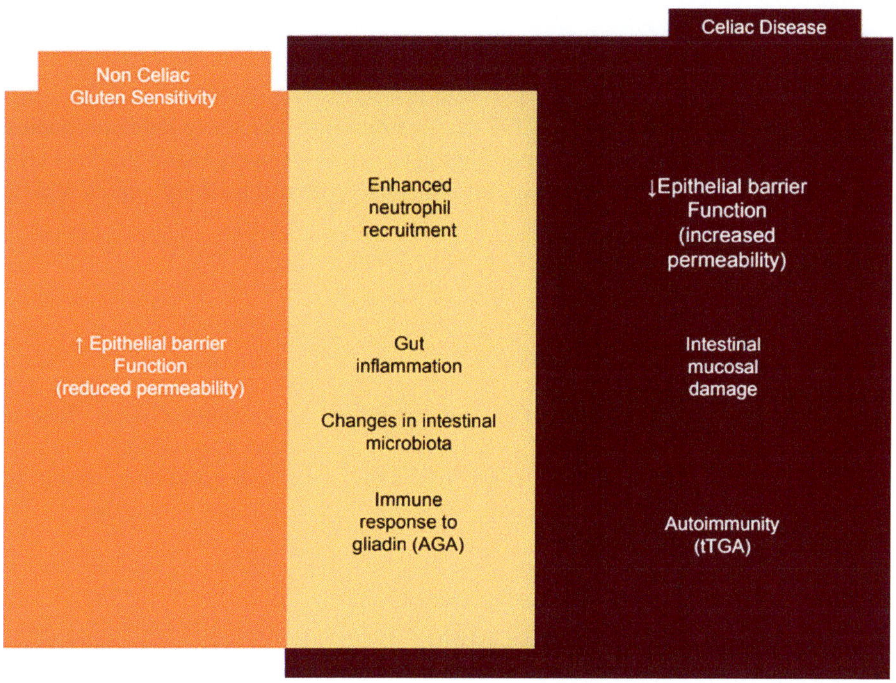

Fig. 3.13 Pathogenesis of non-celiac gluten sensitivity. Increased epithelial barrier function characterises non-celiac gluten sensitivity, while decreased epithelial barrier function and increased mucosal damage by autoimmune mechanisms characterise celiac disease. (Adapted from: Cellular & Molecular Immunology (2013) 10, 383–392)

FODMAPs

As demonstrated by a re-challenge study, FODMAP may be involved in NCGS pathophysiology and symptom manifestation. The GI manifestations were resolved on reduced consumption of FODMAP and vice versa. In only 8% of the subjects, the symptoms were attributed to gluten. Therefore, the NCGS manifestations may all be attributed to components such as FODMAP rather than gluten.

Emerging data show that the fructose challenge is more often associated with NCGS symptoms than gluten.

Amylase–Trypsin Inhibitors

Amylase trypsin inhibitors (ATIs) are protein allergens present in grains and account for 4% of total proteins. These allergens may cause baker's asthma. These may stimulate an innate immune response, intestinal infiltration, activation of myeloid cells, and the release of inflammatory cytokines, which may underlie its role in the pathophysiology of NCGS.

3.41.2 Diagnostic Criteria (Table 3.12)

Nearly half of the patients with NCGS demonstrate anti-IgG anti-gliadin antibodies. Non-celiac gluten sensitivity has no relation to celiac disease genetic markers (HA DQ 2 and DQ 8): 40% in patients with nonceliac gluten sensitivity versus 30% in the general population.

3.41.3 Treatment

Modify dietary habits, including foods with the least or no gluten content. Instead of wheat, rice, corn, millet, and Leguminosae such as quinoa, amaranth, and soybean are recommended. Besides, meat, fish, eggs, fruit, and vegetables should be included in the daily diet to balance the nutrition since these are gluten free.

Table 3.12 Diagnostic criteria for nonceliac gluten sensitivity

Diagnostic criteria for nonceliac gluten sensitivity
• Challenge and de-challenge with gluten ingestion: symptom appearance and disappearance shortly (hours to days) after gluten ingestion and withdrawal, respectively. • Rechallenge with gluten triggers rapid recurrence of symptoms. • Exclusion of celiac disease using negative serology (endomysial and tissue transglutaminase IgA antibodies) and a duodenal biopsy on a gluten-containing diet. • Wheat allergy tests: specific IgE, as well as skin prick tests, must be negative when performed on a gluten-containing diet. • A double-blind placebo-controlled gluten challenge test is necessary for suspected patient.

3.42 Tropical Sprue

3.42.1 Definition

It is an acquired disease that affects inhabitants and tourists to the tropical region, regardless of race and duration spent. Its manifestations include malabsorption (≥ 2 substances), multiple nutritional deficiencies, and mucosal abnormalities of the small bowel in the absence of any other cause.

3.42.2 Epidemiology

- Found typically in the inhabitants of tropical and subtropical regions.
- Southern and Southeast Asia and the Caribbean Islands are primary loci.
- People in Central and South America are seldom affected.
- People in Africa or the Middle East are rarely affected.

3.42.3 Pathophysiology

An intestinal infection could trigger tropical sprue, although its exact aetiology remains unknown. The infection may cause enterocyte injury, intestinal stasis, and likely bacteria overgrowth. Microstructural intestinal damage may result in nutrient malabsorption, causing multiple nutrient deficiencies such as folate, vitamin B12, and iron. Such deficiencies may further deteriorate the mucosal integrity. Abnormal elevations in enteroglucagon may enhance intestinal stasis. Initially restricted to the proximal part of the small intestine, it may encompass the entire intestine throughout the disease. However, the preceding and succeeding organs are rarely affected. The suspected microorganisms in tropical sprue are *Klebsiella*, *E. coli*, and *Enterobacter* species.

3.43 Immunology in IBD

IBD results from a complex interplay of aberrant immune responses in genetically predisposed individuals exposed to environmental risk factors.

Alterations seen in IBD are the following.

1. **Dysregulation at the epithelial barrier**
    - Abnormalities in the intestinal mucus with more significant bacterial contamination.
    - Abnormal intestinal permeability.

- Epithelial cells can act as APCs since these can express some receptors similar to immune cells, particularly in patients with IBD.
- Abnormal handling of unfolded intracellular proteins by epithelial cells (Paneth cells) in response to stressed endoplasmic reticulum has been associated with IBD.

2. **Dysregulation in immune cells**

- Innate immunity:
  - The lamina propria of IBD patients have an inflammatory type myeloid cell that secretes increased amounts of cytokines.
  - Neutrophils in lamina propria indicate recruitment from vasculature by a complex homing process in IBD. It also involves increased expression of adhesion molecules and increased leukocyte binding to vascular endothelial.
- Acquired immunity:
  - Lymphoid cells harbour several activation markers and cytokines.
  - T cells hyper-proliferate and increase cytokine synthesis in response to antigens.
  - Elevated number of mucosal and circulating B cells and plasma cells, autoantibodies indicating abnormal B-cell regulation may also occur in IBD.

3. **Dysregulation in secreted mediators**

- Elevated levels of immunoregulatory and inflammatory cytokines are associated with active IBD.
- **Th 1 cell response**: $CD4^+$ T lymphocytes isolated from patients with CD secrete excessive **IFN-γ and tumour necrosis factor (TNF)-α**.
- **Th 2 cell response**: IL-4, IL-5, and IL-13 secreted by Th2 cells suppress Th1-mediated responses, and **eosinophil infiltration results** in intestinal inflammation.
- **Th 17 cell response**: **IL-23-dependent signalling and secretion of cytokines by Th17 cells** indicate abnormal cytokine secretion earlier associated with Th1 cells.
- **T reg cell response**: Subsets of regulatory $CD4^+$ T cells ($CD4^+$ $CD25^+$ Tregs, Th3 and Tr1 cells) secrete IL-10 and TGF-beta cytokines with inhibitory action. Therefore, these behave as 'suppressor' cells. Regulatory cells typically form in the thymus, while induced ones are found in the gut-associated lymphoid tissue (GALT). There is limited evidence of aberrant T regulatory cell function in IBD.

- Each of these steps mentioned above is therapeutic targets in IBD—biologics, anti-integrins, anti-IL 12/IL 23, and the latest study targets anti-trafficking agents (anti-beta 7—etrolizumab and anti-MAdCAM-1 antibody ontamalimab).

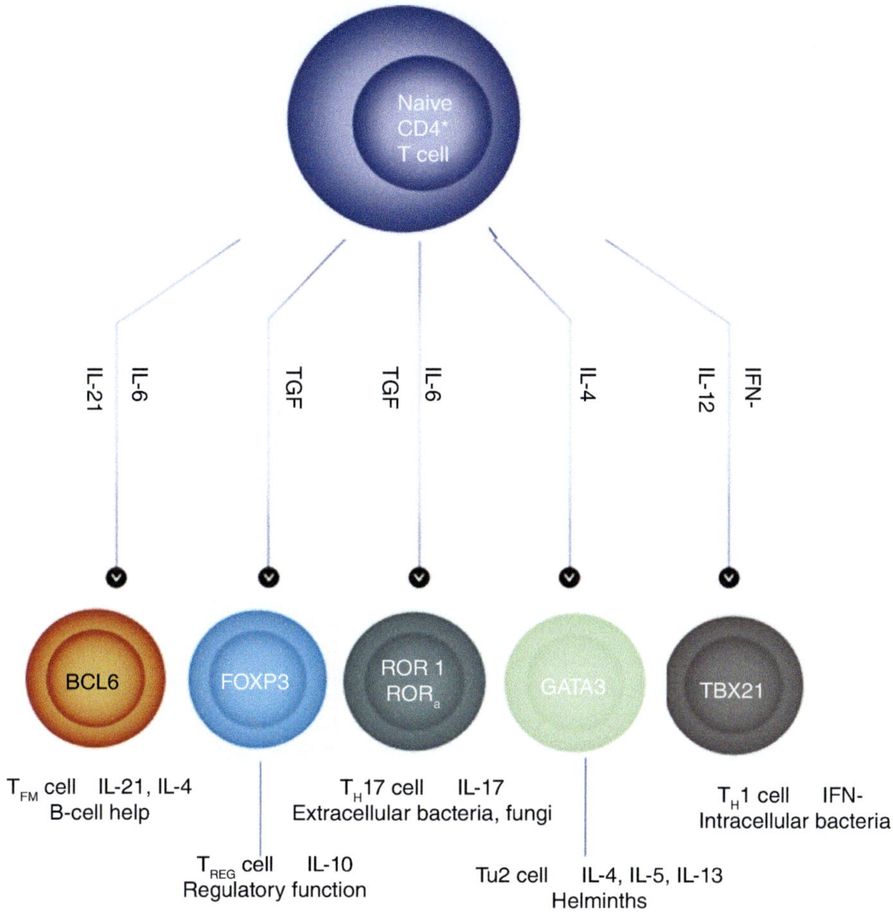

Fig. 3.14 T-cell subsets

3.44 T-Cell Response and Their Role in GI Disorder

There are different types of T cells with distinct patterns of development and function (Fig. 3.14).

T Helper cell response in various GI disorders: The gastrointestinal tract has to accept and assimilate required nutrients from luminal food content while simultaneously antagonising bacterial intrusions via the generation of immune responses. CD4+ T cells are principal mediators of the host's protective and homeostatic responses.

1. **IBD**

   • Abnormal Th1 and Th2 cells may account for the pathogenesis of CD and UC, respectively.

- Recent studies showed an essential role for Th17 cells in the pathogenesis of IBD. Th17 cells synthesise T helper cell subsets such as IFN-γ (Th1) and IL-4 (Th2), leading to the development of pathogenic Th17/Th1 or Th17/Th2 phenotype, which is found to be more pathogenic than the native Th17 cells. These are potential targets for the treatment of inflammatory disease states.
- IL-9 secreted by Th9 cells has been linked to UC progression.
- Microscopic colitis involves Th1 mucosal cytokines with elevations in IFN-γ.

2. **Eosinophilic esophagitis**

- Patients consume food or air allergens followed by APCs presenting them to Th2 cells.
- IL-4 and IL-5 are produced by the activated Th2 cells. These play a role in eosinophil accumulation, terminal differentiation, and proliferation.
- These also produce IgE from B cells, IL-13, and TNF, which support local inflammation.

3. **Celiac disease**

- Commonly gluten tolerance is attributed to anti-gluten $Foxp^{3+}$ regulatory T cells (T reg cells), inhibitory natural killer (NK) receptors.
- However, in celiac disease, HLA-DQ2 or DQ8 on APCs-like dendritic cells present gluten to CD4+ T cells inciting an anti-gluten T-cell response. These release IFN-γ and IL-21, causing epithelial damage.
- IL-15 increases the expression of NK receptors, encourages T-cell receptor-independent cytotoxicity, and inhibits $Foxp^{3+}$ T-reg cell action on IELs.
- IELs transform to acquire an NK phenotype to increase epithelial cell injury implying a role of innate immunity.
- **Production of** gluten-specific antibodies results in dermal manifestations of systemic effects by humoral immunity.

4. **Whipple's disease**
   Studies point to a defective cellular immune response, especially an impaired Th1 response, playing an important pathogenetic role. Decreased IFN production may be the result of reduced monocyte IL-12 production. This results in increased susceptibility to Tropherema Whippeli.

5. **GI motility disorders**
   Th1 and Th2 cytokines act on intestinal smooth muscle contraction via 5-hydroxytryptamine signalling.

- Th1 cytokines reduce the number of L-type Ca^{2+} channels and increase G protein-signalling regulators, resulting in reduced contractility of inflamed intestinal smooth muscle.
- Th2 cytokines increase intestinal smooth muscle contractility via activation of transcription 6 or mitogen-activated protein kinase signalling pathways.

3.45 Ulcers in Small Bowel and NSAID-Induced Enteropathy

Small intestinal ulcers can be solitary or multiple.

1. Solitary ulcers beyond duodenum with unknown aetiology—*nonspecific* or *idiopathic* intestinal ulcers.
2. Systemic diseases associated with small intestinal ulcers—drugs, infections, inflammatory disorder, vasculitides, neoplasms, toxins, Zollinger Ellison syndrome, refractory celiac disease.
3. Non-specific ulcers in the small intestine: chronic nonspecific multiple ulcers of the small intestine (CNSU) or cryptogenic multifocal ulcerous stenosing enteritis (CMUSE).

3.45.1 Non-specific Ulcers in the Small Intestine: Chronic Nonspecific Multiple Ulcers of the Small Intestine (CNSU) or Cryptogenic Multifocal Ulcerous Stenosing Enteritis (CMUSE)

Clinical features

Recurrent abdominal pain and anaemia: common presentation

- Common manifestations include frequent abdominal pain and anaemia
- Acute or chronic GI bleeding: often occult
- Recurrent abdominal pain: small bowel obstruction (intermittent small bowel obstruction—most common presentation)
- Perforation

Physical Examination

- Non-specific abdomen tenderness
- Abdomen distension (ranging from non-specific to abdomen distension due to complete small bowel obstruction)
- Acute abdomen

Laboratory Feature

- Anaemia was seen in 50% of the patients.
- On CECT or MRI, three-fourths of the strictures show moderate while approximately 10% show substantial enhancement with a layered pattern.

Pathology

- The most common site of non-specific ulcers is—the ileum
- Most common site of perforation—jejunum
- Pathologic examination: ulcer size ranged from 3 to 5 cm, located on an anti-mesenteric border with occasional fibrous scar causing narrowing.
- Microscopy demonstrates non-specific chronic inflammation with abrupt end at the ulcer edge, while bowel and vasculature were unremarkable. Resected specimens showed multiple superficial ulcerations limited to mucosa or submucosa and multiple strictures (some ulcerated). Ulcerative lesions appear either linear or circular, with few showing fibrosis of the submucosal layer. Chronic crypt inflammation, pyloric gland metaplasia, or lymphoid aggregates were also associated with ulcerative lesions.

Clinical Criteria for This Diagnosis

1. Constant and occult blood loss from the GI tract (except during bowel rest or postoperative period).
2. Low-grade recurrent small bowel obstruction—presenting as intermittent pain episodes
3. Evidence of typical small intestinal lesions by various investigations, circular or oblique ulcers in alignment, distinct from surrounding healthy mucosa, geographic or linear in shape, numerous cells with less than 4-cm gap from each other, and ulcers not affecting the muscularis propria layer, including scarred ulcers assumed to be the healing stage in cases treated via bowel rest.

Difference from IBD

The CMUSE lacks penetrating features such as fissures, fistula, or abscesses. It also does not show evidence of other GI organs except the small intestine and seldom mesenteric lymphadenopathy or mesenteric infiltration. These features differentiate CMUSE from other IBD, particularly CD.

Treatment

- Segmental intestinal resection.
- Responsiveness to corticosteroids was seen, but dose duration and the inability to taper them because of symptomatic relapse were not specified.

3.45.2 NSAIDS-Related Small Bowel Disease

NSAIDs are among the most commonly prescribed drugs globally, and their gastroduodenal adverse effects are frequently reported. However, diminished GI tract risks of NSAIDs are being increasingly recognised.

Clinical Feature

1. It may occur as an asymptomatic mucosal injury or severe symptoms manifest with considerable ulceration, excessive bleeding, intestinal obstruction, or perforation.
2. NSAID enteropathy may increase intestinal permeability, inflammation, and chronic mild blood loss; macroscopic injury may be obscured. Relevant GI symptoms may be absent, but the patient may manifest anaemia and hypoalbuminemia. Infrequently weight loss, diarrhoea, overt bleeding, and perforation may be noted.
3. Formation of strictures may cause partial small bowel obstruction.

Pathology

Pathologic appearance is nonspecific: solitary or multiple small ulcers to adjoining areas of deep ulcer with stricture formation. Capsule endoscopy reveals an erythematous mucosal layer with erosions, ulcers, and active bleeding with normal intervening mucosa [4].

D/D: other medication or disease-related injury and Crohn's disease ulcers.

Diaphragm disease: chronic use of NSAID may lead to the formation of concentric strictures composed of mucosa and submucosa with or without concomitant submucosal fibrosis.

Pathogenesis

Concomitant inhibition of COX-1 and COX-2 plays an important role in intestinal mucosal injury.

- **Disturbance in the microcirculation**: compromised blood flow to capillaries at the villus tips, endothelial cells become vacuolated, vascular stasis intensifies, and the overlying epithelial cells slough.
- **Topical effects** of NSAIDS: prolonged exposure and enterohepatic circulation lead to **mucosal barrier dysfunction**, which permits entry of luminal contents into the mucosa and worsens the injury. This injury activates neutrophil chemotaxis, which further sustains the nonspecific inflammation and ulceration as a response.

Diagnosis

- EGD, colonoscopy, and barium contrast studies could not detect the cause of blood loss in more than half of the patients with IDA on NSAIDS.

- CE and double balloon enteroscopy—increase detection rates of ulcers and diaphragm dramatically.
- Faecal calprotectin can be used to determine intestinal inflammation as a non-invasive means to diagnose NSAID enteropathy.

Treatment

- Withdrawal of NSAIDs is the most effective management option.
- Use of selective COX-2 inhibitors over non-selective COX inhibitors wherever possible.
- Misoprostol co-therapy has been shown to attenuate damage in both NSAID and aspirin users in studies of short duration.
- Other drugs which have been used in NSAIDS enteropathy—metronidazole and sulfasalazine.

3.46 Pathogenesis of Intestinal Tuberculosis

The bowel is one of the extrapulmonary sites for tuberculosis (Fig. 3.15). Etiologic agents are as follows:

- *Mycobacterium tuberculosis*—most cases; *Mycobacterium bovis*; *Mycobacterium avium* complex

3.46.1 Three Modes of Infection

- Reactivation of quiescent foci
- Primary infection
- Reinfection

3.46.2 Route of Infection

- Swallowed infected sputum
- Ingestion of contagious milk
- Hematogenous spread from an active pulmonary lesion, miliary TB, or asymptomatic bacteraemia during the primary phase
- Adjoining organs

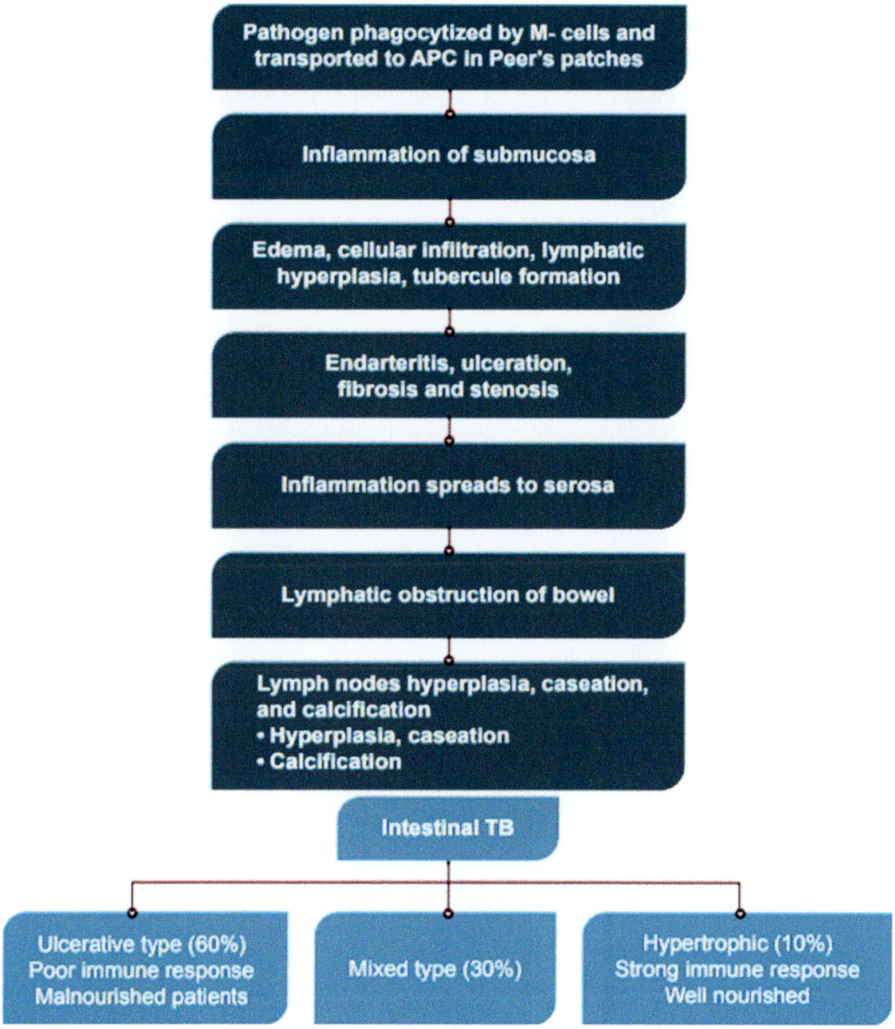

Fig. 3.15 Pathogenesis of intestinal tuberculosis

3.46.3 Site of Involvement

The most common site is the ileocaecal region due to (1) a surge in physiological stasis, (2) accelerated absorption rate, (3) negligible digestive activity, and (4) a myriad of lymphoid tissue.

3.47 Small Intestinal Bacterial Overgrowth

Malabsorption and excessive presence of bacteria in the small intestine are typical features of small intestinal bacterial overgrowth (SIBO). Frequently patients with SIBO do not manifest malabsorption symptoms, such asymptomatic patients may be grouped as 'simple colonization'.

The manifestation may include megaloblastic anaemia and steatorrhea secondary to vitamin B12 deficiency and fat malabsorption, respectively.

- Anaerobic organisms are enabled to utilise vitamin B12 both in its free form and as a complex with IF to produce inactive cobamides. This leads to vitamin B12 deficiency in the host. Moreover, the metabolites formed a contest for the ileal binding sites, consequently reducing the absorption of the available vitamin B12. Thiamine and nicotinamide deficiencies may also occur in patients with SIBO.
- The bacteria synthesise folate. Therefore, SIBO may lead to elevated levels.
- Deconjugation of bile acids may result in malabsorption of fat and fat-soluble vitamins.
- The brush border disaccharidase levels are decreased, possibly leading to carbohydrate malabsorption. Bacterial fermentation of carbohydrates manifests as abdominal discomfort and bloating, which forms the basis of breath tests. Lactose intolerance is not infrequent.

3.47.1 Protein Malabsorption

Protein malabsorption in SIBO may be attributed to the following:

1. Decreased absorption of amino acids and peptides due to mucosal damage.
2. Poor levels of enterokinase, which impedes the functioning of the pancreatic proteases.
3. Enteropathy causes loss of proteins.

Usually, the histology of the small intestine in patients with SIBO is unremarkable. Abnormalities that appear are resolved with antimicrobial pharmacotherapy. These include villus atrophy, cellular infiltration of the lamina propria, and intraepithelial lymphocytosis.

3.47.2 Clinical Features

Associated disease manifestations and symptoms can vary from subtle manifestations to frank features.

Symptoms

- Abdominal discomfort
- Bloating sensation
- Diarrhoea
- Weight loss
- Weakness
- Neuropathy

Clinical Findings

- Vitamin deficiencies (vitamin B12)
- Fat-soluble vitamins (vitamin A, vitamin D, and vitamin E)
- Fat and carbohydrate malabsorption
- Hypoproteinaemia or hypoalbuminemia
- Iron deficiency

3.47.3 Diagnosis

1. **Small intestinal aspirate and culture**
   The presence of $>10^5$ cfu/mL bacterial counts in jejunal aspirate can be considered a confirmation of the diagnosis of SIBO.
   Limitations
   (a) Invasive and time consuming
   (b) The oral flora can contaminate the aspirate and confound the results
   (c) Logistic difficulties along with technical limitations for the culture of the aspirate
2. **As described previously, breath tests** rely on producing hydrogen or radiolabelled carbon dioxide due to metabolism by glucose, lactulose, or xylose bacteria.
   Commonly used are (1) glucose hydrogen breath test and (2) lactulose hydrogen breath test. (3) ^{14}C-xylose and ^{13}C-xylose breath tests. A rise of 12 ppm above the baseline is regarded as a diagnostic of SIBO. Fasting breath hydrogen levels of more than 20 ppm are also considered positive, usually found in patients with celiac disease.
3. Others
   - As measured by gas–liquid chromatography, acetate and propionate are volatile fatty acids in the jejunal fluid.
   - Deconjugated bile salts in small intestinal aspirates.
   - Estimation of cholyl-PABA in the urine and serum bile acids (depends on the presence of bacteroides).

3.47.4 Treatment

- Nutritional rehabilitation; vitamin deficiency should be corrected.
- A diet devoid of lactose may improve diarrhoea.
- Any identified predisposing anatomic or functional abnormality should be resolved, which is an impractical alternative.
- Antisecretory agents should be withdrawn.
- Antimicrobial pharmacotherapy may be effective in SIBO; however, evidence suggests that the superiority of any particular agent over others is lacking.
  - Amoxicillin–clavulanic acid (500 mg t.d.)
  - Ciprofloxacin (250 mg b.d.)—helpful in SIBO with concomitant CD disease
  - Metronidazole (250 mg t.d.)—useful in SIBO with concomitant CD disease
  - Norfloxacin
  - Chloramphenicol (250 mg q.d.)
  - Doxycycline (100 mg b.d.)
  - Neomycin (500 mg q.d.)
  - Tetracycline (250 mg q.d.)
  - Trimethoprim-sulfamethoxazole (1 DS tablet b.d.)
  - Rifaximin (1200 mg o.d.)

Initially, therapy is administered for 2 weeks, followed by a re-assessment of the response. In patients with an anatomic or motility disorder, long-term antibiotic treatment with a rotation of antibiotics every 2 weeks or 2 out of 4 weeks is suggested. Long-term continuous use of single antimicrobial therapy may carry a risk of adverse events, e.g. peripheral neuropathy in patients given metronidazole.

- The somatostatin analogue octreotide stimulates intestinal motor activity. At a low subcutaneous dose (50 μg o.d. for 3 weeks), it was effective in SIBO with scleroderma. However, higher doses (200 μg t.d.) may *cause* SIBO by causing hypomotility.
- Prokinetics are useful in SIBO associated with cirrhosis.
- Probiotic therapy has been tried, but the results are not very promising. Evidence from small studies has found some benefits in paediatric patients with *Lactobacillus plantarum* 299V and *Lactobacillus* GG.

3.48 Short Bowel Syndrome

- Malabsorptive state occurs due to resectioning extensive small or large intestine segments.
- Short bowel syndrome occurs when there is <200 cm, bowel segment is available.
- Survival without enteral or parenteral nutrition is possible if the residual length of the small intestine is >150 cm.
- Surgical spectrum ranges from limited ileocecal resections to extensive small and colonic resections resulting in high jejunostomy.

3.48.1 Causes of Short Bowel Syndrome

1. **Vascular insults**
    - Thrombosis or embolism of SMA
    - Thrombosis of SMV
    - Volvulus/strangulation
2. **Postoperative**
    - Jejunolieal bypass
    - Abdominal trauma
3. **Miscellaneous**
    - Crohn's disease
    - Radiation enteropathy
    - Primary or secondary neoplasms of GIT

3.48.2 Factors Influencing Short Bowel Syndrome

(a) **Extent of resection**
- End jejunostomy with available bowel of less than 115 cm
- Jejunocolic anastomosis with less than 60 cm
- Duodenostomy or jejunoileal anastomosis with less than 35 cm of available bowel

(b) **Site of resection**
Jejunal resection decreased CCK and secretin secretion and gastric hypersecretion

(c) **Ileal resection**
- High stomal output
- Bile acid malabsorption
- Steatorrhea
- Vitamin B12 malabsorption

3.48.3 Colon and Short Bowel Syndrome

1. **Retention of colon: benefits**
    - Improves fluid absorptive capacity: primary function is to absorb 1–2 L of fluid presented to it from the ileum
    - Decreased energy losses
    - Decreased rate of gastric emptying

2. **Retention of colon: maladaptive consequences**
   - Secretory diarrhoea
   - Oxalate renal stones
   - Carbohydrate malabsorption and D-lactic acidosis

Intestinal Adaptation

(a) **Role of diet**

- Animal studies: intestinal adaptation found only when fed orally
- Mechanisms: direct effect on enterocytes
- Trophic hormones
- Pancreatic and biliary secretions
  - Gastrin has a trophic effect on the stomach and the duodenum
  - Cholecystokinin has a trophic effect on the small intestine
  - Enteroglucagon activates ornithine carboxylase, increases polyamine production, and cellular hyperplasia
  - Somatostatin: motility and gastric secretions
  - Neurotensin, EGF, PGE2, and arginine also probably have a role
- Short-chain fatty acids affect the motility of the stomach and ileum by proglucagon and peptide YY

(b) **Glutamine**

- Necessary fuel for enterocytes and colonocytes
- Decreased after resection and increased during intestinal adaptation
- Safeguards the structural integrity of the small intestine

3.48.4 Physiological Factors in Short Bowel Syndrome with End Jejunostomy

- Stoma within about 100 cm from DJ flexure—the net effect is a negative balance in fluids
- Negative balance in sodium if given hypotonic fluids (Na^+ 90 mmol/L stomal output). Balance studies showed a net loss of Na^+ if given fluids with Na 60 mm/L and net absorption if Na^+ was 90 mmol/L.
- Intestinal K^+ balance is generally positive except when the available bowel is less than 60 cm from DJ flexure.
- If jejunostomy output exceeds 3–4 daily, equivalent to 270–360 Na mmol/L, likely to require total parenteral nutrition (TPN).

Three phases are recognised after intestinal resection:

- Stage 1: lasts for 1–2 weeks, requires parenteral supplementation and kept NPO.
- Stage 2: lasts from several months to more than a year; oral intake gradually increases with partial or complete PN.
- Stage 3: relatively regular oral intake is possible if on TPN weaning is difficult, likely to be TPN dependent.

3.48.5 Clinical Manifestations in Short Bowel Syndrome

Diarrhoea: multifactorial

- Gastric hypersecretion—due to hypergastrinemia, responds to H2 blockers or PPI
- Steatorrhea—gastric hypersecretion and bile acid malabsorption
- Carbohydrate malabsorption
- Protein malabsorption—rare

3.48.6 Systemic Manifestations

1. Oxalate renal stones
2. Gall stones
3. D-Lactic acidosis
   - Rare complication noted in patients with safeguarded colon
   - More commonly in overfeeding with a refined carbohydrate diet
   - Carbohydrates metabolised to SCFA and lactate
   - Lower pH in the colon and support growth of acid-resistant Gram-positive anaerobes such as lactobacillus and bifidobacterium
   - D-Lactate produced is metabolised in limited amounts due to a lack of D-lactate dehydrogenase
   - D-Lactate causes metabolic acidosis and typical neurologic sequelae
   - Diagnosis by measurement of D-lactate in the blood (D-lactate >3 mmol/L; N <0.5 mmol/L)

Treatment of D-Lactic Acidosis

- Stop oral intake
- Bicarbonate for acidosis correction
- Antibiotics role debated (neomycin and vancomycin)
- Refined carbohydrates should be replaced with starch

3.48.7 Management of Short Bowel Syndrome

Four Nutritional Outcomes

- Maintenance of regular or modified oral diet
- Use of defined enteral formula diets
- Enteral intake with parenteral fluid and electrolyte supplementation
- Total or partial enteral nutrition with variable amounts of oral intake

Energy Intake

- Energy absorption is two-thirds as much as normal absorption in SBS. Day-time consumption, if inadequate, night-time NG feedings to cover the deficiency
- High energy consumption can result in high outputs making them socially disabled, requiring parenteral supplementation
- If energy absorption is less than one-third of intake—it requires parenteral nutritional supplementation

Protein

- Despite reduced absorption, protein deficiency on oral regimen is rare
- 80–100 g protein per day is sufficient
- Positive nitrogen balance and protein absorption of 60–80% on an oral regime

Fat

- Patients with jejunostomy should consume a usual quantity of fat in the diet
- Patients with complete colon benefit from dietary fat restriction

Carbohydrate

- Carbohydrates should be consumed as glucose polymers (starch)
- High carbohydrate favourable in SBS with an intact colon
- Oligosaccharide diet is associated with osmotic diarrhoea and high chances for D-lactic acidosis
- In the absence of jejunum, lactose should not be restricted
- Lactose in milk (20–25 g per glass) is well-tolerated in patients with end jejunostomy

Magnesium

- Deficiency is common in jejunostomy patients
- Deficiency is subclinical or fatigue, depression, paraesthesia, excitability, or tetany
- Magnesium oxide gelatin capsules (4 mmol, 160 mg) well tolerated, total dose 12–24 mmol/L/day

Calcium and Vitamin D

- In negative balance for calcium, Ca supplementation is recommended (800–1200 mg/day)
- Vitamin D is relatively unimportant
- Low-oxalate diet in patients with intact colon

Vitamin and Mineral Supplements in Short Bowel Syndrome

Vitamin A 1000–5000 IU daily
Vitamin B12 300 μg s.c. monthly (*T. ileal* resections)
Vitamin C 200–500 mg
Vitamin D 1600 vitamin D2
Vitamin E 30 IU daily
Vitamin K 10 mg weekly
Selenium 60–100 μg daily
Zinc 220–440 mg daily (sulphate form)

Gastric antisecretory drugs reduce jejunostomy output and improve absorptive function.

Opiates (codeine, loperamide, and diphenoxylate) should be used regularly an hour before meals for diarrhoea of SBS.

Clonidine stimulates Na^+ and Cl^- absorption and HCO_3^- and Cl^- secretion in the colon and small intestine.

Octreotide reduces jejunostomy effluent volume and sodium loss and does not benefit patients with a positive balance.

Total Parenteral Nutrition and Short Bowel Syndrome

Intestinal failure: Patients who cannot consume an adequate diet orally or cannot absorb sufficient energy should be managed with long-term total parenteral nutrition (TPN).

Home Total Parenteral Nutrition

- Should be compressed to night-time infusion
- Once the goal of fluids was determined, the infusion rate decreased to 2–3 h
- Total parenteral nutrition should be administered via a single-lumen catheter in SVC/IVC to reduce complications
- Tunnelled catheters (Hickman or Broviac) are used for home TPN. Patient education is compulsory prior to discharge

3.48.8 Nontransplant Surgery and Short Bowel Syndrome

The goal of surgical management is to increase nutrient and fluid absorption either by enhancing intestinal transit or by increasing the intestinal surface area.

- Reversed small bowel segments
- Colonic interposition
- Construction of valves
- Recirculating loops
- Tapering and plication
- Sequential intestinal lengthening

Indications for Intestine Transplant

- Life-threatening complications attributable to intestinal failure
- High risk of or evident hepatic failure due to TPN-associated hepatic injury
- Thrombosis in ≥2 central veins
- ≥Two episodes per year of catheter-related systemic sepsis necessitating hospitalisation
- Occurrence of line-related fungemia, septic shock, or acute respiratory distress syndrome
- Recurrent bouts of severe dehydration despite administration of intravenous fluids

3.49 Radiation Enteritis

Radiation therapy is a standard mode of treatment for various abdominal malignancies but can result in bowel injury.

3.49.1 Radiation Dose

Serious injury usually occurs at doses >50 Gy.

Table 3.13 Radiation dose tolerance

Intestinal parts	Minimal tolerance (Gy)	Maximal tolerance (Gy)
Small intestine	45	65
Colon	45	60
Rectum	55	80

Minimal tolerance (TD 5/5): results in 5% of patients developing radiation injury in <5 years.

Maximal tolerance (TD 50/5): results in 25–50% of patients developing radiation injury in 5 years (Table 3.13).

3.49.2 Pathophysiology

Actively dividing cells in the M phase are most susceptible to radiation. Radiation-induced injury is best described in two ways:

1. An acute injury is a function of fractionation of the radiation dose, field size, type, and frequency of treatment. It occurs due to the detrimental effects of radiation on actively dividing crypt cells in the intestine.
2. Chronic radiation injury results from cumulative exposure, which damages the mitotically less active vascular endothelial cells and connective tissue.

This causes biphasic radiation injury.

- Radiation injury compromises the usual replacement of epithelium with new cells from the epithelial crypt cells, generally occurring in 5–6 days. This leads to inconsistent degrees of retraction of villous core cells and expansion of enlarged villous epithelial cells. Thus, the absorbent surface is impaired, leading to diarrhoea resulting in malabsorption of nutrients.
- Based on the extent of loss of mucosal integrity, microulcerations or larger lesions occur. Due to impaired TJs, toxic substances or endotoxins infiltrate the plasma from the intestinal lumen.
- Damage to the vascular endothelium hinders the blood supply increasing the risk of infection and sepsis. Sepsis rarely occurs since radiotherapy is interrupted in case of evident symptoms.
- Cessation of radiotherapy allows crypt mitosis restoration within 3 days, while complete histologic recovery takes longer.
- Chronic effects occurring after 6 months to 2 years are attributed to obliterative arteritis and vessel thromboses, culminating in ischemia or necrosis.
- Thus, acute and chronic radiation injury can cause structural and functional abnormalities in the patient undergoing radiotherapy.

3.50 Microbiota–Gut–Brain Axis

Eamonn M M Quigley, MD FRCP FACP MACG FRCPI MWGO
David M Underwood Chair of Medicine in Digestive Disorders, Chief, Gastroenterology and Hepatology, Houston Methodist Hospital, Houston, TX, USA

3.50.1 Overview of the Gut–Brain Axis

The gut–brain axis is an intricate, two-way signalling system which regulates not only the GIT physiology but also certain cognitive functions. These effects are conducted via neuro-immuno-endocrine mediators that track and modulate the physiology of the gut and central nervous systems. The emotional and cognitive loci of the brain are interlinked with the enteric nervous system; therefore, they mutually influence each other. The psychological behaviour affects GI motility, secretion, immune responses, entero-endocrine signalling, gut integrity, and local reflexes, e.g. stress-induced GI dysfunction, motility, visceral perception, gastrointestinal permeability, and gastrointestinal secretions. This signalling network includes the central nervous, autonomic, enteric, and neurohormonal hypothalamic–pituitary–adrenal axis (HPA). This complex system regulates the activities of various component cells such as immune, epithelial, smooth muscle, and enterochromaffin.

Increasing evidence supports the association of the gut–brain axis with certain diseases. In autism spectrum disorder (ASD), three-fourths of the patients reported abdominal symptoms, and a considerable drop in *Bacteroidetes*/*Firmicutes* ratio and elevations in *Lactobacillus* spp. was noted. Nevertheless, these data have a certain level of ambiguity since a meta-analysis did not find a significant association between the gut–brain axis and ASD. Initiation and persistence of IBS and its symptom severity have been associated with stress.

3.50.2 Microbiota–Gut–Brain Axis

Experimental data demonstrate the role of the gut–brain axis in brain development through the upregulation of genes associated with plasticity and metabolism in various brain regions. The absence of gut microbiota has shown impaired neurogenesis, structural neuronal anomalies, and delayed maturation of nerve cells. Abnormal neurotransmitter levels, their metabolism, and receptor expression are also dysregulated without gut microbiota. Thus, these abnormalities affect the gut sensory-motor physiology cumulatively, resulting in dysmotility and HPA, which produces hyper-responsiveness to stress and a detrimental effect on memory due to reduced expression of brain-derived neurotrophic factors.

The CNS also affects the content and total biomass of gut microbiota. Stress-associated dysregulation of neurotransmitters and neuroendocrine receptors increases the host's susceptibility to inflammatory and infectious stimuli. It results in altered intestinal membrane integrity, motility, immune responses, and abnormal GI secretions, thus leading to various disease conditions.

3.50.3 Microbiota and Neurodegenerative Diseases

Gut microorganisms can modulate the expression of various proteins, signalling factors and receptors, and abnormalities that can result in a neuropsychological disorder's development.

Neurodegenerative diseases occur due to structural and physiological progressive degeneration of either the central nervous system or the peripheral nervous system, or both, e.g. Alzheimer's disease (AD), Parkinson's disease (PD), Huntington's disease, amyotrophic lateral sclerosis (ALS), and multiple sclerosis (MS). These diseases have few overlapping characteristics. These include anomalous folding of proteins, loss of neurons, and inflammation. Considering the interlinked bidirectional association of gut microbiota and the nervous system, it has been argued that microbiota dysregulation has a causal role in some neurodegenerative diseases.

Evidence gathered so far supports the role of the microbiome in the pathogenesis of the following diseases:

1. Alzheimer's disease (AD)
2. Parkinson's disease (PD)
3. Huntington's disease
4. Amyotrophic lateral sclerosis (ALS)
5. Multiple sclerosis
6. ASD
7. Schizophrenia
8. Depression

These are exciting concepts, but it must be stressed that in most instances, evidence is based mainly on animal models and in human studies on association studies where many confounding variables may be operating, and causation cannot be implied. Of these, PD has been extensively studied. The link between PD and gut microbiota was established based on Lewy pathology (LP) presence in the enteric nervous system. Braak's hypothesis suggests an unidentified neurotrophic mediator triggers PD in the gut and olfactory system. It is subsequently disseminated along the vagal nerve up to the substantia nigra. The other crucial neurodegenerative disease, AD, has been hypothesised to be initiated by the copious quantity of lipopolysaccharides and amyloid produced by abnormal gut microbiota.

Moreover, AD is a senile disease; ageing per se alters the intestinal and blood–brain barrier permeability, encouraging LPS, and amyloid passage. It leads to subsequent activation of cytokines and other small proinflammatory molecules, resulting in potentially detrimental effects on the CNS. The microbiome also probably acts as a source of the amyloid plaques, a principal causative factor of AD pathogenesis.

Here, as elsewhere, some common themes have emerged to explain how the microbiome might initiate or perpetuate a central nervous system disease: a disturbed gut microbiome elaborating neuro-active or neuro-toxic molecules, impaired gut barrier function, an inflammatory response to translocated bacterial components generating cytokines that can modify brain morphology and modulate its function all interacting with intrinsic (genetic) and environmental factors. We have a long way to go to unravel this complex mix, but the rewards for new diagnostics and therapeutics could be significant.

3.51 Gut Microbiota and Liver Disease

Eamonn M M Quigley, MD FRCP FACP MACG FRCPI MWGO
David M Underwood Chair of Medicine in Digestive Disorders, Chief, Gastroenterology and Hepatology, Houston Methodist Hospital, Houston, TX, USA

The liver has a crucial role in metabolism and immune functions. It is aptly located to monitor and eliminate the microbes or their components which escape the gut barrier and intrude into the portal circulation. Therefore, hepatic disease, dysfunction, or shunting of portal blood, averting the passage through the liver, can have detrimental effects.

The bidirectional association of the liver and the gut was established almost a century ago, with progressive elucidation in the last few decades. Similarly, the association of microbiota–gut liver axis was identified practically six decades ago. The inflammatory cascade incited by the bacteria has been linked with portal hypertension, complications of chronic liver disease as spontaneous bacterial peritonitis, systemic sepsis, and haemostatic failure.

Recent research has evidence to justify the role of gut microbiota in various liver ailments such as alcoholic liver disease, non-alcoholic fatty liver disease (NAFLD), and TPN/intestinal failure-associated liver disease (IFALD). Evidence also links it with primary biliary cholangitis and primary sclerosing cholangitis, immune-mediated disease conditions.

The interaction of the gut microbiota with the luminal contents, the compromised gut barrier integrity, and immune responses of the gut mucosal layer and the liver per se plays a significant role in the inception and dissemination of the various hepatic diseases.

3.51.1 Changes in Gut Microbiota in Liver Disease

Typically, SIBO and alteration in the faecal microbiome are two abnormalities found in individuals with liver disease or its complications.

Small Intestinal Bacterial Overgrowth (SIBO)

Changes in gut motility and transit and alterations in intestinal permeability subjects with CLD are increasingly susceptible to intestinal stasis and bacterial intrusion from the gut lumen to the portal circulation. Thus, SIBO is frequently found across various hepatic diseases.

The association of SIBO with NAFLD and NASH is known. Besides, it also plays a role in the pathogenesis of steatohepatitis in morbidly obese patients treated with jejunoileal bypass procedures.

The primary problem with SIBO is its description and lack of precision and replicability of invasive and non-invasive techniques such as aspiration and culture or breath hydrogen analysis.

Quantitative or Qualitative Changes in the Microbiota

Advanced analysis of faecal samples has detected changes in liver disease, its severity, and complications such as hepatic encephalopathy (HE), spontaneous bacterial peritonitis, and bacteraemia.

Various mechanisms have been elucidated that indicate the participation of the microbiota in the pathogenesis of NAFLD/NASH:

1. Gut microbiota and their metabolites are involved in the pathogenesis of obesity *and* metabolic syndrome.
2. These microbiotas activate the release of pro-inflammatory cytokines (e.g. TNFα) via TLR and may be involved in disease exacerbation from steatosis to NASH.
3. Microbiota also may interact with inflammasomes which may result in defective/deficient inflammasome sensing, resulting in the translocation of intestinal microbes in the portal circulation with the emergence of high levels of bacterial products (microbial- or pathogen-associated molecular patterns—MAMPs or PAMPs). These changes may underlie the progression of NAFLD to NASH.

The microbiota is qualitatively and quantitatively different from individuals with alcoholism than without. These microbiotas can metabolise ethanol to acetaldehyde and produce ethanol in situ from the fermentation of carbohydrates.

3.51.2 The Roles of Bile Acids

Bile acids are essential for the digestion of fat and fat-soluble vitamins. Additionally, these have local effects on gut motility, sensation, fluid secretion, and permeability. These also have signalling/hormonal actions which modulate activities related to energy expenditure, insulin sensitivity, and lipid metabolism. Bile acids are essential for maintaining intestinal wall integrity and protecting against bacterial infiltration. These functions are performed by activating the farsenoid X receptor (FXR) in the intestinal epithelium.

3.51.3 The Gut Barrier and the Mucosal Immune Response

The gut barrier, which acts as a defence system of the intestine, includes commensal microbiota and the mucus layer, the columnar epithelium per se, the lamina propria including constituent blood and lymph vessels, immune cells, in addition to intrinsic and extrinsic nerve terminals.

Several factors previously described can compromise the gut wall integrity and are linked to the pathogenesis of liver disease, namely ethanol, inflammatory mediators such as IFN-γ and TNF-α, mast cell proteases and neutrophils, and several medicinal agents. As shown in experimental models, translocation of Gram-negative bacteria (overgrown) and its component LPS, endotoxins to the portal system may occur due to impaired gut wall. However, affirmation in humans is difficult due to available assay constraints.

3.51.4 The Immune Response in the Liver

The translocated bacteria and their toxic components can activate the inflammasome complex setting in a cascade of events, including the secretion of pro-inflammatory cytokines, which result in hepatic injury that can eventually progress to fibrosis.

Microbiota-associated immune responses may also be involved with hepatic diseases of autoimmune aetiology. The TLR4 expression, a characteristic of primary sclerosing cholangitis (PSC), is triggered by bacterial LPS. It has also been hypothesised that the loss of immune tolerance to endotoxin in PSC leads to the initial immune-mediated biliary epithelial cell injury and ensuing progression. Similarly, TLR4 may also have a role in primary biliary cholangitis (PBC), characterised by anti-mitochondrial antibodies (AMA), particularly the M2 component of AMA. AMA has demonstrated cross-reactivity with bacterial components, particularly *E. coli* pyruvate dehydrogenase-E2.

3.51.5 Gut Microbiota and Hepatic Carcinogenesis

Progression of NAFLD to HCC has been linked to an evident alteration in the microbiome. Diminished levels of anti-inflammatory species correlate with elevated levels of calprotectin in the faeces; thus, microbiota analysis may aid in the early diagnosis of HCC.

3.51.6 Future Directions

The significance of the microbiome–gut–liver axis in the pathogenesis of important complications of liver disease has been sufficiently established. Recent investigations have divulged its complexities in association with inflammatory/immune-mediated, metabolic, cholestatic, and neoplastic hepatic diseases. Further research needs to focus on the applications of these data in identifying target sites that can be exploited for therapeutic goals.

3.52 Methods of Gut Microbiota Analysis

3.52.1 Assessment Tools for Gut Microbiota

Conventional culture-based techniques can assess gut microbiota. However, many species of gut microbiota are not cultivable. Newer molecular-based methods have helped identify many bacterial species which were not earlier cultivable (Tables 3.14 and 3.15).

3.52.2 Steps in Faecal Microbiome Evaluation

1. Sample collection: It should aim to ensure the integrity and stability of the sample obtained.
2. DNA isolation: Faeces comprise DNA from various sources such as bacteria, food, host, and specific inhibitors. This complex composition makes PCR amplification and NGS procedures highly challenging.

Table 3.14 Methods of assessing gut microbiota

Culture-based methods	Non-culture based methods
• Swab and culture • Jejunal aspiration—culture: fungus and SIBO	• Antigen tests (e.g. stool *H. pylori* antigen) • Breath hydrogen/methane for SIBO • Molecular methods: DNA/RNA/protein or metabolite based

Table 3.15 Culture-independent methods to study host–microbiome interaction

DNA-based approaches	RNA-based approaches	Protein-based approaches	Metabolite-based approaches
Who is there? What can they do? 165 rRNA, 185, ITS gene sequencing	How do they respond? What pathways are activated?	How are they interacting with the host? What proteins are being produced?	What are the chemical outcomes of their activity?
Metagenomics	Metatranscriptomics	Metaproteomics	Metabolomics

Adapted from: Huang, Yvonne & Microbiome, Committee (2018). Environmental Chemicals, the Human Microbiome and Health Risk: A Research Strategy

3. DNA extraction: DNA extraction from faeces aims to obtain suitable quality DNA with precise identification of microbial composition and corresponding abundance. The extraction methods have evolved from traditional Sanger sequencing technology to next-generation sequencing (NGS).

The NGS relies on DNA fragment detection from total DNA isolate to reveal the bacterial composition. The shotgun sequencing method offers the entire dataset of a complete gene pool in the sample. Sequencing of 16S rRNA gene amplicons, covering variable regions of the gene, is the preferred method for analysing bacterial composition given its cost-effectiveness, adequate resolution, and sequencing depth· The 16S rRNA gene is approximately 1500 bp in size and comprises nine variable regions demarcated by conserved regions.

3.52.3 Composition of Faecal Microbiota

- The gut microbiota predominantly comprises Bacteroidetes and Firmicutes.
- A study published in 2011 in Nature found that based on the proportion of Bacteroides, Prevotella, and Ruminococcus, the human populations can be grouped into three distinct enterotypes regardless of the host's ethnicity, geographical location, gender, or age. The study reported that a diet rich in protein and animal fat was associated with a Bacteroides predominant enterotype, while a carbohydrate-rich diet was associated with a Prevotella-predominant enterotype.
- Firmicutes, Proteobacteria, and Bacteroidetes are highly prevalent in adults; however, given a predominantly vegetarian diet, Indians harbour fewer Bacteroidetes than their North American counterparts (Fig. 3.16).

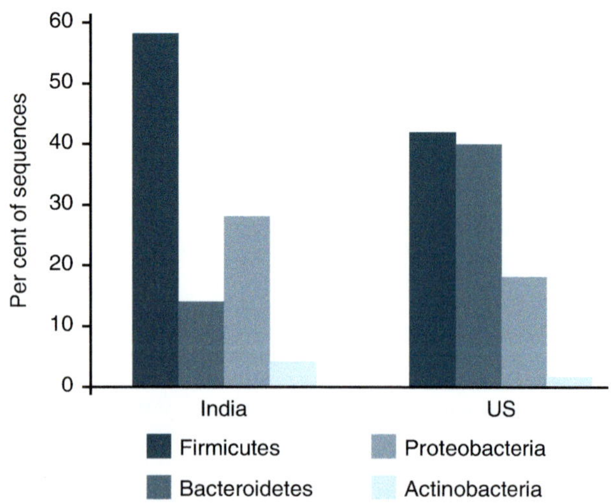

Fig. 3.16 Comparison of phylum level distribution of gut microbiota in healthy Indian and healthy US volunteers determined by cloning the faecal 16S rRNA genes. (Adapted from: Hajela, Neerja & Nair, G & Ramakrishna, Balakrishnan & Ganguly, N. Probiotic foods: Can their increasing use in India ameliorate the burden of chronic lifestyle disorders? The Indian Journal of Medical Research. 2014;139:19–26)

3.53 Intestinal Dysbiosis and the Role of Pre and Probiotics

Intestinal dysbiosis has been associated with various GI disorders (Table 3.16).

3.53.1 Probiotics

- Probiotics are living microorganisms in foods and dietary supplements that improve the host intestinal microbial balance, obstruct adhesion sites on colonocytes (which might improve mucosal barrier function), and improve local immune response.
- A probiotic can either contain a specific non-pathogenic strain of a bacterial species or a mixture of multiple species and strains, most commonly including *Lactobacillus* or *Bifidobacterium* species; seldom, they have fungal antigens as well. VSL#3 contains four strains of *Lactobacillus* (*Lactobacillus acidophilus*, *Lactobacillus delbrueckii* subspecies *bulgaricus*, *Lactobacillus plantarum*, and *Lactobacillus casei*), three strains of *Bifdobacterium (Bifdobacterium infantis*, *Bifdobacterium longum*, and *Bifdobacterium breve),* and one strain of *Streptococcus* (*Streptococcus salivarius* subspecies *thermophilus*) is a common probiotic.
- Current probiotic indications in treating GI diseases are summarised in Table 3.17.

3.53.2 Prebiotics

- Prebiotics are indigestible food components that distinctively activate the growth or activity of single or multiple organisms of the intestinal microbiota, thus, offering significant benefits to the host. These commonly include oligosaccharides, with galacto-oligosaccharide, fructo-oligosaccharide, lactulose, and inulin.

Table 3.16 Diseases associated with intestinal dysbiosis

Mucosal diseases	Non-mucosal disease
• *Clostridium difficile* infection (CDI) • Irritable bowel syndrome (IBS) • IBD • Malignancies: colorectal cancer • Diarrheal diseases	• Chronic fatigue syndrome • Hepatic malignancy and breast cancer • Periodontal diseases • Complications of liver cirrhosis • Autoimmune disorders (T1DM, arthritis) • Obesity and metabolic syndrome • Abnormalities of the gut–brain axis: Autism and other neurological disorders • Allergic conditions

Table 3.17 AGA (2020) summary of recommendations on the use of probiotics

	Disorder	Recommendation	Level of evidence
1	Prevention of antibiotic-associated *C. difficile* diarrhoea in children and adult	• *S. boulardii* or • 2-strain combination of *L. acidophilus* CL1285 and *L. casei* LBC80R or • 3-strain combination of *L. acidophilus*, *L. delbrueckii* subsp. *bulgaricus*, and *B. bifidum* or • 4-strain combination *of L. acidophilus*, *L. delbrueckii* subsp. *bulgaricus*, *B. bifidum*, and *S. salivarius* subsp. *thermophilus*	Conditional
2	*C. difficile* infection	Only in a clinical trial	
3	Ulcerative colitis and Crohn's disease	Only in a clinical trial	
4	IBS	Only in a clinical trial	
5	Acute infectious gastroenteritis	Not recommended	
6	Pouchitis	8-strain combination of *L. paracasei* subsp. *paracasei*, *L. plantarum*, *L. acidophilus*, *L. delbrueckii* subsp. *bulgaricus*, *B. longum* subsp. *longum*, *B. breve*, *B. longum* subsp. *infantis*, and *S. salivarius*	

3.54 Faecal Microbiota Transplant

It aims to restore physiological gut microbiota in diseases with dysbiosis.

Donor selection (Ref.: BSG guidelines on FMT)

- The donors can be selected from among adult family members, intimate partners, friends, or unrelated volunteers.
- Donors are selected based on clinical history (Table 3.18), physical examination, and results of lab investigations (Tables 3.19 and 3.20).

3.54.1 *Types of Delivery of Faecal Microbiota*

FMT can be delivered through a nasogastric tube, colonoscopy, enema, or through capsule (Table 3.21).

3.54.2 *The Procedure of Processing Donor Faeces and Storage*

- **Fresh faecal donation**: The stool sample in clean polythene is obtained within a month of screening from the donor. In the laboratory, it is diluted in normal saline, mixed in a sterile bag by hand stirring, and blenderized. After filtration through moistened 5-layer sterile gauze in a funnel, it will be delivered to the endoscopy suite within 4 h of presentation (Fig. 3.17).

Table 3.18 Recommended donor questionnaire

Donor questionnaire
1. Prior exposure to HIV and/or viral hepatitis and known previous or latent tuberculosis.
2. Risk factors for blood-borne viruses, including high risk, sexual behaviours, drug abuse, tattoo/body piercing/needlestick injury/blood transfusion/acupuncture within the last 6 months.
3. Vaccination with a live attenuated virus within the past 6 months.
4. Pre-existing GI conditions/symptoms (e.g. history of IBD, irritable bowel syndrome (IBS), chronic diarrhoea, chronic constipation, coeliac disease, bowel resection or bariatric surgery), also including acute diarrhoea/GI symptoms within the past 2 weeks.
5. Family history of significant GI conditions (e.g. IBD or colorectal cancer).
6. History of atopy (e.g. asthma, eosinophilic disorders).
7. Any systematic autoimmune conditions.
8. Any metabolic conditions, including diabetes and obesity.
9. Any neurological or psychiatric conditions or known risk of prion disease.
10. History of chronic pain syndromes, including chronic fatigue syndrome and fibromyalgia.
11. History of any malignancy.
12. Medication history within the past 3 months: antimicrobials, proton pump inhibitors, immunosuppression, and chemotherapy.
13. History of receiving growth hormone, insulin from cows, or clotting factor concentrates.
14. History of receiving an experimental medicine or vaccine within the past 6 months.
15. History of travel to tropical countries within the past 6 months.

Table 3.19 Recommended blood investigations for donor

General/metabolic screening	Pathogen screening
• Full blood count with differential • Creatinine and electrolytes • Liver enzymes (including albumin, bilirubin, aminotransferases, gamma-glutamyl transferase and alkaline phosphatase) • C reactive protein	• Hepatitis A IgM • Hepatitis B (HBsAg and HBcAb) • Hepatitis C antibody • Hepatitis E IgM • HIV-1 and HIV-2 antibodies • HTLV-1 and HTLV-2 antibodies • Treponema pallidum antibodies (TPHA, VDRL) • Epstein–Barr virus IgM and IgG[a] • Cytomegalovirus IgM and IgG[a] • Strongyloides stercoralis IgG[a] • Entamoeba histolytica serology

[a] Epstein–Barr virus and cytomegalovirus testing is only recommended where there is the potential that the faecal microbiota transplant prepared from that donor will be administered to immunosuppressed patients at risk of severe infection

- **Storage of faeces**: Faecal material is converted to 250-mL aliquots in a similar standardised method and stored at −80 °C and delivered on dry ice on request.
- **Capsules**: Faecal microbiota capsules can be obtained by concentrating the slurry. Subsequently, the faecal solution is transferred into a capsule holding 650 mL and then sealed by a second capsule made of acid-resistant, hypromellose material. Each donor stool is converted into 30 capsules, acting as a single therapeutic dose. These capsules can be stored at −80 °C (−112 °F) before use for a maximum of 6 months.

Table 3.20 Recommended stool screening for donors

- *C.* difficile PCR
- Campylobacter, Salmonella, and Shigella by standard stool culture and/or PCR
- Shiga toxin-producing *E. coli* by PCR
- Multi-drug resistant bacteria, at least CPE and ESBL[a]
- Stool ova, cysts, and parasite analysis, including for Microsporidia
- Faecal antigen for Cryptosporidium and Giardia
- Acid-fast stain for Cyclospora and Isospora
- *H.* pylori faecal antigen
- Norovirus, rotavirus, PCR

[a]While carbapenemase-producing Enterobacteriaceae (CPE) and extended-spectrum beta-lactamases (ESBL) are the only multi-drug resistant bacteria that are recommended to be screened for universally, consider testing for other resistant organisms [including vancomycin-resistant Enterococci (VRE) and/or methicillin-resistant *Staphylococcus aureus* (MRSA)] based on risk assessment and local prevalence

Table 3.21 Types of faecal microbiota transplant—advantages and disadvantages

Method	Advantages	Disadvantages
Oral/nasogastric/ nasoduodenal/ nasojejunal	• No sedation necessary • Cost-effective	• Discomfort with tube placement • Risk of vomiting and aspiration • Evaluation of mucosa or biopsy sample collection not feasible
Colonoscopy	• Enables evaluation of mucosa and obtaining biopsy samples • Highest efficacy for treatment of recurrent *Clostridium difficile* infection (rCDI)	• Invasive • Generally, it requires sedation • Standard risks of colonoscopy (e.g. discomfort, perforation, bleeding) • Expensive
Enema	• Less invasive • It can be administered in-office setting or at home • No sedation necessary • Cost-effective • Easily reproducible	• Donor stool does reach the entire colon and is limited to the left-sided colon • Less effective than other routes for rCDI
Capsules	• Less invasive • It can be administered in an office setting • No sedation necessary	• Expensive but eliminates the cost of endoscopy • Less effective than colonoscopy for rCDI • Large capsule burden • Risks of vomiting and aspiration

3.54.3 Clinical Utility of Faecal Microbiota Transplant

Faecal microbiota transplant has been studied in the following conditions:

1. *C. difficile* infections
2. IBD
3. Corticosteroid non-responsive alcoholic hepatitis
4. NASH

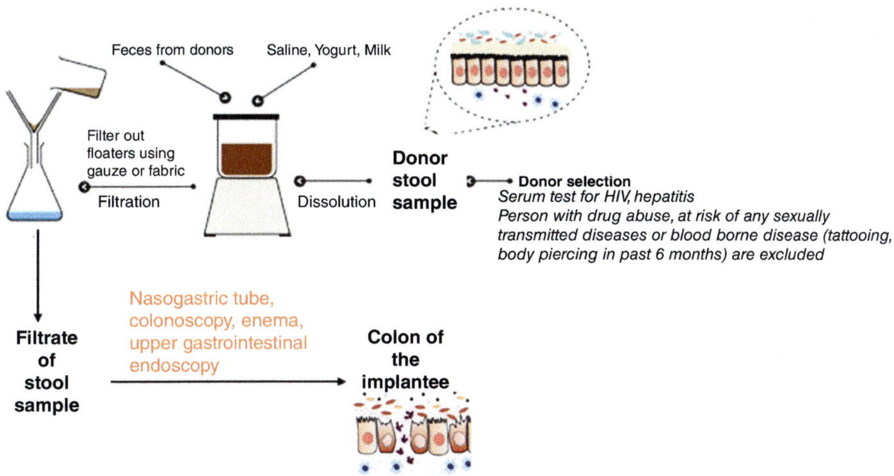

Fig. 3.17 Optimal processing of faeces. (Adapted from: Wu et al., Optimization of faecal sample processing for microbiome study: The journey from bathroom to bench. Journal of the Formosan Medical Association (2019) 118, 545–555)

FMT for Recurrent *C. difficile* Infection

- Patients with mild-moderate rCDI respond most to FMT if they respond to antibiotics that reduce the *C. difficile* load by the fourth day. Antibiotics are withdrawn at 24–48 h before FMT.
- FMT transferred by colonoscopy requires bowel preparation or administration of an enema to enhance colon visualisation. The standard dose of FMT is ~50–100 g diluted to 250–500 mL.
- FMT can also be administered from the upper GIT through capsule ingestion.

3.54.4 Efficacy

BSG guidelines currently recommend FMT as a treatment solely for recurrent *C. difficile* infection and not for any other gastrointestinal or non-gastrointestinal disease.

3.54.5 Follow-Up and Side Effects

- The common side effects of FMT are bloating and loose stools for the first 24 h, which generally subside by 1–2 weeks and are replaced by formed stools. Initial follow-up assessment is performed 3–7 days post-FMT, followed by again at 4–8 weeks.

- Stool testing is considered if ≥ 3 episodes/day of diarrhoea occur after a few weeks of FMT. It should be noted that *C. difficile* toxin PCR returns positive for a month after FMT.

3.55 Duodenal Mucosal Resurfacing

Principle: Insulin resistance is an important underlying factor in the pathogenesis of metabolic syndrome (diabetes, hypertension, dyslipidaemia, obesity, and NAFLD).

3.55.1 Role of the Small Intestine in Glucose Metabolism

It was noticed that Roux en Y gastric bypass (RYGB) caused improvement in glycaemic control immediately post-surgery before any significant weight loss could occur. It was postulated that the bypass of nutrients from proximal intestinal segments was possibly the reason for the early metabolic benefit.

Presumably, duodenal mucosal resurfacing improves insulin sensitivity and, consequently, hepatic glucose production.

3.55.2 Role of Intestinal Permeability and Endotoxemia in Insulin Resistance

- Accumulating evidence has shown increased intestinal permeability, systemic endotoxemia, and chronic low-grade systemic inflammation in obese subjects and patients with type 2 diabetes mellitus (T2D), which may be associated with the development of insulin resistance.
- RYGB surgery is associated with a reduction in endotoxemia in patients with obesity and T2D. Change in microbiome composition correlates with improved peripheral insulin sensitivity in patients with metabolic syndrome. Roux en Y surgery helps restore healthy microbiota and improve insulin sensitivity.

3.55.3 Increase of Insulin Secretion Mediated via Incretins Secreted from the Small Intestine

- Glucagon-like peptide 1 (GLP-1) and GIP are incretins secreted from L and K cells in small intestines.

- Administration of exogenous GLP-1 has significant antidiabetic effects. It is hypothesised that L-cell stimulation may cause increased endogenous secretion of GLP-1 and help in glucose haemostasis.

3.55.4 Glucose Haemostasis Is Modulated by Bile Acid and Its Receptors in the Intestine and Liver

- Bile acids improve glucose metabolism and insulin sensitivity via specific bile acid signalling pathways.
- Bile acids modulate energy expenditure and glucose and lipid metabolism via the nuclear FXR in the liver and distal small intestine and G-protein-coupled receptor 5 (GPCR) in the terminal ileum.
- Takeda GPCR-5 induces direct secretion of GLP-1 from L cells of the small intestine, which can modulate glucose metabolism on stimulation by bile acids.

3.55.5 Procedure

The duodenal mucosal resurfacing (DMR) procedure requires the administration of general anaesthesia or deep sedation with propofol.

- An initial screening gastro-duodenoscopy is performed to eliminate the risk of contraindications to the procedure. With the help of argon plasma coagulation or placement of an endoscopic clip, the position of the papilla of Vater was marked on the contralateral duodenal wall.
- A guidewire should be placed ahead of the ligament of Treitz.
- DMR catheter should be progressed along the guidewire using the endoscope.
- DMR comprises submucosal expansion (to create a protective layer of saline between the mucosa/submucosa and the duodenal proper muscle layer) followed by stepwise circumferential hydrothermal ablation at 90 °C for 10 s over 9–10 cm of the post-papillary duodenum.
- Usual antidiabetic medicines are continued for 24 weeks.

3.55.6 Benefits

A multi-centre open-label study has shown.

1. Diabetes control (reduction in HbA1C) after 1 year.
2. Improve dyslipidaemia (reduction in LDL and triglyceride and elevation of HDL).
3. Improvement in LFT after 1 year.

3.56 Endoscopic Therapy for Obesity

3.56.1 Indications

Indications for endoscopic therapy are listed in Table 3.22.

3.56.2 Endoscopic Bariatric Interventions

1. Gastric '**restrictive**' procedures
   - Intragastric balloons—Orbera, ReShape Duo, Spatz adjustable balloon system, Obalon gastric balloon
   - Other space-occupying devices—transpyloric shuttle, full sense device
   - Aspiration therapy—a tube and aspire assist device
   - Gastroplasty techniques—endoscopic sleeve gastroplasty, POSE
2. Small intestinal **'malabsorptive'** procedures
   - Duodenojejunal bypass
   - Gastroduodenojejunal bypass
   - Duodenal mucosal resurfacing

Table 3.22 Indications for endoscopic treatment for obesity

Procedure	Goal
Early intervention	Weight loss or stabilisation in early-stage obese patients yet ineligible for conventional surgery
Bridge to surgery	Attenuating the obesity-related operative risk for various bariatric and non-bariatric surgeries
Metabolic	Managing comorbid illness (e.g. diabetes)
Revision	Repairing unsuccessful bariatric surgical procedures

3.57 Enteral Stents

3.57.1 Indications

1. Malignancy of the pancreas, duodenum, and stomach causes gastroduodenal obstruction.
2. Benign gastroduodenal obstruction (peptic ulcer disease) who are poor surgical candidates.

3.57.2 Technique

- The patient is positioned in the left lateral or prone position (better for stent visualisation) rather than the supine position (unless the patient is intubated for airway protection) to avoid aspiration.
- Gastroduodenal stent placement is performed under endoscopic and fluoroscopic guidance. A therapeutic upper endoscope, colonoscope, or duodenoscope is used with a working channel large enough to accommodate the stent delivery system.
- During endoscopy, the stricture length may be estimated accurately by advancing the endoscope through the stricture or by injecting contrast material with fluoroscopy.
- A guidewire is advanced across the stricture into the distal bowel under fluoroscopy.
- A through-the-scope stent of appropriate size, generally 4 cm longer than the size of the stricture, is then advanced over the guidewire and deployed under endoscopic and fluoroscopic guidance. At least a 2 cm length of the stent should be flared at both ends of the stricture to attain an appropriate 'waist'; otherwise, overlapping stents may be needed to traverse the length of the stricture fully.
- Peritoneal carcinomatosis is a relative contraindication to stent placement, although limited recent data suggest that these patients have similar outcomes to those without carcinomatosis.
- If biliary obstruction is present or impending, placing a biliary stent before gastroduodenal stenting is prudent to avoid difficult biliary access later.

3.57.3 Efficacy

- Gastroduodenal SEMS placement has demonstrated 95% technical success and 85–95% clinical success with adequate decompression of outlet obstruction and patients' ability to tolerate at least a soft diet.
- Several studies have shown that stent placement is usually associated with better short-term outcomes, such as a faster ability to tolerate oral intake and shorter

Table 3.23 Enteral stents

Stent	Composition	Delivery system	Unconstrained outer diameter	Features
Ultraflex (Boston Scientific)	Nitinol	22F	25 mm body, 30 mm flare	Over-the-wire proximal suture release for left-sided colonic obstruction; indicated for palliation and as a bridge to surgery for malignant colonic neoplasms
WallFlex (Boston Scientific) duodenal	Nitinol	10F	22 mm body, 27 mm flare	Low-profile 10F through-the-scope/the over-the-wire delivery catheter, reconstrainable up to 70% deployment
Wallstent (Boston Scientific) (colonic and duodenal)	Elgiloy	3.3 mm	20 mm, 22 mm	Through-the-scope deployment possible; reconstrainable before full deployment

duration of hospital stay. In contrast, surgery is associated with better long-term outcomes and a low reintervention rate. Therefore, stent placement is a reasonable alternative to palliative surgery in patients with a short-life expectancy, generally less than 6 months.

3.57.4 *Adverse Events*

- Mild—abdominal discomfort, mild fever, and occasional vomiting without obstruction.
- Major (in the first week)—bleeding, perforation, stent migration, severe pain, fever, and jaundice.
- Major (late)—fistula formation, stent obstruction, late perforation or bleeding, biliary obstruction, and stent migration.

3.57.5 *Types of Enteral Stent and Characteristics (Table 3.23)*

3.58 Evaluation of Small Bowel: Use of Capsule, CT Enterography, and CT Angiography

3.58.1 *Video Capsule Endoscopy*

- Used as a diagnostic tool for small intestinal imaging.
- Provides higher magnification and higher resolution as compared to traditional endoscopy.

- Primarily used for detecting obscure GI bleeding, but can also be used for diagnosis of IBD, celiac disease, and small-bowel neoplasia.

Obscure Gastrointestinal Bleeding

- Capsule endoscopy may detect superficial small intestine lesions that barium studies miss.
- Superior diagnostic yield of capsule endoscopy as compared with that of push enteroscopy plus small-bowel follow-through study.
- Capsule endoscopy can also detect lesions within reach of conventional methods (e.g. colonoscopy and upper endoscopy). Capsule endoscopy can also identify previously undetected sources of obscure GI bleeding in the colon and stomach.
- A sensitivity of 95% and a specificity of 75%.
- The most common lesions detected include telangiectasias and ulcerations.
- Does not allow immediate intervention. However, capsule endoscopy may be used to plan subsequent interventions.

IBD

- Crohn's disease affects the small bowel in most individuals; in 30% of patients, the disease is limited only to the terminal ileum.
- Conventional studies are limited in diagnosing proximal ileal and jejunal lesions. Thus, VCE helps diagnose IBD, localise the lesion, and assess the extent of the disease.
- Diagnostic yield of capsule endoscopy with that of other modalities (e.g. barium studies, colonoscopy with ileoscopy, computed tomography [CT] enterography or enteroclysis, and small-bowel magnetic resonance imaging [MRI]) found capsule endoscopy to be superior in the diagnosis of a recurrence in non-stricturing small-bowel Crohn's disease.

Small-Bowel Polyps and Neoplasia

- Small-bowel neoplasia occurs in 75% of Peutz–Jeghers syndrome (PJS) patients and 90% of patients with familial adenomatous polyposis (FAP).
- Patients undergoing CVE for other reasons, if found to have a significant polyp, can prompt operative interventions in patients with FAP and PJS (a retrospective study found 8.9% of patients had small bowel tumours, out of which 48% were malignant).

Celiac Disease

- Capsule endoscopy has been considered an alternative diagnostic tool (to duodenal biopsy) because its magnification is sufficient for imaging the villi and detecting villous atrophy, with capsule endoscopy having a sensitivity of 87.5% and specificity of 90.9%.

  Contraindication to capsule endoscopy: small bowel obstruction/stenosis.

3.58.2 CT Enterography

- It is a modification to 'standard' abdominopelvic CT examination to examine the small bowel in detail specifically. It combines neutral (low-density) oral contrast with 'enteric phase' CT to optimise contrast resolution between mucosa and lumen, thereby maximising the chance of identifying abnormalities arising from the small bowel wall.
- They are broadly categorised into CT enterography (where patients drink oral contrast) and CT enteroclysis (luminal contrast is introduced via a nasojejunal tube).
- Positive oral contrast agents (containing iodine or barium) are not routinely used for CT enterography because they obscure mucosal enhancement, intraluminal haemorrhage, and assessment of subtle mural disease. Positive oral contrast should be avoided in obscure gastrointestinal bleeding because the contrast can obscure the bleeding site. However, it can be used occasionally to help establish fistula patency or the exact site of mechanical obstruction because it will track the flow of luminal contents.

Assessment of CT Enterography

1. Enhancement pattern
   (a) Target appearance with mural stratification—benign conditions like—vasculitis, CD, and venous thrombosis with oedema/ischemia.
   (b) Homogeneous and mild enhancement—a chronic disease like CD/radiation-related enhancement.
   (c) Homogenous hyper-enhancement—active CD.
   (d) Heterogeneous enhancement—neoplasms.
2. Length of bowel involvement
   (a) Focal—neoplasms, endometriosis, small bowel diverticulitis, foreign body perforations, and small bowel ulcers.

Table 3.24 Characterisation of small intestinal mural thickening

Size of the mural thickening	Mild	Hypoalbuminemia, infectious, mild ischemia, or Crohn's disease
	Moderate	Crohn's disease, intestinal ischemia, intramural haemorrhage, vasculitis, lymphoma, or early adenocarcinoma
	Severe	Neoplasms, including lymphoma, vasculitis, Crohn's disease, or intramural haemorrhage
The symmetry of bowel thickening	Symmetrical	Benign conditions and a few cases of lymphoma
	Asymmetrical	Crohn's disease, TB, adenocarcinoma, and GIST
Site of abnormal mural thickening	Proximal small intestine	Adenocarcinoma and celiac disease
	Distal small intestine	Lymphoma, carcinoid, and Crohn's disease

(b) Segmental—Crohn's disease, lymphoma, infectious enteritis, and ischemia: SMA and SMV related.
(c) Diffuse: hypoalbuminemia, low-flow intestinal ischemia, vasculitis, graft-versus-host disease, and infectious enteritis.

3. Mural thickening and symmetry

   Size, pattern, and area of involvement can provide vital clues regarding the aetiology of mural thickening (Table 3.24).

Disease-Specific Features on CT Enterography

1. Crohn's disease includes mucosal hyper-enhancement, mural thickening and stratification, transmural ulceration, mesenteric inflammation, engorgement of vasa recta, and strictures associated with upstream dilatation. The mucosa and serosa enhance avidly, but the intervening bowel wall enhances to a varying degree due to oedema or intramural fat.

   Complications of Crohn's disease may be due to transmural ulceration, resulting in abscesses or the formation of fistulae between bowel segments and other organs.
2. Small bowel tumours: account for <5% of gastrointestinal tumours (Table 3.25).
3. GI bleed

   - Superior to MRI enterography for the investigation of chronic blood loss.
   - It can be considered complementary to CVE.
   - Radiation exposure remains a concern and should be avoided in young patients in non-emergency settings. Whereas in elderly patients risk-to-benefit ratio favours CT.

Table 3.25 Imaging characteristics of small intestinal tumours

Tumours	Adenocarcinoma (at least 25%)	Carcinoid tumour (25–41%)	Lymphoma	Gastrointestinal stromal tumour (benign or malignant)
Location	Proximal small bowel proximity to the ampulla	Frequently in the ileum	Multiple sites, frequent in the ileum	Frequent in the ileum or jejunum
Imaging characteristics	Different shapes, soft tissue masses with heterogeneous attenuation, duodenal ones can be papillary or polypoid; more distal ones are mostly annular	Polypoid or carpet-like with mesenteric desmoplastic reaction may have calcified deposits in the mesentery, likely hypervascular or necrotic liver metastases	Exoenteric mass with adjoining lymphadenopathy and aneurysmal dilation	Pedunculated or exoenteric is usually homogeneous, but large tumour scans play necrotic areas

- Definite role in young patients with bleeding from suspected Meckel's diverticulum. Technetium 99 pertechnetate scintigraphy is the modality of choice for investigating paediatric patients with gastrointestinal haemorrhage and a suspected Meckel's diverticulum, with a sensitivity of 85%, a specificity of 95%, and an accuracy of 90%.

Specific indications for CT enterography are listed in Table 3.26.

3.58.3 CT Angiography (CTA)

As a result of improvements in imaging protocols and scanner technology, it is possible to acquire exquisite CTA images of the bowel, mesentery, and mesenteric vasculature (including second-order and third-order branch vessels) and create 3D reconstructions of extraordinary detail.

Procedure

- Intravenous (IV) contrast: 100–120 mL of non-ionic contrast is typically injected rapidly (4–5 mL/s).
- Using the dual phase technique—arterial phase images are acquired using a bolus trigger (usually at roughly 30 s), whereas venous phase images are acquired using a fixed scan delay of roughly 60 s.
- Arterial phase images are critical for evaluating the mesenteric vasculature and identifying it.

Table 3.26 Indications for CT enterography

Small bowel disorder	Crohn's disease	Small bowel tumour	Occult gastrointestinal bleeding	Partial small bowel obstruction	Coeliac disease
Indication	Assessment of extent, severity, and luminal/extraluminal complications	Detection, characterisation, and staging	Detect and characterize vascular/neoplastic origin	Locate the site and detect the cause	Assessment of complications, particularly lymphoma
Advantage	Allows simultaneous diagnosis of extraluminal disease and complications. Preferred examination in suspected small bowel stricture	Allows simultaneous detection and staging	Improved visualisation of lesions compared with conventional CT and fluoroscopy, lesion-directed planning	Improved visualisation of mucosal lesions compared with conventional CT	CT enterography may show a typical mucosal pattern and lymph node involvement
Disadvantage	Less sensitive in detecting early mucosal abnormalities than capsule endoscopy Radiation concerns preclude CT enterography in paediatric patients. Repeated examinations in young adults, ultrasound or MR enterography, are preferred in these situations	Less sensitive than capsule study for detection of small mucosal lesions	Unlike nuclear medicine techniques, CT enterography does not allow prolonged imaging times necessary to detect intermittent bleeding	Conventional barium follow-through study is more sensitive in assessing adhesional obstruction than CT enterography	Unlike endoscopic techniques, it does not depict subtle mucosal changes

- Hyper-enhancing small bowel tumours and assessing subtle abnormalities in small bowel wall enhancement.
- Venous phase images are the most important for assessing the parenchymal organs of the abdomen, identifying hypovascular small bowel tumours, and assessing small bowel wall thickening.

In acute presentation, oral contrast of any kind should typically be avoided. In a sub-acute or chronic presentation, where there is greater importance placed on the need for good bowel distension—standard CT enterography protocol (neutral oral contrast agents) + IV contrast should be used.

Indications

Acute Gastrointestinal Bleeding

CTA

Reasonable sensitivity at blood loss of 0.35 mL/min, additional diagnostic information like ischemic versus infection versus inflammatory can also be acquired. Even in cases where active extravasation is not visualised, CT offers the opportunity to diagnose various other abnormalities that may explain the patient's bleeding, such as ischemia, IBD, small bowel tumour, and arteriovenous malformation.

Tagged RBC Scans

Detects bleed as slow as 0.1 mL/min with an additional ability to detect intermittent bleed.

Limitations: poor spatial resolution, exact site, and cause cannot be determined.

Conventional Angiography (DSA)

The effective modality for treatment.

Limitations: Sensitive only for rapid bleeding (>1 mL/min), limited contrast resolution, inability to identify intermittent bleeding.

Acute Mesenteric Ischemia

Divided into four major categories.

(a) **Acute arterial occlusion**: a most common cause of mesenteric ischemia (up to 70%).

  - Most common sources—cardiac from AF, myocardial clot post-MI, and atherosclerotic plaque ulceration.
  - A clot in proximal SMA—entire small bowel ischemia.

Table 3.27 Imaging characteristics of mesenteric vascular occlusions

Characteristic	Arterial occlusion	Venous occlusion
Superior mesenteric	Embolus	Thrombosis
Bowel thickening	No	Yes
Bowel wall	Paper thin	Hypodense due to oedema or hyperdense due to haemorrhage
Mucosal hyperenhancement	No	Yes
Bowel dilation with infarction	Yes	It may or may not
Mesenteric findings	Slight stranding, haemorrhage, or oedema	Significant inflammation with fluid and haemorrhage

- Smaller emboli can extend more distally into SMA branches and affect just a short segment of the bowel.
- Rarely arterial occlusion can also result from SMA thrombosis with extensive atherosclerotic disease.

- **Imaging**—abnormal hypo-enhancement with thinning of the bowel wall over time with a lack of adjacent mesenteric fat stranding, oedema, and haemorrhage (Table 3.27).

(b) **Acute venous occlusion**: 10% of all cases.

- Typically related to thrombosis of the SMV.
- Cause—underlying hypercoagulable state, drug-induced hypercoagulability, e.g. oral contraceptives, status post complex hepatobiliary surgery, and septic thrombophlebitis caused by an underlying GI tract infection.

Imaging—significant mesenteric oedema, fat stranding, haemorrhage, and markedly thickened bowel wall with significant mucosal hyperaemia. The thickened wall can be hypodense or hyperdense.

(c) **Nonocclusive mesenteric ischemia** (NOMI): diminished blood flow to the small bowel in hypotension.

Imaging—does not have a consistent imaging pattern and can resemble either arterial or venous occlusion or may appear as nonspecific bowel wall thickening.

(d) **Complicated bowel obstructions**: causes ischemia because of strangulation of the arterial and venous blood supply to the bowel, with ischemia most likely in cases of either severe obstructions or closed-loop bowel obstructions with twisting and occlusion of the vascular pedicle supplying the bowel.

Imaging—often has features of both arterial and venous occlusion because the twisting of the vascular pedicle can interrupt the arterial and venous blood supply to the bowel.

Regardless of the cause of the obstruction, the presence of pneumatosis and portal venous gas is virtually diagnostic of the presence of transmural bowel infarction, with specificity approaching 100% for the presence of ischemia.

Vasculitis

Vasculitis represents inflammation and necrosis of the blood vessels, with different types of vasculitis showing involvement of vasculature (small, medium, and large).

Small bowel involvement is seen in HSP, lupus vasculitis, Behcet disease, and PAN.

(a) **PAN**—medium vessel vasculitis—mesenteric vessel involvement is seen in up to 50% of patients.

    Imaging—aneurysms at arterial branch points, as well as the development of bowel ischemia, GI haemorrhage, and a bowel perforation.

(b) **HSP** is a small vessel vasculitis—bowel involvement (~60%), with the small bowel being the most common.

    Imaging—severe inflammation or ischemia, with evidence of bowel wall thickening, ulceration, mesenteric oedema, haemorrhage, ascites, and reactive lymphadenopathy. Young patients can present with intussusception.

(c) **Lupus vasculitis** is an autoimmune disease that can affect any portion of the bowel because of immune complex deposition or result in ischemia caused by these patients' hypercoagulability and consequent venous thrombosis.

    Imaging—small bowel ischemia or severe inflammation, with bowel wall thickening, mucosal hyperaemia, and mesenteric inflammation.

(d) **Behcet disease** is a small vessel vasculitis occurring in young men with a strong tendency to involve the distal ileum. It can mimic Crohn's disease/malignancy.

    Imaging—bowel wall thickening can be severe, mass-like, and irregular, mimicking the appearance of a discrete mass.

Other syndromes that can be diagnosed with CTA include the following:

- Superior mesenteric artery syndrome
- Median arcuate ligament syndrome
- Mesenteric artery dissection
- Chronic mesenteric ischemia

Further Reading

Arnott ID, Lo SK. The clinical utility of wireless capsule endoscopy. Dig Dis Sci. 2004;49(6):893–901.

Arumugam M, et al. Enterotypes of human gut microbiome. Nature. 2011;473:174–80.

Hartmann D, Schmidt H, Bolz G, Schilling D, Kinzel F, Eickhoff A, et al. A prospective two-centre study comparing wireless capsule endoscopy with intraoperative enteroscopy in patients with obscure GI bleeding. Gastrointest Endosc. 2005;61(7):826–32.
Ilangovan R, Burling D, George A, Gupta A, Marshall M, Taylor SA. CT enterography: review of technique and practical tips. Br J Radiol. 2012;85(1015):876–86.
Kitiyakara T, Selby W. Non-small-bowel lesions detected by capsule endoscopy in patients with obscure GI bleeding. Gastrointest Endosc. 2005;62(2):234–8.
Leighton JA, Sharma VK, Hentz JG, Musil D, Malikowski MJ, McWane TL, et al. Capsule endoscopy versus push enteroscopy for evaluation of obscure gastrointestinal bleeding with 1-year outcomes. Dig Dis Sci. 2006;51(5):891–9.
Panek M, Čipčić Paljetak H, Barešić A, et al. Methodology challenges in studying human gut microbiota—effects of collection, storage, DNA extraction and next generation sequencing technologies. Sci Rep. 2018;8:5143. https://doi.org/10.1038/s41598-018-23296-4.
Triester SL, Leighton JA, Leontiadis GI, Gurudu SR, Fleischer DE, Hara AK, et al. A meta-analysis of the yield of capsule endoscopy compared to other diagnostic modalities in patients with non-stricturing small bowel Crohn's disease. Am J Gastroenterol. 2006;101(5):954–64.
van Baar ACG, et al. Endoscopic duodenal mucosal resurfacing for the treatment of type 2 diabetes mellitus: one year results from the first international, open-label, prospective, multicentre study. Gut. 2020;69:295. https://doi.org/10.1136/gutjnl-2019-318349.
van Baar ACG, et al. The duodenum harbors a broad untapped therapeutic potential. Gastroenterology. 2018;154:773. https://doi.org/10.1053/j.gastro.2018.02.010.

Chapter 4
Large Intestine

Piramanayagam Paramasivan

4.1 Embryology, Development, and Malformations of the Colon, Rectum, and Anal Canal

4.1.1 Formation of Gut Tube and Its Divisions (Days 21–50)

The gut tube is formed when the yolk sac is retracted into the embryo and nipped off as the horizontal germ layers of the early embryo fold laterally and cephalocaudally. The tube has an endodermal lining from which the epithelial layer is derived.

The gut tube remains in contact with the yolk sac (midgut) and splits the gut tube into foregut, midgut, and hindgut regions.

Blood Supply The foregut is innervated by the coeliac artery, the midgut by the superior mesenteric artery, and, lastly, the hindgut by the inferior mesenteric artery.

Development of Midgut

At week 5, the midgut slings from the dorsal abdominal wall and connects with the yolk sac through the vitelline duct. It undergoes rapid elongation and loop formation. The proximal part of the midgut forms the duodenum, jejunum, and a segment of the proximal ileum, and the tail end forms the distal part of the ileum, caecum, appendix, ascending colon, and proximal two-third part of the transverse colon.

1. Physiological umbilical herniation

   Due to the rapid growth of the liver and relatively small abdominal cavity, the intestinal loops herniate into an extra-embryonic cavity.

P. Paramasivan (✉)
Department of Medical Gastroenterology, Apollo Hospitals, Chennai, India

G. Choudhuri et al. (eds.), *Gastroenterology and Hepatology*,
https://doi.org/10.1007/978-981-99-9287-4_4

2. Rotation of midgut

   It rotates around the axis of Superior mesenteric artery (SMA) in the counter-clockwise direction for 270°. During rotation, elongation and coiling of small intestinal loops take place, whereas only elongation of colonic loops takes place.
3. Return into the abdominal cavity

   At the 10th week, herniated loops return to the abdominal cavity. The caecal bud is the terminal part to re-enter, lying in the right superior quadrant just inferior to the right lobe of the liver. This later sinks into the right iliac fossa, positioning the right colon and hepatic flexure on the right side of the abdomen.

Development of Hindgut

The terminal part of the hindgut ends in a cavity called cloaca (which is continuous with allantois, a remnant of yolk-sac—that regresses later to form the superior part of the bladder).

Uro-rectal septum—a wedge of mesoderm that grows posteriorly towards the cloacal membrane, splitting the cloaca into primitive urogenital sinus towards the anterior side and the anorectal canal towards the posterior side. The cloacal membrane ruptures into amniotic cavity. Thus, the caudal part of the lining of the anus is derived from the ectoderm and the cranial part from the endoderm.

Therefore, the caudal part is innervated by blood vessels from the internal iliac artery and the cranial part from branches of IMA.

4.1.2 Congenital Abnormalities of the Colon, Rectum, and Anal Canal

1. Caecal and colonic volvulus: The inability of the ascending colon to attach to the post-abdominal wall can lead to a mobile colon which can twist upon itself, causing caecal and colonic volvulus. Incomplete attachment can lead to the formation of retro colic pockets, which can later lead to retrocolic hernia.
2. Omphalocele: Containing colon as a result of a failure of reduction of physiological hernia.
3. Malrotation: When the rotation of 90° takes place (instead of 270°), the large intestine and the caecum are the initial segments to be retrieved into the abdominal cavity and are positioned to the left side, forming a left-sided colon (Incomplete rotation). When a rotation of 90° takes place clockwise (reverse rotation), it leads to a transverse colon lying behind the duodenum and SMA.
4. Failure of the uro-rectal septum to split the cloaca into urogenital and hindgut space leads to an anomalous link between the two resulting in the formation of a recto-urethral fistula or recto-vaginal fistula.
5. Recto-anal atresia: Presents as a narrow tube or fibrous remnant connected to the perineal surface—a result of interrupted blood supply to the hindgut region,

probably due to misexpression of genes during epithelial–mesenchymal signalling.
6. Imperforate anus: Failure of the cloacal membrane to open into the amniotic cavity.
7. Congenital megacolon (Hirschsprung's disease): Lack of parasympathetic ganglia in the intestinal wall, which usually originate from the neural crest cells that translocate from the neural fold to the intestinal wall. Only the rectum is involved in 80% of the cases; the defect can extend up to the midpoint of the sigmoid. Around 10–20% of patients can have involvement up to the transverse/right side of the colon.

4.1.3 Anatomy of Anal Canal

- The length of the anal canal, which originates at the anorectal junction and terminates at the anal verge, is approximately 3.5–4.5 cm. It is usually shorter in females. The anterior wall is marginally smaller than the posterior. Owing to the positional orientation of the anal sphincter, an oval slit instead of a circular one is formed in the anteroposterior plane.
- It is positioned at an angle with respect to the rectum due to the pull of the puborectalis.
- The anal canal is bordered by the coccyx on the posterior side, the perineal body and vagina in women and the urethra in men on the anterior end, and the ischiorectal fossa along with its components on either of the lateral sides. The anococcygeal ligament connects the anal canal on the posterior side to the coccyx.

4.1.4 Musculature of Anal Canal

- The anal canal is circumscribed by the internal and external anal sphincter muscles.
  - The internal anal sphincter is made up of an inner circular smooth muscle layer of the rectum that terminates 1.5 cm below the dentate line fairly cephalad to the external sphincter, thus creating the intersphincteric groove. The internal anal sphincter is enclosed by the levator ani muscle and superficial external sphincter muscle at the anterior and the posterior end, respectively. The internal anal sphincter is perfused by the terminal branches of the superior rectal vessels and branches of the inferior rectal vessels. The internal anal sphincter is innervated by the sympathetic fibres originating in the lower two lumbar segments and parasympathetic fibres in the second to fourth sacral spinal segments. These regulate the contraction and relaxation of the sphincters, respectively.

- The external anal sphincter is perfused by the terminal branches of the inferior rectal vessels. It is innervated by the inferior rectal branch of the pudendal nerve that begins in the anterior segments of the second to fourth sacral nerve roots.

4.1.5 *Mucosal Lining of Anal Canal*

- The epithelium that lines the anal canal is different at various segments. The epithelium of the anal canal proximal to the dentate line is columnar epithelium as is at the rectum and contains several tubular glands or crypts. Squamous epithelium without hair and glands lines the mucosa distal to the dentate. The transitional epithelial lining between these two comprises cuboidal cells scattered with columnar epithelium.
- The anal canals are vertical folds in the upper part of the anal canal and are 6–10 in number. These contain a terminal radicle of the superior rectal artery and vein. The subepithelial tissues expand into three '**anal cushions**' in the left-lateral, right-posterior, and right-anterior sections of the canal where the vessels are maximum in size. The anal canal is sealed by cushions to avoid incontinence, flatulence, and fluids. Internal haemorrhoids are formed as a result of their enlargement.
- The columns may form small crescentic folds at the lower ends called the **anal valves**, between which lie **anal sinuses** that are small recesses. The dentate or pectinate line comprises anal valves and sinuses.

4.1.6 *Rectum*

- The rectum is 12–15 cm in length. Along the plane of the third sacral vertebra, it extends from the sigmoid colon and ends at the proximal end of the anal canal. In the true pelvis, it is placed in the curve of the sacrum, and the posterior surface is almost completely extraperitoneal. The peritoneum encloses the upper 30% of the rectum on its anterior and lateral aspects, while the middle third is covered only on the anterior side.
- 'Pouch of Douglas' or the 'pelvic cul-de-sac' is the anterior peritoneal space that acts as a location for the drop metastases from abdominal tumours forming the Bloomers shelf that can be detected on digital rectal assessment.

Rectal Folds (Valves of Houston)

- The rectum commonly has three permanent semilunar transverse folds called the valves of Houston, most evident in rectal distension.

- Of the two forms of transverse folds, one is formed of the mucosa, a circular muscle layer and section of the longitudinal muscle, and is marked by an indentation, while the other only consists of the circular layer.
  - The uppermost fold at the initial part of the rectum could be on the left or right side, or it infrequently circumscribes the rectal lumen.
  - The middle fold that lies superior to the rectal ampulla is the largest and has a conspicuous circular muscle. It projects from the anterior and right walls just below the level of the anterior peritoneum.
  - The last fold lies with a gap of 2.5 cm below the left side of the middle fold. Infrequently a fourth fold may form 2.5 cm above the middle fold. These structures are called folds instead of valves since they perform no relevant valve-like functions.

4.2 Solitary Rectal Ulcer Syndrome

4.2.1 Macroscopically

The term 'solitary rectal ulcer syndrome' (SRUS) is a misnomer since ulcers are prevalent in merely 40% of patients, while a solitary ulcer is found in only 20% of patients, and the remaining vary from hyperaemic mucosa to broad-based polypoid lesions.

- Appearance: The shallow ulcerating lesions are surrounded by hyperaemic mucosa. These may appear as polypoid lesions (25%), patchy mucosal erythema (18%), and multiple lesions (30%).
- Location: It is usually found 5–10 cm from the anal verge on the anterior wall of the rectum.
- Size: ranges from 0.5 to 4 cm in diameter, but normally it is 1–1.5 cm.

4.2.2 Differential Diagnosis (on Endoscopy)

- Inflammatory bowel disease (IBD)
- Infections (amebiasis, lymphogranuloma venereum, and secondary syphilis)
- Malignancy
- Chronic vascular insufficiency (chronic ischemic colitis)
- Endometriosis
- Colitis cysticaprofunda
- Drug-induced (e.g. ergotamine tartrate-containing suppositories)
- Stercoral (pressure) ulcer
- Trauma

- Idiopathic

4.2.3 Histology

There is a thickening of the mucosa with extension and deformation of the glands, especially at their bases. Histological features include the following:

- Fibrous expunging of the lamina propria, distortion and enhanced thickness of the muscularis mucosa, and re-generative changes, with the disorientation of the cytoarchitecture. The lamina propria undergoes fibromuscular obliteration resulting in hypertrophy and disorganization of the muscularis mucosa with few inflammatory cells.
- The whitish appearance of the submucosa on proctosigmoidoscopy is attributed to dense fibrosis.

4.3 Adenomatous Polyps

Gastrointestinal (GI) polyposis is the manifestation of multiple protruding tissue masses throughout the GIT and is a distinct entity clinically. These are predominantly genetic and increase the risk of colon malignancy.

4.3.1 Epidemiology

Factors affecting the frequency of adenomatous polyps:

1. There is a large variability in the prevalence of colon adenomas. However, populations at greater risk for colon cancer tend to have higher frequency and vice versa.
2. The elderly population is at higher risk of colonic adenomas (40%).
3. The relative risk is 1.5-fold greater in males than in age-matched females.
4. A positive family history of colorectal cancer (CRC) increases the risk of adenomatous polyps.

4.3.2 Histology

The adenomatous polyps are pedunculated or sessile benign neoplasm. The predominant glandular pattern assigns a polyp its histologic type. They may be as follows:

Table 4.1 Classification of adenoma

Types	Proportion (%)	Degree of dysplasia (%)		
		Mild	Moderates	Severe
Tubular	80	88	8	4
Tubulovillous	8–16	58	26	16
Villous	3–16	41	38	21

- Tubular: Most frequent, complex mesh of bifurcating adenomatous glands.
- Villous: Long, finger-like projections are created by bifurcations of adenomatous glands from the surface to the core of the polyp.
- Tubulovillous (villoglandular): Combination of tubular and villous histologic subtypes.

According to the World Health Organization (WHO), adenomas are classified either as tubular or villous if ≥80% of the glands are of the branching or villiform tubule type, respectively (Table 4.1).

- Mild dysplasia: 70–86% of adenomatous polyps
- Moderate dysplasia: 18–20%
- Severe dysplasia (carcinoma in situ): 5–10%
- Invasive carcinoma: 5–7%

Higher grades of dysplasia are frequent in adenomas of greater size and villous content and adenomas with a higher probability of severe dysplasia consisting of foci of invasive cancer.

4.3.3 Significance of Adenomatous Polyps

- The clinical significance of adenomatous polyps lies as precursors of almost all sporadic CRCs.
- Although the risk of malignancy, which is reported only in 10% of patients, depends on size, histology, and extent of dysplasia, these factors are interdependent (Table 4.1).

4.4 Serrated Polyps

Most common non-adenomatouspolyps.

Types based on histology:

1. Sessile serrated polyp/adenomas (SSP/As)
2. Traditional serrated adenomas (TSAs)
3. Hyperplastic polyps (HPs)

4.4.1 Sessile Serrated Polyp/Adenoma (SSP/As)

- <1% of all polyps and 1–11% of adenomas.
- The occurrence of high-grade dysplasia and cancer is between 5% and 16%.
- Associated with both synchronous and metachronous cancers as well as MSI cancers.
- Flat, sessile.
- Found in the right colon.
- Enclosed with a mucus cap.

When sporadic, SSP/As is frequent in the proximal colon. Most SSP/As appear to progress to colorectal carcinoma via a distinct molecular pathway that involves increased methylation of CpG islands. The CpG island methylation pathway (CIMP) leads to the downregulation of DNA mismatch repair genes, resulting in microsatellite instability (MSI).

Genetic mutations: *BRAF* mutations and extensive DNA methylation.

High-resolution white-light images and NBI describe SSA as a cloud-like surface with vague demarcations, contorted shapes, and dark spots inside the crypts. These characteristics adequately help to differentiate SSA from HP with 93% precision.

4.4.2 Traditional Serrated Adenoma (TSA)

- Behave like conventional adenomas
- Pedunculated
- More often in the distal colon
- Associated with K-ras mutations

Management

Resection is the most preferred management method since adequate differentiation of SSP/A or TSA from an HP is challenging, and these are highly susceptible to malignant transformation to invasive carcinoma. Diminutive SSP/As can be removed by biopsy forceps, but larger polyps should be removed by snare polypectomy.

4.4.3 Hyperplastic Polyps

Features

- Small.
- Sessile.
- More common in the distal colon.

Histopathology

- Colonic crypts are lengthened.
- Epithelial cells have a papillary configuration.
- The epithelium comprises well-differentiated goblet and absorptive cells.
- Mitoses and DNA synthesis occur only at the base of the crypts, and organized cell maturation is restored.

Prevalence

Hyperplastic polyps were found in 9–10% of asymptomatic patients >50 years.

Treatment

HPs usually remain small, sessile, and asymptomatic. Resection is advisable even if malignancy transformation is unlikely since these cannot be differentiated from neoplastic or serrated polyps by gross examination. This is supported by clinical evidence, which found that in patients with the absence of symptoms, distal HP had a ~25% and ~4% risk of proximal and advanced proximal neoplasia, respectively. This also supports colonoscopy if sigmoidoscopy detects a lesion.

4.5 Anatomy and Physiology of Faecal Continence Mechanism

An anatomically healthy and functional anorectal unit is required for sustaining normal continence of bowel contents (Fig. 4.1).

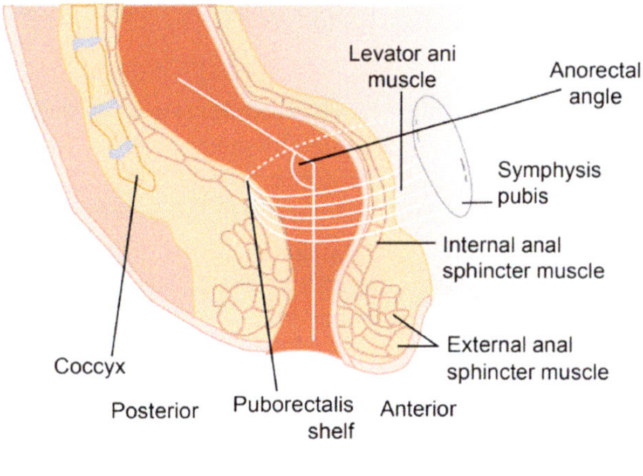

Fig. 4.1 Muscles involved in anal continence

Rectum The rectum is a muscular elongated structure consisting of a unique arrangement of sequential longitudinal muscles with interlaying circular muscles which allows the rectum to act as a reservoir for stools as well as a pump for defecation.

Anus The anus is a muscular elongated tubular structure 2–4 cm in length that, at rest, forms a right angle with the axis of the rectum which changes to acute (~70°) on voluntary squeeze and obtuse during defecation (~110°–130°).

The anal sphincter consists of the internal and external anal sphincter (IAS and EAS); these are 0.3–0.5 cm and 0.6–1.0 cm thick, respectively. The IAS is composed of a circular smooth muscle layer and the EAS of levator ani muscles.

Structurally these sphincters are distinct. The IAS consists primarily of slow-twitch, fatigue-resistant smooth muscle. The IAS produces mechanical activity with a frequency of 15–35 cycles/min as well 1.5–3 cycles/min, which ultraslow waves. The IAS contributes 70–85% of the resting anal sphincter pressure but only 40% of the pressure after acute distention of the rectum and 65% during constant rectal distention. The balance pressure is contributed by the EAS or puborectalis, or both. Therefore, the IAS is primarily responsible for sustaining anal continence at rest.

The anus remains closed by the tonic activity of the IAS, anal mucosal folds, and expansive vascular cushions. The barrier is reinforced during voluntary squeeze by the EAS. These barriers are enhanced by the puborectalis muscle, which creates a flap-like valve that generates an anterior pull and supports the anorectal angle.

Nerves The anorectum is adequately innervated by sensory, motor, and autonomic nerves and the enteric nervous system. The primary one is the pudendal nerve that originates in the second, third, and fourth sacral nerves (S2, S3, and S4) and innervates the EAS.

In all likelihood, the sensation of rectal distention is passed along the S2, S3, and S4 parasympathetic nerves, which lie the pelvic splanchnic nerves and act independently of the pudendal nerve.

Reflexes

- Recto-anal inhibitory reflex: Reduction in anal resting pressure caused by rectal distension. It is transmitted by the myenteric plexus via NO and VIP. It facilitates the discharge of flatus/stools.
- Sampling reflex: It refers to the transient relaxation of IAS leading the stools to reach the anal canal. It helps to differentiate between flatus and stools.
- Recto anal contractile reflex: It is transmitted by pelvic and pudendal nerves and is responsible for the prevention of flatus/stool incontinence by causing contraction of EAS.

4.6 Short-Chain Fatty Acids

Fat is an ester of fatty acids and glycerol present in animals, vegetables, oil, or waxy substances. Fatty acids are aliphatic monocarboxylic acids. Natural fatty acids commonly have an unbranched chain of 4–28 carbons with an even number of carbon atoms, that is, 8, 10, 12, 14, etc. These may have double- or triple-carbon bonds and are grouped as unsaturated fatty acids, while those without such bonds are called saturated fatty acids.

4.6.1 Classification of Fatty Acids

Conventionally fatty acids have been classified as follows:

- Short-chain fatty acids (SCFA) are fatty acids with an aliphatic chain of <6 carbons.
- Medium-chain fatty acids (MCFA) are fatty acids with aliphatic tails of 6–12 carbons.
- Long-chain fatty acids (LCFA) are fatty acids with aliphatic tails >12 carbons.
- Very-long-chain fatty acids (VLCFA) are fatty acids with aliphatic tails >22 carbons.

What are short-chain fatty acids?

SCFA include acetic acid, propionic acid, butyric acid, isobutyric acid, valeric acid, isovaleric acid, caproic acid, lactic acid, and succinic acid.

Production of SCFA

The various SCFA acetic acid, butyric acid, and propionic acid are formed via the bacterial fermentation of various polysaccharides and proteins in the colon by saccharolytic bacteria and proteolytic bacteria, respectively. Carbohydrates are a preferred substrate over proteins due to their conversion to toxic metabolites such as ammonia, sulphur-containing compounds, indoles, and phenols. Carbohydrates are fermented in the proximal colon, while proteins in the distal colon. Apart from SCFA, the other by-products of fermentation are carbon dioxide, methane, and hydrogen gases along with heat.

Faecal SCFA production is in the order of acetate > propionate ≥ butyrate in a molar ratio of approximately 60:20:20, respectively, and seldom varies except with extreme dietary alterations. The uniformity in molar ratio in the proximal and distal colon is attributed to the differential absorptive capability of the gut mucosa in different segments of the colon. In the proximal colon quantum of bacteria, the extent of fermentation and proliferation is proportional to substrate quantity. Cecum and proximal colon are the principal sites of colonic fermentation, whereas the distal colon is devoid of carbohydrates and water. The cumulative SCFA in the proximal colon is greater than the distal colon, 70–140 mM versus 20–70 mM. The proximal colon has the lowest pH, which increases distally.

4.6.2 Importance of SCFA in Colon

- SCFAs raise energy mainly from carbon sources such as undigested dietary fibre.
- Approximately 5–15% of the total calorie requirements of humans are met by SFCAs.
- Butyrate is the primary source of energy for colonocytes.
- The composition of the colonic microflora is influenced as they lower the colonic pH (e.g. acidic pH lowers pathogenic clostridia), reduce the solubility of bile acids, enhance absorption of minerals (indirectly), and lower the absorption of ammonia (NH_3) by the protonic dissociation of ammonia (NH^{4+}) and other amines.
  - Acetate produced in the colon is rapidly absorbed and less metabolized in the colon. Acetate is used for lipogenesis and cholesterol synthesis.
  - Propionate has lipid-lowering action by inhibiting hepatic cholesterol synthesis.
  - Butyrate may have an anti-carcinogenic and anti-inflammatory action.
  - Butyrate affects the intestinal barrier.
  - Butyrate plays a role in satiety and oxidative stress.

Absorption of SCFAs

- SCFAs are weak acids primarily occurring in the anionic, dissociated form in the colonic lumen. These are taken up by colonocytes either via passive diffusion of the undissociated (lipid soluble) form, SCFA/HCO^{3-} exchange, or active transport of the dissociated form by SCFA transporters [monocarboxylate transporter isoform 1 (MCT1) and sodium-coupled monocarboxylate transporter (SMCT1)]. Colonic epithelium metabolizes the majority of the absorbed SCFAs resulting in poor concentrations in portal circulation. Further, the liver draws out a large of the balance SCFAs, lowering the venous systemic serum concentrations.

Source of SCFAs in Colon

The fermentation of resistant starch, dietary fibre, unabsorbed sugars, raffinose, stachyose, polydextrose, and modified cellulose produces colonic SCFAs. The human small intestine does not absorb 5–20% of dietary starch, which acts as a source of SCFAs. The majority of fibre-containing foods contain about ~30% soluble (e.g. pectins, gums, mucilages, and some hemicelluloses) and insoluble fibres (e.g. lignins, cellulose, and some hemicelluloses). Insoluble fibres resistant to bacterial fermentation form the major part of the stool bulk. Soluble fibres are better fermented by colonic microflora and mainly contribute to SCFA production and minimally in increasing faecal bulk.

The following clinical conditions have a potential role for SCFAs:

- Colonic cancer
- Ulcerative colitis (UC)
- Constipation
- Hepatic encephalopathy
- Osteoporosis
- Coronary artery disease

4.7 Faecal Occult Blood

- Colorectal cancers and polyps can bleed intermittently in small quantities, which may not be obvious to the naked eye. However, stool tests can be used to identify small quantities of blood in stools and alert to the possibility of neoplastic lesions in the colon. So, stool occult blood tests were used for screening for colorectal neoplasia.
- Stool tests for occult blood can detect <50 mg haemoglobin/g of stool (normal adult stools have <2–3 mg haemoglobin/g of stool).

4.7.1 Degradation of Haemoglobin in GI Tracts: Basis of Faecal Occult Blood Tests

- RBCs in the digestive tract are broken down into hematin and globin. Hematin is partly reabsorbed by the intestine (15%), and the rest (85%) are converted into faecal porphyrins by colonic bacteria. Globin is broken down by pancreatic proteases into peptides and amino acids. Each of these steps allows us to detect blood or its substrates in faeces.
- RBCs can be tracked by 51Cr-labelled RBCs scan. On degradation, the hematin component can be identified by using Guaiac-based tests. Porphyrins, the breakdown product of hematin, can be quantified by fluorescence method. Globins can be detected by immunological tests.
- Of these, 51Cr-labelled RBC scan and faecal porphyrin levels are the quantitative tests. Guiac-based tests to detect hematin and immunological tests to detect globin are qualitative tests. Since quantitative tests are cumbersome, only qualitative tests are used in clinical practice.

4.7.2 *Type of Faecal Occult Blood Tests*

There are two types of stool tests used to detect blood:

1. Tests which detect haeme component of haemoglobin—Guaiac-based tests
2. Test which detects globin fraction of haemoglobin—faecal immunochemical tests

Guaiac Test

The stool sample should be collected in a dry container. A wooden applicator should be used to apply a small sample of stool to the two windows: A and B, on the guaiac-coated card and placed in an envelope. Three consecutive day stool samples are collected and mailed to the lab. In the lab, a developer solution (hydrogen peroxide) is applied to the opposite side of stool impregnated card. If the card turns blue within 30 s, the test is positive.

Principle of Guaiac Card Test

$$H_2O_2 + \text{Indicator in reduced state}(\text{Guaiac}) \xrightarrow[\text{Haeme}(\text{in stool})]{} \text{Quinone}(\text{blue}) + H_2O$$

The card for the faecal occult blood test contains guaiac (alpha guaiaconic acid) coated paper. A developer solution (stabilised hydrogen peroxide) is added to the card. In the presence of haeme in the stool sample, hydrogen peroxide oxidises the guaiac to a blue-coloured quinone.

Dietary Restriction and Modification of Medicines Before the Guaiac Test

- Dietary restriction—For 3 days prior and 3 days during stool collection: avoid red meat, vegetables like radish, broccoli, cabbage, turnip, carrot, cucumber, cauliflower, mushrooms, and grape—as they are sources of vegetable peroxidases.
- Avoid aspirin, anticoagulants, and NSAIDs-alternate causes for bleeding.
- Avoid antioxidant-containing food like citrus fruits/juices and high doses of oral vitamin C (>250 mg/day).

Clinical Significance of Faecal Guaiac Test

- In Western nations, 3–5% of asymptomatic individuals above 45 years of age test positive for faecal occult blood on an unrestricted diet. With diet restriction, ~2% of samples test positive.

- On further evaluation of positive stool occult blood test, 10–15% will be found to have colonic carcinoma, and an additional ~15–30% will have benign colonic adenomas/polyps.
- Most of the carcinomas detected by stool occult blood tests are Dukes cat A or B1 (early stage) with >80% 5-year survival after resection. Most symptomatic colonic carcinomas are advanced and have lower 5-year survival.

Faecal Immunochemical Test (FIT)

- FITs use polyclonal anti-human haemoglobin antibodies to detect human haemoglobin in stool samples. They are more specific as there is no cross-reaction with animal haemoglobin or vegetable peroxidases noted with the Guaiac test. So, no dietary restriction is required for this test.
- FIT identifies undegraded human haemoglobin. Bleeding from the upper GI source (small bowel or upper GI tract) is usually degraded by pancreatic proteases by the time they are passed in stools. On the contrary, bleeding from colonic sources have less time to be degraded by proteases and so are identified by FIT. So, positive FIT is more specific for lower GI bleed.
- FIT has replaced Guaiac-based faecal occult blood tests in US Multisociety task force colorectal cancer screening guidelines due to increased sensitivity and specificity and patient adherence. Annual FIT and colonoscopy are initial tests for CRC screening in patients aged more than 50 years.

Clinical Significance

- FIT tests have replaced guaiac tests in screening programs in Europe with good acceptability by patients. Around 20–40% reduction in mortality has been observed in patients who had undergone screening with FIT as compared to those who did not undergo screening tests.
- Newer stool tests (utilising FIT and multi-target DNA) have increased sensitivity in identifying carcinomas and advanced adenomas. These tests are expensive and are yet to be integrated with screening programs.

4.8 Antibiotic-Associated Diarrhoea

Antibiotic-associated diarrhoea can vary from mild disease to severe disease in the form of pseudomembrane colitis (Fig. 4.2).

Risk factors for antibiotic-associated diarrhoea are the following:

- Antibiotics in the last 3 months (penicillins, cephalosporins, clindamycin, and quinolones)

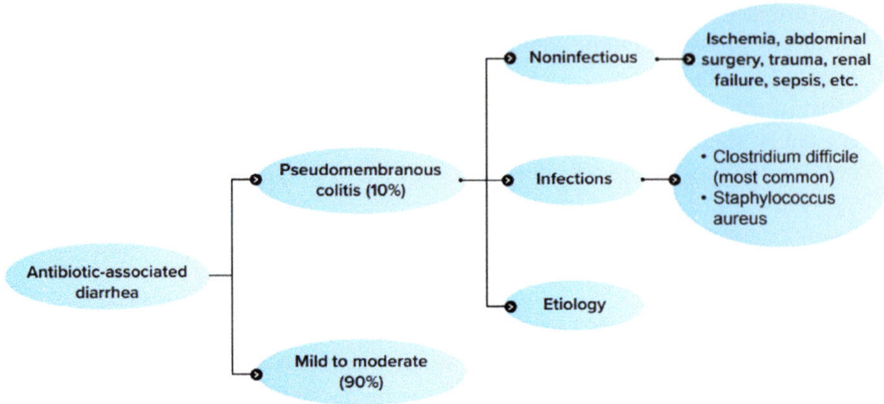

Fig. 4.2 Antibiotic-associated diarrhoea

- Hospitalization
- Old age
- Comorbidities
- Nasogastric tube in situ
- Proton pump inhibitors (PPIs)
- Cytotoxic chemotherapy
- IBD

The pathogenesis of antibiotic-associated diarrhoea is explained in Fig. 4.2.

4.9 Colorectal Cancer

Colon cancer is the third most prevalent malignancy globally and is increasing in incidence across the world. The rise in incidence is partly attributable to poor dietary habits and lifestyle.

4.9.1 Risk Factors

Most colon cancers are sporadic. There are genetic and environmental risk factors for colorectal cancer. (Table 4.2). The genetic and molecular alterations are listed in (Tables 4.3 and 4.4).

Table 4.2 Risk factors of colon cancer

Genetic		Environmental (modifiable lifestyle factors) contributes to 15–70% of colon cancers	
Higher risk in patients with first-degree relatives who have had colon cancer (higher if relative diagnosed <50 years of age or if >2 relatives have cancer)	Contributes to 15–20% of all colon cancer	Alcohol Smoking Increased BMI Type 2 DM IBD	The benefit of lifestyle modification can be augmented by regular ingestion of aspirin, NSAID, statin, or hormone replacement therapy
Hereditary colorectal cancer syndrome: • Lynch syndrome—defect in DNA mismatch repair gene • Familial adenomatous polyp—mutation in APC gene • Peutz–Jegher's syndrome • Mutations associated with mut Y DNA glycosylase gene (MUTYH) gene	Contributes to 5–10% of colon cancers	Increased risk with consumption of red meat	Decreased risk with higher intake of fibre, calcium, and vitamin D Decreased risk with daily physical activity

Table 4.3 Genetic alterations noted in colon cancer carcinogenesis

Tumour suppressor genes	Function	Molecular lesion	Frequency (%)
APC	Wnt signalling pathway	Inactivating mutation	40–70
SMAD 4	TGF and BMP pathway	Inactivating mutation	25
TP53	Regulates cell cycle, DNA repair, and apoptosis	Inactivating mutation	50
Proto-oncogenes	*Function*	*Molecular lesion*	*Frequency (%)*
BRAF	MAPK signalling pathway	Activating mutation	25
ERBB2	EGF-MAPK signalling pathway	Amplification	30
GNAS	G protein signalling	Activating mutation	25
KRAS	MAPK pathway	Activating mutation	40
NRAS	MAPK pathway	Activating mutation	2
MYC	Regulates proliferation and differentiation	Amplification	
PIK3 CA	Regulates PI3K-AKT pathway	Activating mutation	20

Table 4.4 Molecular alterations noted in colon cancer carcinogenesis

Alteration	Molecular lesion	
Chromosomal instability	Aneuploidy	70%
CpG island methylator phenotype (CIMP pathway)	Methylation of >20% loci from a select panel of markers	15%
Microsatellite instability	Unstable microsatellite repeats	15%
Mismatch repair genes	Loss of protein in immunohistochemistry	1–15%

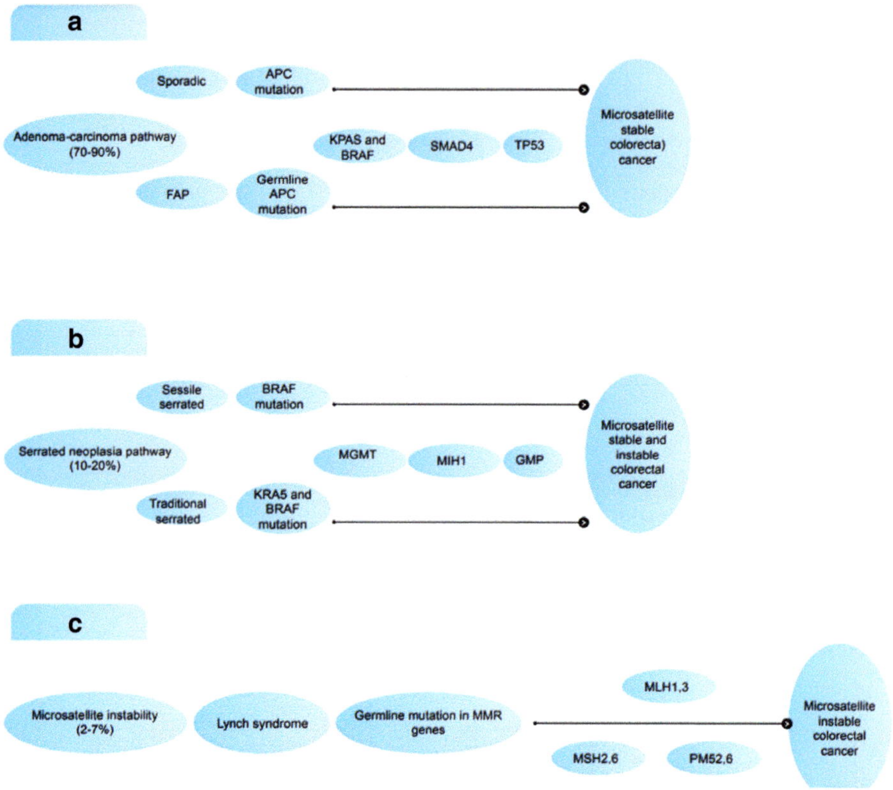

Fig. 4.3 Molecular pathways of colorectal cancer development

Pathogenetic Mechanisms of Colorectal Cancer Carcinogenesis

There are three main pathways (Fig. 4.3):

1. "Traditional" pathway or adenoma-carcinoma sequence (chromosomal instability sequence)
2. Serrated adenoma pathway
3. Microsatellite instability pathway

4.9.2 Implications in Prognosis and Treatment

The pathogenesis and prognosis of colon cancer are depicted in Fig. 4.4.

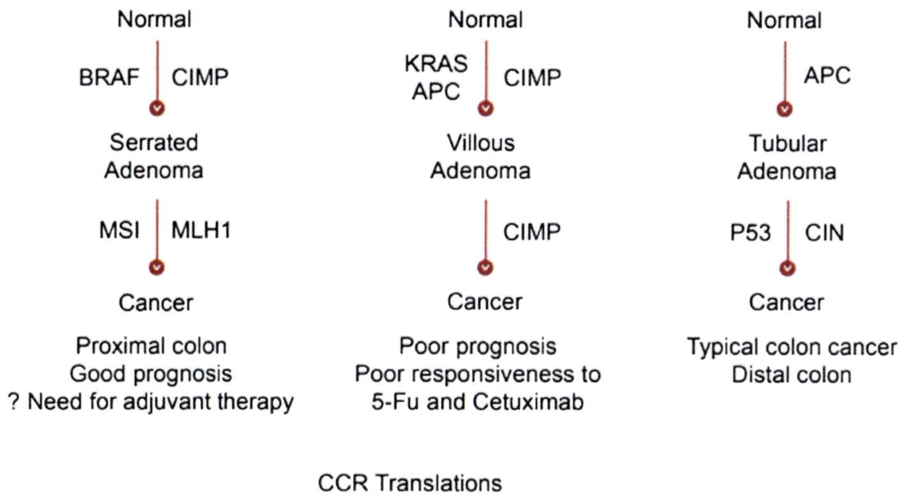

Fig. 4.4 Pathogenesis of colon cancer and prognosis

4.10 Pathogenesis of Amebiasis

4.10.1 Epidemiology

- About 10% of the world population is chronic *Entamoeba* carriers, but only 10% of them have pathogenic *Entamoeba histolytica*, while the rest of them have nonpathogenic *E. dispar*.
- In the majority of cases, amoebic infections are asymptomatic and self-limiting. A mere fraction of 1% disseminates to the hepatic region but is the principal extraintestinal site.

4.10.2 Lifecycle

- *Entamoeba histolytica* is a food-borne and water-borne infection transmitted via cysts. Trophozoites are released into the terminal ileum, which translocates and colonizes the mucus layer of the large intestine.
- Normally, trophozoites exist as luminal commensals, proliferating via binary fission and meeting their energy requirements using resident microflora and nutrients from the host.
- Some parasites form mature cysts in the descending colon and their faecal–oral dissemination results in the propagation of the life cycle.

4.10.3 Pathogenesis

Adhesion to Colonic Epithelium and Mucosal Invasion

The *E. histolytica* trophozoites attach to colonic epithelial cells via galactose/*N*-acetyl galactosamine-specific lectin on the colonic cell surface, which then invade the colonic epithelium through two mechanisms.

1. Incorporation in the host cell membrane of amoeba pores which are membrane-active peptides: At least, three small peptides with pore-forming in lipid bilayers have been identified.
2. Apoptosis: *Entamoeba histolytica* trophozoites can cause apoptosis of mammalian cells.

Submucosal Invasion

After amoebic invasion through the mucosa, there is an extension to submucosa:

1. Interaction of amoebic trophozoites and the extracellular matrix protein fibronectin activates signalling cascades in the parasite and leads to actin reconfigurations that modify the adherence and motility of the parasite.
2. The *histolytica* cysteine proteinases (CP) cause proteolysis of the extracellular matrix proteins, aiding the trophozoite entry.
3. Colonic epithelium imposes an inflammatory immune response against parasites and causes chemotaxis to the site of amoebic invasion. *Entamoeba histolytica* causes neutrophil lysis, which leads to the release of mediators, which cause diarrhoea and adjacent epithelial tissue damage. This also elucidates the appearance of hardly any neutrophils adhering to amoebic trophozoites in human biopsies.
4. The initial lesion is in the caecum or sigmoid colon, which is possibly due to the gradual passage of luminal contents and prolonged mucosal contact time that amoeba gets facilitating invasion. Untreated superficial lesions may extend into the deeper layers of the submucosa and muscularis, creating the typical flask-shaped ulcer.

Vascular Invasion

Once amoebas reach the vascular or lymphatic circulation, these are disseminated initially to the liver and later to the other organs.

1. Dissemination to locations outside the intestine occurs if the anaerobic trophozoites survive the oxygen-rich milieu in the vasculature and escape the immune cells and other humoral factors.

   E. histolytica has many ways to escape natural immune mechanisms:

(a) Parasite CP cleaves human IgA.
(b) Trophozoite CP can also disable circulating IgG.
(c) Masking the host-immune system identified surface receptors of the amoeba in membrane-bound vesicles.
(d) Gal-lectin has epitopes of the complement inhibitory protein, CD59 which averts trophozoite lysis via membrane attack complexes.
(e) *Entamoeba histolytica* deactivates the anaphylatoxins C5a and C3a to escape the complement effect.

Development of Liver Abscess

Little is known about how parasite emigrates from sinusoids to liver parenchyma to establish the infection. Sinusoidal endothelium is fenestrated, which facilitates trophozoite migration to the space of Disse and may enter into the liver parenchyma.

1. Detection of trophozoites in the liver triggers an acute inflammatory cascade that leads to a large number of neutrophil recruits surrounding them. Neutrophil degranulation worsens hepatocyte destruction, and subsequent progression causes variation in immune cell infiltration.
2. The second bout of cells has pronounced influx of macrophages with corresponding granuloma development, which culminates into the production of collagen production and fibroblast proliferation. The liver abscesses have the development of central dense necrotic masses. Trophozoites at the inner side of the granuloma lyse the fibrous component and cause cytolysis, thus migrating to adjoining sites and developing multiple granulomas. In due course, these lesions coalesce to form a single large mass typical of an amoebic hepatic abscess.

4.11 Microscopic Colitis

Microscopic colitis (MC) is marked by prolonged watery diarrhoea but unremarkable endoscopic, radiological findings of colon and histopathologic abnormalities in the colonic biopsy. It includes two typical forms (collagenous colitis and lymphocytic colitis), several atypical patterns, and nonspecific colitis.

4.11.1 Epidemiology

- Most common in the fifth or the sixth decade
- Male–female ratio is about 1:3
- The lymphocytic: collagenous colitis ratio is about 3:1

4.11.2 Clinical Features

- Chronic large bowel watery, non-bloody diarrhoea, with nocturnal symptoms
- Diarrhoea responds to fasting
- Waxing and waning course
- Pain in the abdomen
- Weight loss—minor

Frequently associated conditions: thyroid disease, rheumatoid arthritis, celiac disease, hypertriglyceridemia, and IBD.

4.11.3 Pathogenesis

Many factors have been implicated in pathogenesis, but there is no unifying hypothesis for pathogenesis.

- Genetic factors: Family history of IBD, celiac disease, or collagenous colitis noted in 12%.
- Luminal factor
  - Diversion ileostomy resolves the typical histopathologic changes in collagenous colitis. Ileostomy closure has a risk of symptom recurrence and histopathologic alterations.
  - Liquid meal replacement providing nutrition in the form of basic elements reduces diarrhoea in patients with MC.
- Luminal infection
  - Microscopic colitis manifests as acute watery diarrhoea with mucosal lymphocytosis without crypt distortion, epithelial destruction, or collagen deposition that occurs for several years. Findings consistent with acute inflammation on biopsy or a history indicate the occurrence of an acute infection in several patients with MC.
  - A subset of MC patients respond to antibiotic therapy.
  - *Yersinia enterocolitica*, *Campylobacter jejuni*, and *Clostridium difficile* have also been reported in patients with different types of MC.
- Drugs: NSAIDs, aspirin, lansoprazole, ranitidine, and sertraline.
- Bile acids: Concurrent bile acid malabsorption in MC patients.
  - Autoimmunity: MC has been linked with other autoimmune diseases such as thyroid disease, celiac disease, diabetes mellitus, or arthritis that has been suggested as an autoimmune process.

4.11.4 Histopathologic Types of Microscopic Colitis

1. **Classical lymphocytic colitis**: Lymphocytic colitis is characterized by the following:
    (a) Increased (>20 IELs/100 colonocytes) intra-epithelial lymphocytes (IELs) with surface epithelial damage (a hallmark of the disease)—IELs are predominantly present in the surface epithelium than in the crypts.
    (b) Normal crypt architecture with increased cryptal IELs.
    (c) A greater number of inflammatory cells infiltrate in the lamina propria include lymphocytes, plasma cells, and eosinophils.
2. **Classic collagenous colitis**: The subepithelial collagen band is >10 μm thickness and is considered diagnostic (healthy 4–5 μm), with its overlying epithelium usually damaged.
3. **Cryptal lymphocytic colitis**: The features match that of lymphocytic colitis with IELs in the crypts rather than the surface epithelium.
4. **Paucicellular lymphocytic colitis**: Features match that of MC along with additional ones like patchiness and a lower density of surface IELs, such that the appearance does not allow categorization as lymphocytic colitis.
5. **Microscopic colitis with giant cells**: This atypical form of MC is characterized by the atypical presentation, including the presence of multinucleated giant cells without granuloma formation on histological analysis.
6. **Pseudomembranous collagenous colitis**: Presence of morphologic characteristics of collagenous colitis with pseudo-membranes.
7. **Microscopic colitis with granulomatous inflammation**: This is characterized by a conspicuous granulomatous reaction.
8. **Microscopic colitis with IBD-like histologic features**: MC does not manifest architectural distortion, unlike IBD. Infrequently some features of IBD, such as active crypt inflammation, surface erosion or ulceration, Paneth cell metaplasia, and crypt architectural irregularity, may exist in patients with MC subtypes, either lymphocytic or collagenous colitis.
9. **Microscopic colitis, not otherwise specified (NOS)**: It describes changes in appearance showing elevated lymphoplasmacytic infiltration in the lamina propria, devoid of a significant rise in IELs or a thickened subepithelial collagen band.

4.11.5 Treatment of MC

Antimotility agents are initiated. If not responsive, oral ASA formulations are started. If not responsive, cholestyramine is started. If no response, to consider budesonide.

4.12 Pathogenesis of Irritable Bowel Syndrome (IBS)

IBS typically manifests as recurrent abdominal pain/discomfort associated with alteration in stool frequency or consistency, devoid of any structural identifiable cause.

The varying factors involved in the pathophysiology are the following:

1. Gastrointestinal motility

   Abnormalities include the following:

   (a) Greater frequency and inconsistency of intestinal contractions.
   (b) Delayed passage of bowels in constipation-predominant IBS.
   (c) Elevated motor response to CCK and food consumption in diarrhoea-predominant IBS.

2. **Visceral hypersensitivity**

   (a) Distention—Balloon distension studies show a lower threshold for pain in IBS patients as compared to those without.
   (b) Bloating—Although unrelated to the amount of intestinal gas, nearly half of the patients have a sizeable rise in abdominal girth due to bloating.
   (c) This is mediated by the following:

   - The enteric nervous system is mediated by specific GI hormones like serotonin, kinins,
   - By central brain modulation, or
   - Increase in spinal cord excitability due to activation of *N*-methyl-D-aspartate receptor (NMDA).

3. **Intestinal inflammation**

   (a) Inflammatory cells—A higher number of lymphocytes and mast cells in the jejunum, terminal ileum, and colon of IBS patients, which release substances enabled to activate the enteric nervous system to generate abnormal motor and visceral responses.

4. **Post-infectious IBS**

   (a) Idiopathic bile acid malabsorption—It occurs due to enteric infections leading to diarrhoea-predominant IBS.
   (b) Increase in enteroendocrine cells/lymphocytes—Elevated serotonin levels heighten GI motility and visceral hypersensitivity.
   (c) Antibiotic use—The use of antibiotics for GI or other infections increases the susceptibility of the host to the development of functional bowel conditions.

5. **Alteration in faecal microbiota**

   (a) The faecal microbiota in patients with IBS varies with its predominant subtype and is significantly different from otherwise healthy subjects.

(b) These patients may, therefore, be managed with probiotics, which enhance the gut microbiota.

6. **Bacterial overgrowth**

   (a) Breath tests show abnormal breath hydrogen levels with carbohydrates due to reduced oro-cecal transit time.
   (b) Elimination of overgrowth attenuates symptoms.

7. **Food sensitivity**

   (a) Food allergy—Elimination of foods triggering a surge in IgG titres alleviate gastrointestinal symptoms.
   (b) Carbohydrate malabsorption—FODMAPs in patients with IBS worsen symptoms and compromise intestinal integrity and trigger inflammation
   (c) Fructose intolerance due to carbohydrate malabsorption causes flatus, pain, bloating, belching, and altered bowel habits.
   (d) Gluten sensitivity—The presence of serum IgG antigliadin antibodies and expression of HLA-DQ2 (associated with celiac disease) may help in forecasting the prognosis of a gluten-free diet in patients with diarrhoea-predominant IBS.

8. **Genetics**

   (a) Data from studies of twins are contradictory, with some demonstrating higher (2–22% in monozygotic twins) and others poorer (1–9% in dizygotic twins) concordance rates for IBS.

9. **Psychosocial dysfunction**

   (a) Psychosocial problems related to anxiety and insomnia are independent risk factors for the development of IBS, usually mediated by corticotropin-releasing factor (CRF) as a response to stress.

4.13 Recent Understanding of the Pathophysiology of Colonic Diverticular Disease

- Diverticula are the commonest structural abnormality appearing as a sac-like protrusion in the wall of the colon.
- Its prevalence is proportional to age. It was detected primarily (90% cases) in the sigmoid or left colon in the Western patient population, while in the Asian population, it dominates the ascending or right colon (75–85% of patients)
- A quarter of individuals progress to symptomatic diverticulosis or colonic diverticular disease. Its spectrum consists of symptomatic uncomplicated (SUDD) to symptomatic complicated states such as acute diverticulitis or diverticular haemorrhage, largely found in Eastern populations.

- Progression to diverticulitis is reported in merely 1% with more than a decade of diverticulosis, with ~12% of patients experiencing a serious complication (perforation, abscess, or fistula).

4.13.1 Pathophysiology

The aetiology of diverticular disease is likely to be multifaceted, and the mechanisms depend on the displayed symptoms. For example,

- Connective tissue abnormalities (altered elastin crosslinking) increase the susceptibility of the host to experiencing asymptomatic diverticulosis.
- Gut microbial dysbiosis or medication use are aggravating factors that increase the probability of complicated diverticular disease.

4.13.2 Faecal Stasis and Faecal Impaction

- Faecal stasis and faecal impaction can result in the formation of a fecolith, as found in resection specimens, which might obstruct a diverticulum resulting in bacterial stasis and local trauma. This can cause ischemia, micro-perforation, inflammation, and infection.

4.13.3 Chronic Inflammation

- Along with the overt acute inflammation that occurs in acute diverticulitis, chronic low-grade inflammation in individuals with diverticulosis increases the risk of diverticulitis and SUDD.
- Several risk factors are common for diverticulitis and lifestyle diseases such as CVD and T2D, such as a low-fibre diet, high red meat consumption, obesity, smoking, and sedentary behaviour since these increase low-grade systemic inflammation.
- Refractory endoscopic and/or histological inflammation after relief of acute diverticulitis increases the risk of recurrent diverticulitis.

4.13.4 Alterations in the Intestinal Microbiota

- Some lifestyle risk factors influence the composition and function of the intestinal microbiota, and these may, therefore, increase the risk of diverticulitis.

- Levels of SCFA-producing bacteria (*Clostridium* spp.) are found to be low, and those producing pro-inflammatory mediators (*Marvinbryantia* spp. and *Subdoligranulum*) were found to be high in patients with a history of acute diverticulitis compared with healthy controls.

4.13.5 Neuromuscular Alterations

The combined effect of elevated intracolonic pressure and frailty of the colon wall may underlie the development of diverticulosis.

- Poor consumption of dietary fibre may lead to elevated intracolonic pressure, while colon weakness is linked to age-related changes.
- Abnormal connective tissue and collagen metabolism have been noted in patients with diverticulosis. A colonoscopy reveals hypertrophy of the longitudinal and circular muscle layers leading to the thickening of the colonic folds and a decrease in the lumen calibre, which is together termed myonecrosis coli. Additionally, the level of MMP, which degrades the tissue, is higher in patients with than without diverticulosis.
- The enteric nervous system shows abnormal changes such as reduced glial cell counts, nerve cells, and intestinal pacemaker cells, along with abnormal levels of neurotransmitters and their receptors are also observed in patients with diverticulosis.
- These patients also demonstrate an increased frequency of colonic responses to eating and high-amplitude contractions in affected colonic segments.

4.13.6 Genetics

A genetic basis for diverticulosis and diverticular disease has been suggested based on certain findings.

- The higher prevalence of diverticulosis and the left versus right colon distribution in Western versus Eastern parts of the global population may indicate genetic predisposition in the former.
- Early onset and extensive diverticular disease are frequent in several inherited connective tissue disorders like Marfans and Ehler–Danlos syndrome.
- A couple of familial aggregation studies in a large population subset showed that siblings of patients with diverticular disease are at a 50% higher risk of hospitalization compared to those without. Further, the risk is greater in monozygotic twins compared to dizygotic.
- A large genome-wide association study (GWAS) in more than 27,000 patients with diverticular disease revealed 40 loci with significant associations with

diverticular disease. None of these were common among diverticular disease, IBD, or IBS.
- *PHGR1*, a gene that plays a role in gastrointestinal epithelial cell function, was found to be particularly related to diverticulitis.

Thus, the current understanding of diverticulitis is that of multifactorial alterations at microscopic, macroscopic, and molecular levels, from genetic predisposition to ageing-related changes in gut motility and musculature complicated with lifestyle and dietary factors and underlying chronic inflammatory condition.

4.14 Rectal Manometry and Its Utility in Faecal Incontinence and Defecation Disorder

- High-resolution manometry is recommended in patients who present with faecal incontinence due to obstetric, neurologic, or iatrogenic injury to the sphincters or the nervous supply to the anorectum.
- Resting anal sphincter pressure primarily denotes IAS function, while voluntary anal squeeze pressure denotes EAS and puborectalis function.
- The IAS and EAS weakness lead to low resting and squeeze pressures resulting in faecal incontinence.
- The length of maintained squeeze pressure gives an estimate of the sphincter muscle fatigue.
- In the event of a sudden surge in the intra-abdominal pressure, as during a cough, the EAS contracts reflexively to preserve continence. It can be assessed by an abrupt rise in rectal pressure. This reflex is mediated via the spinal cord. In patients with a spinal cord lesion anterior to the conus medullaris, this reflex is retained regardless of inadequate or absence of voluntary squeeze. However, in patients with a lesion of the cauda equina or sacral plexus, the reflex and voluntary squeeze are missing.
- During straining, there is relaxation of the sphincters along with the contraction of the rectum and abdominal muscles. It helps in diagnosing and evaluating patients with dyssynergic defecation. They are classified into four types:
  - Type I: Inconsistent increase in anal sphincter pressure parallels the increase in intra-abdominal pressure.
  - Type II: The patient is incapable of producing a pushing force, but there is a contradictory surge in the anal sphincter pressure.
  - Type III: The patient can raise sufficient intra-abdominal pressure; however, anal sphincter relaxation is missing or defective.
  - Type IV: The patient is incapable of producing a pushing force in addition to missing or defective anal sphincter relaxation.
- Rectal balloon distention using air or water is used to assess the sensory responses and compliance of the rectal wall. A progressive increase in the volume of water

in the balloon helps to determine the verge for first sensation (normal is 30–50 ml), first impulse to excrete (80–100 ml), and urgency to excrete.
- In patients with rectal hyposensitivity, these cut-offs for sensation are greater than normal, while the volume of water/air required to stimulate anal sphincter relaxation is lower in patients with incontinence than without.
- In patients with colitis, low spinal cord lesion, and concomitant diabetes and incontinence, rectal compliance is poor compared to those with a high spinal cord lesion.
- Healthy individuals can excrete a 50-ml water-filled balloon or a silicone-filled artificial stool from the rectum in less than 60 s. Those with faecal incontinence expel easily. Nevertheless, patients with faecal seepage and elderly with faecal incontinence due to faecal impaction show weak evacuation due to asynchronous excretion.
- Manometric tests of anorectal function are also helpful for the assessment of prognosis post-pharmacotherapy, biofeedback therapy, or surgery.
- Biofeedback is performed as an initial treatment in patients with faecal incontinence who have sphincter weakness with or without neuropathy and patients with dyssynergic defecation and rectal hyposensitivity. It is performed with the patient receiving visual cues from the manometry instrument. Sphincter strength can be improved by asking the patient to squeeze within less than 2 s after balloon inflation. The rectal sensation can be improved by asking the patient to perceive a lower volume of balloon distension. Coordination can be improved by making the patient strain along with anal relaxation. Normally 4–6 training sessions are recommended. Biofeedback helps in the improvement of symptoms in 40–85% of patients.

4.15 Immunotherapy in GI Cancer

4.15.1 Rationale for Immunotherapy

There exists a dynamic interplay between tumour cells and the host's immune system. Schreiber et al. have described the immunoediting process whereby this interplay has been described in three stages:

1. Elimination phase: Where the host's innate and adaptive immune system identifies and clears tumour cells.
2. Equilibrium phase: Tumour cells accumulate genetic mutations to overcome the host's immune surveillance.
3. Escape phase: Tumour cells can escape the host's immunological surveillance and proliferate.

4.15.2 Mechanisms of Immuno-Editing

This includes alterations induced by tumour cells, tumour microenvironment, and alterations in the host's immune response.

- **Tumour cells down-regulate immune activation** by secreting immune-suppressing soluble factors (like TGF-β and IL-10), activating negative co-stimulatory signals involved in antigen presentation to T cells (like PD-L1), and down-regulating MHC-I expression on tumour cells.
- **The tumour microenvironment** plays a key role in the pathogenesis of pancreatic and colon cancer. Tumour microenvironment includes regulatory immune cells, matrix proteins, and cancer-associated fibroblasts. These secrete and promote tumour cells by secreting tumour-promoting factors which are involved in tumour invasion and angiogenesis and increased activation of Treg cells which downregulate immunity.
- Among **alterations in host immune cells**, modifying the immune thresholds by cytotoxic T-cell lymphocyte-associated antigen-4 (CTLA-4) and programmed death-1 (PD-L1) are important mechanisms. Immune checkpoint molecules play a key role in T-cell activation by either upregulating or downregulating co-stimulatory or co-inhibitory pathways.
  - CTLA-4 is an inhibitory receptor found on T cells, which binds to CD80/CD86 on dendritic cells. On binding, inhibitory pathways in T cells are activated and lead to downregulation of immunity.
  - PD-L1 is expressed commonly in T cells, B cells, monocytes, and NK cells. It is a cell surface co-inhibitory receptor, which on activation downregulates T-cell immune response. They bind to programmed death ligands 1 (PD-L1) and PD-L2, which are secreted during inflammation. Tumour cells increase PD-L1 expression and downregulate T-cell activation.

4.15.3 Immunotherapeutic Strategies

Immunotherapeutic strategies can be either active or passive.

- Active immunotherapeutic strategies involve activating the host's immune response. The use of cytokines, cancer vaccines, and checkpoint inhibition are types of active immunotherapeutic strategies.
- Passive strategies include the use of monoclonal antibodies and adoptive cellular therapy (ACT), which involve the translocation of tumour-specific T cells into the host tumour to eradicate the tumour cells.

Among the active immunotherapeutic strategies, cytokine-based therapy and cancer vaccines have not been successful. The breakthrough therapy has been

Table 4.5 Immunotherapy in GI cancers-current status

Checkpoint inhibitors		
Anti CTLA-4		
Ipilimumab	FDA approved for the pharmacotherapy of metastatic colorectal cancer as combination therapy with nivolumab, which expresses microsatellite instability (MSI-H) or deficiency of mismatch repair (dMMR)	MSI-H or d MMR is noted only in ~6% of metastatic colorectal cancer. So current use is limited to this subgroup
Tremelimumab	Under investigation for metastatic colorectal cancer	
Anti PD-1		
Nivolumab	(1) FDA approved for the pharmacotherapy of metastatic colorectal cancer either alone or as a combination therapy with nivolumab, which expresses microsatellite instability (MSI-H) or deficiency of mismatch repair (dMMR) (2) FDA approved as the second-line agent for HCC	
Pembrolizumab	(1) In 2017, FDA approved the use of any pretreated solid malignancies expressing microsatellite instability (MSI-H) or deficiency of mismatch repair (dMMR) (2) NCCN-category 2b recommendation for pretreated HCC	36% objective response rate in pretreated colorectal cancer expressing MSI-H or dMMR
Anti PDL-1		
Avelumab	Under trial in gastroesophageal cancer	
Durvalumab	Under investigation in colorectal cancer	
Atezolizumab	Under investigation in colorectal cancer	

the discovery of checkpoint inhibitors: anti-CTLA-4 (ipilimumab) and anti-PD-1 (nivolumab and pembrolizumab). These molecules block the downregulation of T-cell response by checkpoint molecules and activate the host immune response to tumour antigens. They were first approved for use in metastatic melanoma. The current status of checkpoint inhibitors in GI cancers is tabulated in Table 4.5.

Among the passive immunomodulatory strategies, monoclonal antibodies have been in widespread use. Another interesting concept is the use of adoptive cellular therapy, whereby T cells isolated from the patient are activated against tumour antigen in the laboratory and reinfused into the patient. A modification of this is Chimeric antigen receptor therapy (CAR). The T cells are engineered to display chimeric antigen receptors on their surface, which are independent of MHC receptors. CAR cells can engage their target independently of APC and MHC classes. Thus, CAR therapy is beneficial in cancers with downregulation of MHC class I. There is increasing evidence of CART therapy in haematological malignancies, while the role in GI malignancy is under evaluation.

4.16 Role of Anti-TNF in IBD

4.16.1 Indications for TNF Inhibitors

- Crohn's disease
  - Immunomodulator refractory disease
  - Fistulizing disease
  - Steroid-refractory or dependent disease
  - Severely active disease with a risk of poor prognosis
- UC
  - Immunomodulator refractory disease
  - Steroid-dependent active colitis
  - Steroid-refractory severe colitis

4.16.2 Approaches

'Top-down therapy'.

In patients with severe disease phenotype

- Start anti-TNF treatment in patients with less than 2 years of disease duration demonstrates better efficacy (REACT study).
- Therapy with both IFX and an immunomodulator in biological and immunomodulator-naive patients offers better clinical and endoscopic outcomes than either therapy alone (SONIC study).

4.16.3 Screening for Infections Before Initiation of TNF Inhibitor Treatment

Tests recommended:

1. HBV serology, HCV, HIV, EBV, CMV (optional)
2. M tuberculosis screen—IGRA, CT chest, and Mantoux test
3. Other bacteria, viruses, and parasites—blood, urine, and stool cultures

4.16.4 Contraindications for Anti-TNF ('STOIC')

- Sepsis/abscess/active infection
- Tuberculosis

- Optic neuritis (demyelination)
- Infusion reaction (anaphylaxis)
- Cancer and congestive heart failure

4.16.5 Summary of Anti-TNF Inhibitors Available for the Treatment of IBDs

Infliximab—Infliximab is a chimeric monoclonal antibody constituting human and murine sequences in a 3:1 ratio. Its induction dose is 5 mg/kg intravenously administered at 0, 2, and 6 weeks. Patients who respond adequately are continued on maintenance therapy for remission at the same dose, administered usually every 8 weeks.

Adalimumab—Adalimumab is a recombinant human monoclonal antibody that binds to TNF-α. Induction therapy includes 160 mg adalimumab administered subcutaneously, followed by a 50% reduction in dose at weeks 2 and 4, and, subsequently, 40 mg weekly maintenance dose.

Certolizumab pegol—Certolizumab pegol is a humanized monoclonal antibody Fab fragment linked to polyethene glycol that improves its plasma half-life with a consequent reduction in dosing frequency and immunogenicity. The recommended dose for induction in CD is 400 mg, administered subcutaneously every 2 weeks for three doses, followed by maintenance therapy every 4 weeks.

4.16.6 Vaccination

- Patients should be vaccinated for influenza and pneumonia yearly and 5-yearly, respectively.
- Usual vaccines for infections such as diphtheria, tetanus, hepatitis B, and HPV should also be administered.
- Vaccines that contain live infectious pathogens should not be administered a minimum 4 weeks before chemotherapy; these include vaccines against oral polio; rotavirus; measles, mumps, and rubella; yellow fever; and varicella should be avoided at least 4 weeks before immunosuppression.

4.17 Anti-integrins in IBD

Anti-integrin agents block the interactions between white blood cells (WBCs) and the intestinal vasculature, averting the infiltration of inflammatory cells and mediators in IBD.

4.17.1 *Natalizumab*

Natalizumab is a chimeric recombinant human IgG4 antibody that blocks the α_4 subunit in $\alpha_4\beta_7$ and $\boldsymbol{\alpha_4\beta_1}$ integrins on WBCs that interact with **vascular cell adhesion molecules**.

Indication: It has been approved by the FDA only and not EMEA in moderate-to-severe CD patients refractory or intolerant to conventional treatment or TNF inhibitor therapy.

Adverse effect: It has been associated with progressive multifocal leukoencephalopathy (PML) due to the JC virus (1 in 1000), which can be fatal.

4.17.2 *Vedolizumab*

Vedolizumab is also a humanized monoclonal IgG1 antibody against $\boldsymbol{\alpha_4\beta_7}$-integrin that attenuates the adhesion of WBCs to the vascular endothelium by blocking the interaction between $\alpha_4\beta_7$-integrin and **MAdCAM-1** expressed on GI blood vessels and lymph nodes. Vedolizumab is gut selective and so inhibits lymphocyte trafficking only in the intestine.

Indication: Vedolizumab is approved by the FDA and the European Medicines Agency for the treatment of moderate-to-severe UC and CD adult patients refractory to the first-line therapy, including steroids, immunosuppressive agents, or TNF antagonists.

Dosing and administration: The induction dose for treatment of patients is 300 mg intravenously at 0-, 2-, and 6-weeks, then every 8 weeks thereafter. For patients who flare while on maintenance therapy, the dose can be escalated by decreasing the vedolizumab dosing interval (i.e. 300 mg intravenously every 4 weeks).

GEMINI 1 study showed the benefit of using Vedolizumab (vs. placebo) in patients with UC; GEMINI 2 trial showed the benefit of Vedolizumab (vs. placebo) in patients with CD.

The VARSITY trial found that Vedolizumab was more effective at week 52 than adalimumab in inducing clinical remission and mucosal healing in patients with moderate-to-severe active UC. However, this statistical difference disappeared when the outcome assessed was steroid-free remission. These drugs did not raise any safety concerns and were well tolerated.

4.17.3 *Etrolizumab*

Etrolizumab is a humanized monoclonal IgG1 antibody against the β_7 subunit of $\alpha_4\beta_7$ and $\alpha E\beta_7$ (blocks E cadherin). It is a lymphocyte trafficking inhibitor (phase II trial ongoing).

4.18 Anti-IL12/IL23 Molecules in IBD

- The GWAS has recognized the IL-12/IL-23 pathway as an important one in the pathogenesis of IBD.
- In the intestinal mucosa of patients with CD, increased production of the proinflammatory cytokine IL-12 stimulates the production of IFN-γ and differentiation of Th1 cells.
- IL-12 is a heterodimeric cytokine with two covalently linked subunits, p40 and p35, synthesized by various immune-inflammatory cells.
- Similarly, IL-12 receptor (IL-12R) is a heterodimeric protein consisting of two units, IL-12Rβ1 and IL-12Rβ2. On interacting with its substrate IL-12, Janus kinase 2 (JAK2) and tyrosine kinase 2 (TYK2) are activated, followed by signal transducer and activator of transcription (STAT) 4 activation. This series of events is necessary for the induction of IFN-γ and Th1 differentiation.
- IL-23 regulates the inflammatory responses within peripheral tissues. Nonetheless, sometimes an uncontrolled expression of IL-23 may, such as in the case of IBD, trigger and maintain innate T-cell-mediated intestinal inflammation.

4.18.1 Ustekinumab

- It is a monoclonal antibody targeting the p40 subunit resulting in the inhibition of IL-12 and IL-23 effects.
- Patients are administered ustekinumab at a single dose of 130 mg/kg or 6 mg/kg intravenously of body weight. Maintenance dosing is 90 mg subcutaneously every 8 weeks. Common adverse events include infection and nasopharyngitis. Trials have shown efficacy in CD and UC (UNITY and UNIFI trials).
- Other medications, such as briakinumab and risakizumab, are still in the trial stage.

4.19 Role of Janus Kinase (JAK) Inhibitors in IBD

Janus kinase is named after Janus (two-faced God in Roman mythology), as they have both catalytic and pseudo-kinase activity. JAK inhibitors have been used in multiple autoimmune disorders like rheumatoid arthritis and psoriasis. Its role in IBD is evolving.

Table 4.6 Current status of JAK inhibitors in IBD

Compound	Selectivity	Crohn's disease	UC
Tofacitinib	Pan JAK	Development halted	Approved for moderate-to-severe UC remission; 10 mg twice daily Achieved remission in 18% at 8 weeks and maintained remission in 34% at 52 weeks
Filgotinib	JAK1	Phase 3 trial	Phase 2 trial
Upadacitinib	JAK 1	Phase 3 trial	Phase 2 trial
Peficitinib	Pan JAK	Phase 2 trial	Phase 2 trial

4.19.1 Mechanism of Action

JAK family consists of four distinct members: JAK1, JAK2, JAK3, and tyrosine kinase 2.

4.19.2 JAK Inhibitors and Role in IBD (Table 4.6)

4.19.3 Side Effects of JAK Inhibitors

- Herpes Zoster and other opportunistic infections.
- Thromboembolic events.
- Increase of LDL cholesterol levels: Check lipid profiles after 8 weeks and 6 monthly.
- Cytopenias: CBC at baseline, 4–6 weeks after initiation and 3 monthly thereafter.
- Increased liver enzymes at baseline and 4 weeks later and 3 monthly thereafter.
- Increase in serum creatinine: Check at baseline and 12 monthly.

4.20 SIP 1 Modulators: Ozanimod and Etrasimod

Sphingosine 1-phosphate is a phospholipid-signalling molecule that targets and couples with cell surface (G protein) receptors called sphingosine 1-phosphate (S1P) receptors. They regulate multiple and diverse cellular responses that play a role in immunity, heart rate, smooth muscle contraction, and endothelial barrier function. There are five types: S1PR1, S1PR2, S1PR3, S1PR4, and S1PR5.

S1P is a cell-surface protein that regulates lymphocyte migration out of lymphoid tissues. It has been found that synthetic S1P agonists can bind with the S1P receptors causing sustained internalization within the cell, thus acting as functional antagonists. They modulate lymphocyte migration from lymphoid tissue, reduce peripheral lymphocytes, and prevent their participation or recruitment in the inflammatory processes.

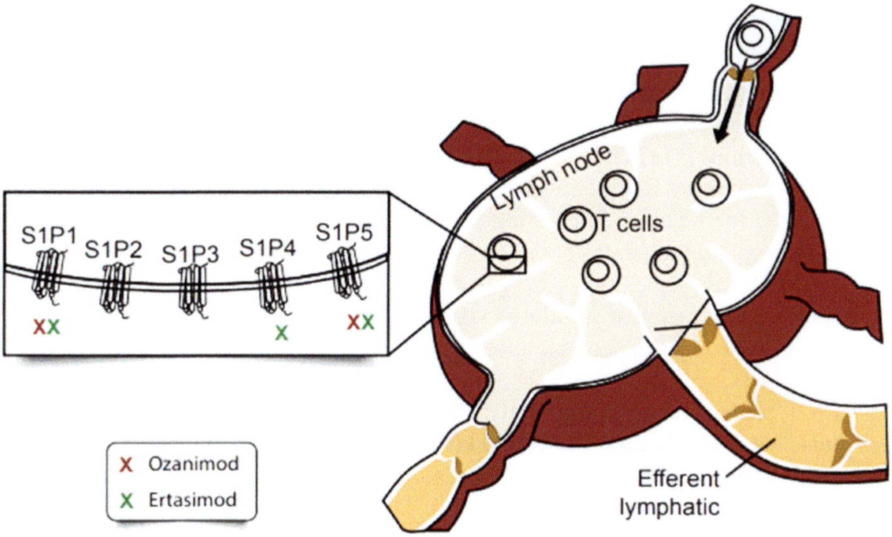

Fig. 4.5 Target sites for synthetic S1P agonists

Synthetic S1P agonists are an emerging class of small molecules that are being tried in a range of immune-mediated inflammatory disorders (IMID) and are now in use as treatment options for Multiple Sclerosis (MS) and IBD, particularly UC.

The first molecule in the class of S1P modulators, OZANIMOD (ZEPOSIA, 0.23, 0.46, 0.92) S1P 1 and 5 modulator, has been approved and in use in the United States, European Union, and Australia for MS and moderate to severe UC (Fig. 4.5).

One of the several adverse effects of Ozanimod was bradycardia and AV conduction blocks. Although most settled after a while, a cardiac evaluation and monitoring of HR after the initial dose. Also, starting with a smaller dose and gradual escalation is recommended.

ETRASIMOD is the next-generation SIP modulator that acts as an agonist of S1PR 1, 4, and 5 receptors (Fig. 4.5). It has completed phases 2 and 3 clinical trials in moderate-to-severe UC and is both efficacious and safe for both the short and the long term. The dose is 2 mg daily.

Adverse effects reported are bradycardia, cardiac conduction defects, blood pressure changes after the first dose (much less than with Ozanimod), headache, cough, macular oedema, and diarrhoea.

Based on the benefit/risk ratio, its overall efficacy and safety even in the long term (OASIS study), and its oral route of administration, it is emerging as a potential alternative to injectable biologicals in the treatment of IBD.

Ozanimod, Etrasimod, Laquinimod, and Obefazimod are on the growing list of synthetic SIP receptor modulators; the first two are now approved for clinical use, while the latter are in early clinical trial stages.

4.21 New Molecular Targets and Therapies for IBD: RIPK1

Due to the huge unmet need for an effective oral treatment for moderate-to-severe IBD, new targets and agents are being discovered and developed frequently.

RIPK1 inhibitor (ECLITASERTIB) is a new molecule with a different target.

The full name of RIPK1 is receptor-interacting protein kinase 1. To be more specific, it is the receptor-interacting **serine–threonine** protein kinase 1, an intracellular protein that regulates both cell death and inflammation, both of which are at play in IMIDs.

There are two types of regulated cell death in which the RIPK1 plays a role: **apoptosis** (the RIPK!-dependent apoptosis or programmed cell death) and **necroptosis** (Fig. 4.6).

RIPK1 protein has two terminals: the C terminal regulates non-inflammatory apoptosis, while the N-terminal-coded kinase activity causes necroptosis. If necroptosis (inflammation and cell death) plays a central role in inflammatory diseases, it appears promising to target the N-terminal of this protein to suppress chronic inflammation and cell loss of the mucosa.

RIPK1 is a key mediator of pro-inflammatory pathways within the cell, activating cell-death signalling from several pathways, including TNF, TLR, and IFN receptor pathways. Inhibition of the N-terminal could provide an effective reduction in inflammation.

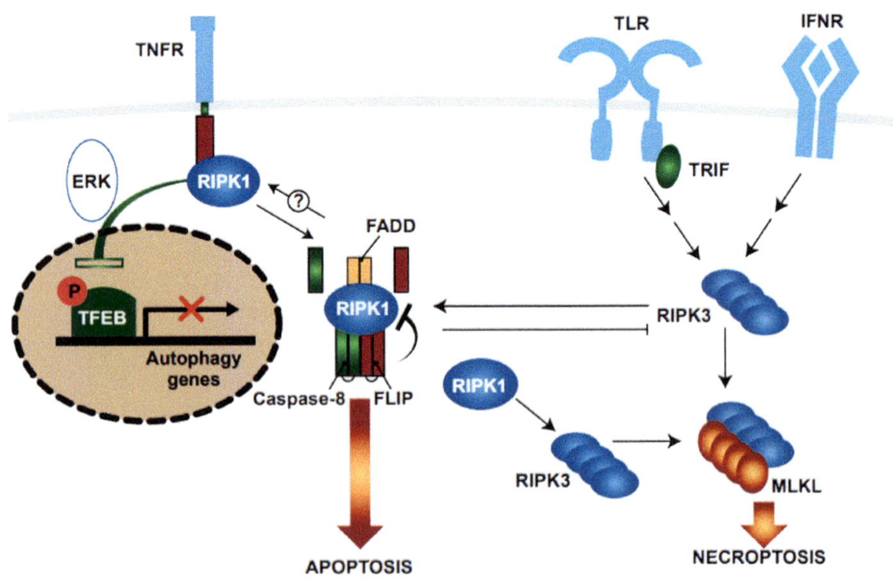

Fig. 4.6 Role of RIPK1 in regulated cell death

Inhibition of this kinase can be beneficial in necroptosis of surface cells too. Hence, it has been tried in psoriasis and cutaneous lupus, both of which show inflammation and necroptosis of skin cells.

The hallmark mechanism in IBD is quite similar: chronic inflammation has a key causative role, perpetuated by damage-associated molecular patterns called DAMPs. This also leads to necroptosis and loss of epithelial cells.

Inhibition of RIPKI with its dual mechanism of action with a selective antagonist seems promising in the treatment of IBD, both UC and Crohn's disease. Early clinical studies have been encouraging. A large global study is underway to establish its efficacy and safety in UC.

4.22 Curcumin in IBD

- Curcumin (diferuloymethane) is the pharmacologically active ingredient of the turmeric plant (*Curcuma longa*), a commonly used food additive in Indian cuisine. It has pleiotropic anti-inflammatory effects and has been studied in various disorders like psoriasis, arthritis, and IBD.
- Curcumin is poorly absorbed with rapid elimination limiting bioavailability. Semisynthetic and synthetic curcumin derivatives have been developed to overcome bio-availability issues for systemic treatment. Synthetic curcumin compounds with 6–7 times higher anti-cyclooxygenase and antioxidant activity have been manufactured.
- However, the luminal concentration of curcumin in GIT is high and is potentially useful for the treatment of inflammatory disorders of GIT like IBD.
- Experimental studies exploring its anti-inflammatory mechanism have suggested inhibition of the NF-κB pathway, p38 mitogen-activated protein kinases (MAPK) activity, and decreased pro-inflammatory Th1 cytokine response. This results in the suppression of inducible nitric oxide synthase and poor neutrophil recruitment.
- In humans, randomised controlled trials in IBD with curcumin as an adjunct to standard treatment or placebo for UC have been limited by heterogeneity due to different doses and preparation and have not been comparable. A recent meta-analysis of selected randomised controlled trials had not shown favourable effects of curcumin in inducing remission or maintenance in UC when results were analysed by intent to treat basis. Another meta-analysis showed better odds of attaining clinical remission in UC when curcumin was added to mesalamine.
- There is no clarity yet on the type of curcumin compound, dosage, and modality of administration—oral tablet or rectal enema for the treatment of IBD.
- Results of ongoing clinical trials with standardised protocol in adult and paediatric IBD patients are awaited.

4.23 The Doctor's Dilemma: When and How to Investigate the Individual with Common Gastrointestinal Complaints

Eamonn M M Quigley, MD FRCP FACP MACG FRCPI MWGO
David M Underwood Chair of Medicine in Digestive Disorders, Chief, Gastroenterology and Hepatology, Houston Methodist Hospital, Houston, TX, USA

Surveys from all around the world indicate that GI symptoms such as heartburn, constipation, dyspepsia/indigestion, abdominal pain, and diarrhoea are remarkably common—so common, indeed, that their occasional occurrence can be regarded as normal and often related to dietary indiscretion and/or the stresses of modern life. We also now know that the various functional syndromes that are comprised of various clusters of these symptoms; functional heartburn, functional dyspepsia, IBS, and chronic idiopathic constipation are also highly prevalent and were found to affect 40% of adults in the first global study on the prevalence of functional gastrointestinal disorders (FGIDs). These numbers illustrate the modern doctor's dilemma—how to evaluate the sufferer with these non-specific symptoms, which are statistically likely to be benign, but which could, in theory, be the harbingers of a fatal disease. Let us take IBS, the most common gastrointestinal problem seen by primary-care physicians and gastroenterologists alike, as an example and explore how one might efficiently and effectively evaluate functional symptoms. It lacks a diagnostic biomarker, and, therefore, definite diagnosis is challenging even with a plethora of investigations. These will not only be an expensive affair for the patient but also not worth it due to the high risk of missed diagnosis of malignancy or an equally serious condition.

4.23.1 IBS: Narrowing the Differential Diagnosis

Most IBS descriptions include abdominal pain as the principal manifestation, and thus a differential diagnosis is unlikely. However, attention to key facts in history can help enormously to narrow the differential diagnosis. Considering the definition that IBS is chronic, all causes of acute abdominal pain should be ruled out. Also, the abdominal pain recurs. It is not continuous but rather pulsating. Additionally, the site of pain is generally the lower quadrants in the Western population, while in Asians, pain in the upper quadrant is common. Most importantly, the abdominal pain in IBS is related to bowel movement, stool form, frequency, and consistency.

The context is all-important—IBS is a condition of young adults; cancer is rare in this age group. The typical IBS sufferer is a 20–40-year-old female with chronic lower abdominal pain, which is exacerbated or alleviated by bowel action and is accompanied by an alternating bowel habit. Abdominal bloating with concomitant distension is frequently noted in patients with IBS although not underscored in various diagnostic criteria.

Clinical history is an important consideration in understanding the exact meaning of the patient's symptoms (diarrhoea or only increased stool frequency or constipation) at presentation and their effect on their quality of life.

4.23.2 *Investigation*

The most frequently used diagnostic criteria for IBS are the Rome consensus which has stood the test of time, and there is adequate clinical evidence demonstrating reasonable credibility.

It is essential to identify indications that suggest a definite diagnosis of IBS or some other GI disease. The diagnostic accuracy can be enhanced with the recognition of additional characteristics such as anxiety, depression, and somatization, along with common laboratory investigations of erythrocyte sedimentation rate, full blood count and C-reactive protein. Symptoms such as nocturnal diarrhoea, age at symptom onset, weight loss, gastrointestinal bleeding, anaemia, fever, and a family history of colon cancer are warning signs to be considered to enhance diagnostic accuracy.

The literature is certainly reassuring, but the individual clinician will usually face a dilemma and anxiety of missing IBD, celiac disease, and especially CRC.

4.23.3 *How Not to Miss Alternate Diagnoses*

Colon Cancer

In a cohort study of more than 500 patients ~50 years of age, only one case of colon cancer was detected. However, interestingly, this patient did not manifest a single alarm symptom. Therefore, screening is underscored in patients with alarm symptoms over 50 years of age and those who have not had a previous screening.

Inflammatory Bowel Disease

IBS-type symptoms may precede a diagnosis of IBD by several years and especially of CD. Faecal calprotectin or lactoferrin levels can distinguish between IBD and IBS and, therefore, commonly used to screen individuals with diarrhoea-predominant IBS to rule out suspected IBD.

Microscopic Colitis (MC)

MC predominantly affects middle-aged to elderly females; unlike IBS, pain is not a characteristic feature; persistent nocturnal diarrhoea is often associated with concomitant weight loss. These differentiating factors should provide clues to clinicians to recommend colonoscopy with biopsies.

Celiac Disease

There is adequate epidemiological evidence to demonstrate that celiac disease has a worldwide distribution with its incidence in a particular region dependent on the prevalence of risk-enhancing genotypes and exposure to gluten. At such locations, serological screening is widely recommended.

Idiopathic Bile Acid Diarrhoea

More than 30% of patients with diarrhoea-predominant IBS may have idiopathic bile acid diarrhoea.[24] However, the availability of laboratory investigations, namely the ^{75}selenium homotaurocholic acid test (^{75}SeHCAT) or measurements of bile acids in faeces, is limited. In such scenarios, a more cost-effective approach would be to perform an empiric trial of a bile-acid sequestrant.

Small Intestinal Bacterial Overgrowth (SIBO)

The unavailability of reproducible SIBO tests and inadequate evidence to clarify its association with IBS make alternatives to manage these far from available.

Dyssynergic Defecation/Difficult Defecation

Defecatory dysfunction should be evaluated, considering its high prevalence among patients with constipation-predominant IBS. Clinical history and a digital rectal examination may provide hints towards IBS with defecatory dysfunction.

Disaccharidase Deficiency/Intolerance

Lactose intolerance causes acute onset of bloating, distension, and diarrhoea, which is confirmed by the symptom relief obtained by excluding milk and milk products, fructose, and other fermentable oligosaccharides. Among the other genetic causes is a congenital sucrase–isomaltase deficiency that manifests early in life; however, this rare disorder can also present in adulthood, depending on the type of mutation. The manifestations include IBS-type symptoms of abdominal discomfort, diarrhoea, bloating, and distension.

4.23.4 Conclusions

The predilection to assess each individual with common GI symptoms or symptom clusters until all possibilities have been ruled out should be discouraged. A scrupulous clinical and family history and comprehension of the context with reference to age, gender, location, and duration of manifestations should be considered. These should guide the advice on investigations which should be targeted rather than non-selective. The patient should be evaluated thoroughly for warning signs. Combining easily available cost-effective tests with a few specific ones along with trial-run responses to empirical treatment can enhance diagnostic accuracy. Invasive testing should be guided by clinical presentation.

4.24 Genetics of Colorectal Cancer

Genetic anomalies that increase the susceptibility to the development of CRC can be typically grouped as follows:

1. Variations in proto-oncogenes
2. Depletion of tumour suppressor gene activity
3. Abnormal modifications in genes associated with deoxyribonucleic acid (DNA) mismatch repair (MMR)

4.24.1 Genomic Instability

The loss of genomic stability can encourage the development of CRC by aiding the multiple tumour-associated mutations (Table 4.7). In CRC, genomic instability may occur in different forms, each with a different aetiology.

4.24.2 Chromosomal Instability

- It is the most frequently reported type of genomic instability in CRC. It is an efficient system giving rise to the physical disappearance of a wild-type copy of a tumour-suppressor gene, such as *APC*, *P53*, and SMAD family member 4 (*SMAD4*). These genes are protective against malignant mutations.
- Unlike other malignancies, CRC does not usually entail amplification of gene counts or gene reconfigurations.

4.24.3 DNA-Repair Defects

- In a subset of patients with CRC, the repair genes for base–base mismatches in DNA are deactivated. This deactivation may be either due to inherited mutation or acquired, such as in hereditary nonpolyposis colon cancer (HNPCC or Lynch syndrome) or in tumours with methylation-associated suppression of a gene that encodes mismatch repair protein, respectively.

Table 4.7 Genes altered in colon cancer

Type of instability and syndrome	Type of defect	Genes involved	Phenotype
Chromosomal instability Loss of heterozygosity at multiple loci	Somatic	Loss of heterozygosity at APC, TP53, and SMAD4	Characteristic of 80–85% of sporadic CRCs, depending on the stage
DNA mismatch-repair defects Hereditary nonpolyposis colon cancer	Germ-line	MLH1, MSH2, MSH6 germline gene mutations 9–14	Multiple primary CRCs, accelerated tumour progression and increased risk of endometrial
Sporadic CRC with mismatch-repair deficiency	Somatic	MLH1 somatic methylation 15–17	Colorectal cancer with increased risk of poor differentiation, more commonly located in the right colon, less aggressive clinical behaviour than tumours without mismatch repair deficiency
CpG island methylator phenotype-methylation target loci	Somatic	Target loci MLH1, MINT1, MINT2, and MINT318-23	15% of CRCs, with most showing mismatch repair deficiency from loss of tumour MLH1 expression
Base excision repair defect-MYH-associated polyposis	Germ-line	MYH24-26	Development of 15 or more colorectal adenomas with increased risk of CRC

DNA deoxyribonucleic acid, *CRC* colorectal cancer

4.24.4 Aberrant DNA Methylation

Epigenetic silencing of genes, carried out by aberrant DNA methylation, is an underlying gene inactivation mechanism in patients with CRC.

4.24.5 Growth Factor Pathways Affected in Colorectal Cancer

1. **Aberrant Regulation of Prostaglandin Signalling**
   The activation of growth factor pathways is frequent in CRC.
   (a) A crucial step in the development of an adenoma is the stimulation of prostaglandin signalling.
   (b) High levels of COX-2 are found in more than 60% of CRCs, and there is a depletion of 15-PGDH in 80% of colorectal adenomas and cancers.

2. **Epidermal Growth Factor Receptor**
   (a) Epidermal growth factor (EGF) protein increases intestinal cell proliferation by interacting with its receptors and activating MAPK and PI3K signalling.
   (b) Signalling through the EGF receptor (EGFR) is an important clinical finding in a subgroup of CRCs.
3. **Vascular Endothelial Growth Factor**
   (a) Certain angiogenic pathways have been shown to worsen prognosis and increase fatality in patients with CRC.
   (b) Treatment of patients with advanced CRC with the anti-VEGF antibody bevacizumab prolonged survival by an additional 4.7 months compared to the conventional therapy.
4. **Stem-Cell Pathways**
   (a) Stem cell therapy is a potential alternative for the prevention and treatment of CRC, which needs further evaluation.

4.24.6 Strategies of Screening for Colorectal Cancer

Principles of Screening

The primary prevention of CRC involves recognizing various genetic, biological, and environmental factors that increase the risk of the disease by modifying tumour development.

While secondary prevention aims at detecting existing preneoplastic and early neoplastic lesions and treating them aggressively.

USPSTF (US Preventive Services Task Force) recommends screening persons aged 50–75 years with an average risk of CRC, but routine screening is not recommended for adults 76–85 years of age and none for those >85 years of age.

Screening strategies for various risk groups are summarised in Table 4.8.

Table 4.8 Summary of screening strategy for patients at risk of colorectal cancer

Category	Test	How frequently
For average-risk individuals	FOBT Sigmoidoscopy Colonoscopy	Starting at 50 years Annual Every 5 years Every 10 years
For patients with a family history of CRC	Colonoscopy	Starting at 40 years of age or 10 years earlier than the youngest case in the family, repeat once every 5 years
For patients with genetic syndromes predisposing to colorectal cancer		
Familial adenomatous polyposis	Sigmoidoscopy	Every year after 10–12 year If the genetic test was positive, colectomy
Lynch syndrome	Colonoscopy	Every 1–2 years starting from 20 to 25 years or 10 years earlier than the youngest affected member
IBD (UC and Crohn's disease)	Colonoscopy	Every 1–2 years, 8 years after the onset of pancolitis, and 15 years after the onset of left-sided colitis
For persons with adenoma on colonoscopy		
<2 tubular adenomas with low-grade dysplasia	Colonoscopy	5–10 years after polypectomy
3–10 adenomas or villous morphology or high-grade dysplasia or adenoma >1 cm	Colonoscopy	3 years after polypectomy
>10 adenomas	Colonoscopy	<3 years after polypectomy
Piecemeal removal of sessile adenoma	Colonoscopy	2–6 months after polypectomy
For persons undergoing curative resection for colorectal cancer	Colonoscopy	Intraoperatively or within 3–6 months after resection. If normal, perform at 1 year, 3 years, and 5 years thereafter

4.25 Chemoprevention of Colorectal Cancer

Administration of a synthetic or naturally derived agent to reduce the risk of developing cancer, impeding the time of cancer onset, or the carcinogenesis process, is defined as chemoprevention.

Because the natural history of CRC is long-drawn-out, clinical studies frequently aim at averting colorectal adenomas that precede CRC.

Various agents have been studied in the chemoprevention of CRC, and the results are summarised in Table 4.9.

Table 4.9 Chemoprevention of colorectal cancer

Agent	Primary target	Mechanisms	Highest level of clinical evidence	Effect	Significant concerns
Aspirin	COX1 and COX2 (irreversible inhibition)	Inhibit prostaglandin synthesis, platelet activation, and inflammation	RCT	20% risk reduction in the development of adenoma; 40% risk reduction in advanced adenoma at 3 years	Elevated risk of bleeding
COX-2 inhibitors	COX2 (reversible inhibition)	Inhibit prostaglandin synthesis, platelet activation, and inflammation	RCT	Celecoxib and rofecoxib showed dose-dependent ~30–60% reduction in risk of advanced adenomas at 3 years	Elevated cardiovascular risk
Statins	HMG-CoA reductase (reversible inhibition)	Disruption of the mevalonate pathway with downstream effects on membrane integrity, cell signalling, protein synthesis, and cell cycle progression	RCTs with cardiovascular effects as the primary endpoint	A case-controlled study showed some benefits—*equivocal*	Myalgias, rhabdomyolysis (rare), elevated transaminases (rarely severe)
Long-chain omega-3 polyunsaturated fatty acids	Membrane phospholipids component	Anti-proliferative, apoptotic and antiangiogenic properties	RCT	Reduced number of polyps in FAP	Unpleasant smelling breath and sweat, nausea, diarrhoea
Vitamin A	Combines with retinol-binding protein, a plasma-specific transport protein	Regulates nuclear receptors that suppress tumour formation induces apoptosis, and enhances immune function	RCT	*Equivocal results*; do not support routine use	Hepatic damage, arthralgia, alopecia, headaches, vomiting, and skin desquamation

(continued)

Table 4.9 (continued)

Agent	Primary target	Mechanisms	Highest level of clinical evidence	Effect	Significant concerns
Vitamin C	Cofactor in collagen formation and tissue repair	Reduces oxidative stress enhances the immune system	RCT	*Equivocal results*; do not support routine use	Nausea, vomiting, diarrhoea, and headaches
Vitamin E	Present in cell and organelle membranes	Suppresses lipid peroxidation in cell membranes, decreases oxidative stress, and impedes carcinogen production	RCT	*Equivocal results*; do not support routine use	Elevated risk of haemorrhagic stroke
Beta-carotene	Functions as a provitamin A	Conflicting evidence of pro- and anti-oxidant properties	RCT	No benefit	Increased risk of lung cancer among smokers and individuals exposed to asbestos with beta-carotene supplementation
Selenium	Important for the synthesis of certain proteins	Antioxidant effects of certain proteins are due to selenium as an essential component but not selenium per se	RCT	No benefit	Gastrointestinal distress, alopecia, and nail discolouration
Folic acid	Coenzyme or cosubstrate in single carbon transfers in the synthesis of nucleic acids and amino acid metabolism	Proposed mechanism is through effects on DNA replication, repair, and methylation through the one-carbon metabolic pathway	RCT	RCT failed to show any beneficial effect	Mask vitamin B12 deficiency or exacerbate progression of neoplastic lesions

Calcium	Incorporated into the skeleton	Bile acid-binding capacity, direct effect on calcium-sensing receptors on colonocyte	RCT	20% risk of recurrence of adenomas	Nausea, vomiting, constipation, bone pain, kidney stones, confusion, and palpitations
Vitamin D	Regulates gene transcription via binding to vitamin D receptors located in cell nuclei	Inhibits proliferation and angiogenesis; induces differentiation and apoptosis	RCT	No benefit	Dose-limiting hypercalcaemic effects
DFMO (difluoromethylomithine)	Ornithine decarboxylase (irreversible inhibition)	Inhibits polyamine synthesis which is important for cell survival	RCT	~70% reduction in new adenomas at 3 years when used in combination with sulindac	Ototoxicity
Erlotinib	EGFR tyrosine kinase inhibitor (reversible inhibition)	Inhibits EGFR signalling	RCT		Gastrointestinal side effects and rash
Curcumin	Inhibits reactive oxygen-generating enzymes, protein kinase C, and EGFR	Anti-inflammatory activity induces apoptosis	Experimental studies	In murine models, utility in preventing adenoma	Gastrointestinal side effects at higher doses

4.26 Artificial Intelligence in Gastroenterology

4.26.1 Terminology

Artificial intelligence is the capacity of a computer to carry out a function that requires intelligence, including 'cognitive' skills that might mirror the human mind, such as the skill to 'learn'.

Machine learning is a stream of computer science in which a system can be enabled to 'learn' with data in the absence of detailed programming. A computer system can create an algorithm for data prediction based on identified patterns.

Deep learning indicates machine-learning algorithms that use multiple nonlinear processing methods for feature extraction and transformation. Feature extraction is the choosing of variables that have power in forecasting a target, while transformation is applying this data into a form that can be definitively used for the development of a model.

Neural networks are specific sections within machine learning that mimic densely interlinked neurons to detect patterns to forecast a result. Many of the current AI healthcare models apply the neural network paradigm.

4.26.2 Utility of AI in Gastroenterology

1. **Analysis of Malignant and Premalignant Lesions**
   (a) Enhance the diagnostic utility of endoscopy for recognizing premalignant lesions.
   (b) Prognostication of patients with GI cancer.
   (c) Estimate the patients' risk for metastases.
   (d) Recognises patients with an absolute response to neoadjuvant chemoradiation for locally advanced rectal cancer with high accuracy.
2. **Analysis of Inflammatory and Other Non-malignant Lesions**
   (a) Identify patients with IBD, ulcers, celiac disease, lymphangiectasia, and hookworm infestation on endoscopic images.
   (b) Prediction of response to treatments in patients with IBD—including response to biologics.
   (c) Predict the progression of IBD.
3. **Gastrointestinal Bleeding**
   (a) AI has been used in the detection of small intestinal haemorrhages based on images obtained during capsule endoscopy and can detect
      - Patients at risk for recurrent bleeding.
      - Patients presumably needing treatment.
      - Risk of mortality.

4. **Liver and Pancreatobiliary Disorders**

   In hepatology, AI has been used:

   (a) To detect fibrosis and progression to cirrhosis.
   (b) To recognize patients with oesophageal varices.
   (c) To identify patients with cirrhosis whose life expectancy is less than a year with the highest accuracy to group them as candidates for liver transplantation.
   (d) To match liver donors with recipients, which could prove to be a powerful decision-making tool.

4.26.3 Limitations

- Currently not applied in routine clinical practice.

4.26.4 Ethical Challenges

- Legal liabilities.

4.27 Next-Generation Sequencing in Gastroenterology

4.27.1 Terminology

- Next-generation sequencing (NGS), also called high-throughput sequencing, deep sequencing, and second-generation sequencing, uses sequencing of multiple DNA fragments performed simultaneously.
- Unlike Sanger sequencing, the speed of sequencing and quantity of DNA sequence data obtained with NGS is cumulatively greater at cost-effective rates.
- The template is prepared using DNA sourced from various cells, which is sometimes modified to get rid of non-coding segments. The sample is prepared by purification, amplification, and fragmentation and subsequently physically isolated.
- Sequence data are produced on these small fragments, and the digital findings are computationally arranged against a previously mapped 'reference' genome.

4.27.2 Clinical Use

The clinical significance of NGS lies in its capacity to detect multiple types of genomic variations, including nucleotide substitutions, small insertions and deletions, copy number alterations, and chromosomal regrouping.

NGS technology can explain the mechanism of pathogenesis and enhance diagnostic and therapeutic testing.

4.27.3 NGS in Identifying Pathogenetic Mechanisms

- NGS also has the potential to identify novel somatic mutations linked to acute myeloid leukaemia, melanoma, mesothelioma, small-cell lung cancer, and prostate cancer.
- Using pairs of tumour tissue from liver metastases and normal tissues from peripheral blood can help to recognize the mutations in genes causing tumour progression. Around 38 pathways with 280 gene mutations have been identified for CRC.

4.27.4 NGS in Diagnostic and Therapeutic Testing

1. ESMO guidelines recommend testing exons 2, 3, and 4 of KRAS and NRAS and BRAF exon 15, crucial for guiding therapeutic strategies in patients with CRC.
2. NGS has also been used to record MMR abnormalities and MSI. The monoclonal antibody pembrolizumab is indicated for the treatment of adult and paediatric patients with unresectable or metastatic solid tumours that have been classified as MSI-high (MSI-H) or MMR deficient (dMMR). This is an important achievement since it is the first-ever chemotherapeutic agent approved regardless of the site primary cancer. This applies to CRC and other solid organ tumours.

4.28 Non-coding RNA in GI Cancer

- A non-coding RNA (ncRNA) is a functional RNA molecule that is transcribed from DNA but does not code any proteins (Fig. 4.7).
- Their function is to regulate gene expression at the transcriptional and post-transcriptional levels and aid in heterochromatin formation, histone modification, DNA methylation targeting, and gene silencing.
- Those ncRNAs part of epigenetic processes are of two types the short ncRNAs (<30 nucleotides) and the long ncRNAs (>200 nucleotides).
- The three major classes of short non-coding RNAs are microRNAs (miRNAs), short interfering RNAs (siRNAs), and piwi-interacting RNAs (piRNAs).

Fig. 4.7 Non-coding RNA

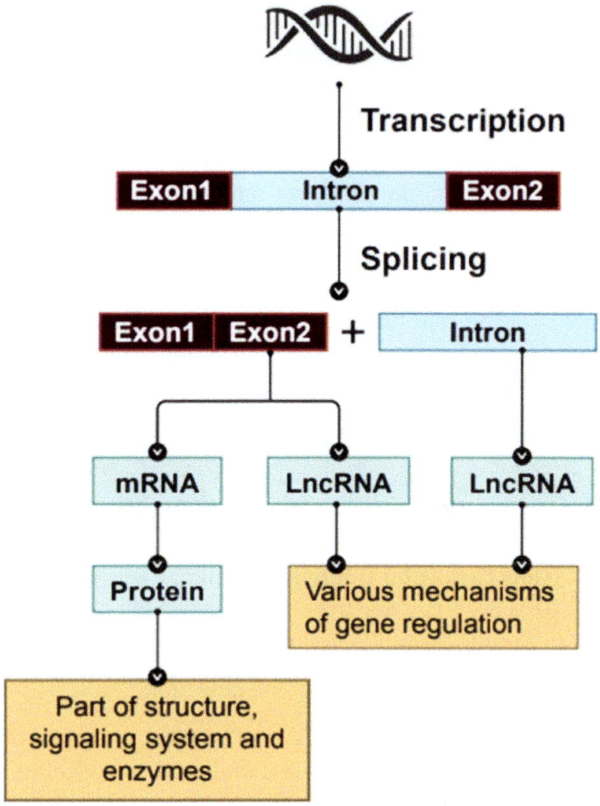

4.28.1 Role of Non-coding RNAs in Cancer

Non-coding RNAs are involved in

- Maintaining proliferative signalling, averting growth suppression, conferring replicative immortality, angiogenesis, and helping invasion and metastasis.
- Has been associated with genomic instability, and mutations.
- Help in evading immune destruction.
- Emerging hallmark: neurogenesis and axonogenesis.

4.28.2 Role of ncRNAs Therapeutics in GI Cancers

- Choosing an overexpressed, oncogenic ncRNA (ncRNA inhibition).
- Restoring an ncRNA with tumour suppressive characteristics if it is downregulated (ncRNA mimetic).

4.29 Role of Advanced Endoscopic Techniques in the Assessment of IBD

Endoscopy is an important investigation method to diagnose and determine the extent of IBD. On follow-up, it is used for the assessment of therapeutic response by monitoring the extent of inflammation and for surveillance for colorectal carcinoma.

The role of high-definition white light endoscopy, dye-based chromoendoscopy, optical endoscopy, and EUS in the assessment of IBD is expanding. Newer technologies like autofluorescence imaging (AFI), FUSE, endocytoscopy, and confocal laser endomicroscopy are the latest additions to the armamentarium used for the evaluation of IBD.

4.29.1 Commonly Used Endoscopic Imaging Techniques

1. **High-Definition White Light Endoscopy (HD-WLE)** is the current standard in GI endoscopic practice that overrides the standard definition (SD)-video endoscopy with >1 million pixels per image.
   (a) It improved the adenoma detection rate versus SD video endoscopy.
2. **Chromoendoscopy (CE)**—The most commonly used imaging technique for IBD screening and is mentioned in the latest international guideline (SCENIC).
   (a) Dye-Based CE—Using topical methylene blue and indigo carmine.
      - Good relativity of UC inflammation severity with conventional histology.
      - Better than SD endoscopy for adenoma detection both in IBD patients and the general population.

   (b) Dye-less digital chromoendoscopy.
      This technology may make use of
      - Optical filters, so-called optical chromoendoscopy, like the narrow-band imaging technique (NBI, Olympus, Japan) or
      - Virtual chromoendoscopy or virtual video post-processing in real time (i-Scan, PENTAX, and FICE, Fujinon, Japan).
      - NBI colonoscopy
        – More precise for diagnosing the degree of inflammation in patients with quiescent UC (as compared to HD-WLE)—increased vessel density in inflamed areas.
        – Poorly differentiates neoplastic versus non-neoplastic mucosa and is inferior to WLE or CE for inflammation and pseudopolyp histology prediction.

- i-Scan
  - It offers a precise assessment of inflammation compared to histological analysis.
  - It provides a real-time diagnosis of both gastric and duodenal CD, facilitating an earlier diagnosis of upper GI CD and facilitating targeted biopsies for histological affirmation.
  - It is equal to dye-based CE and superior to SD—endoscopy in adenoma detection.

3. **EUS**

   During the diagnosis and staging of IBD, delineating the intestinal layers may be critical in doubtful cases to distinguish between UC and CD.

   (a) Rectal EUS is used for estimating the severity of IBD and for perianal fistula and abscess characterization.
   (b) EUS enables differentiation of UC from CD; wall thickness corresponds adequately with activity and histology. The presence of submucosal and muscularis propria thickening along with pericolic nodes was in favour of Crohn's disease.

4. **Capsule Endoscopy**

   In cases where the findings of bidirectional conventional endoscopy and enteroscopy were ambiguous, capsule endoscopy can be used for
   (a) Diagnosis of uncertain IBD cases.
   (b) Elucidation of site and extent of inflammation in CD.
   (c) Monitoring prognosis.

4.29.2 Advanced Endoscopic Imaging Techniques

1. **Full-Spectrum Endoscopy (FUSE®)**

   This is a new high-definition endoscope which has additional lateral camera lenses at the colonoscope tip apart from the conventional forward-viewing camera to enhance the picturization of lateral walls and folds. This lens configuration enables a 330° field of view of the mucosa in contrast to 170° with the conventional one.

   (a) Studies in IBD patients have reported greater detection of dysplastic lesions compared to the conventional forward-viewing colonoscopy.

2. **Endocytoscopy (EC) (Olympus, Japan)**

   It works on the optical principle of contact light microscopy. It delivers in vivo real-time ultra-magnified (1390×) microscopic images of the intestinal mucosal surface. Its findings correspond adequately with those of histological analysis in estimating disease activity in patients with IBD. An endocytoscopy system score (ECSS) has been validated.

3. **Fluorescence and Autofluorescence Endoscopy**

   It relies on the visualization of fluorescence light (wavelength: 500–630 nm) released by either exogenous or endogenous fluorophores.

   (a) Fluorescence endoscopy with 5-aminolevulinic acid (5-ALA) and WLE show similar dysplasia detection rates in IBD.
   (b) Autofluorescence imaging was better than WLE in recognizing inflammation.
   (c) Comparative studies have shown that autofluorescence is better than NBI and WLE in neoplasia detection in UC.

4. **Confocal Laser Endomicroscopy (CLE)**

   Two types of CLE systems:
   (a) The integrated endoscopic system (iCLE, from PENTAX, Tokyo, Japan)—attached to the distal tip of the endoscope—not commercially available now.
   (b) The probe-based system (pCLE, from Mauna Kea Technologies, Paris, France) can pass through the working channel of standard endoscopes.

   Endomicroscopy has high resolution and tissue penetration of approximately 250 μm and can, therefore, reveal the following:
   (a) Mucosal architecture—alterations in crypts such as form, density, integrity, lumen distortion, leakage, and goblet cell density.
   (b) Microvascular modifications such as increased vascularity, vascular integrity, and leakage. Based on these parameters, endomicroscopic inflammation criteria for a Crohn's colitis activity score and UC severity score have been confirmed.

   It can also be used
   (a) For forecasting an IBD relapse by elucidating the process of cell shedding and quantifying the loss of local epithelial barrier function; estimating the risk of relapse in quiescent CD and UC by underscoring the fluorescence leakage and micro-erosions as risk factors for inflammation and based on crypt structural and microvascular criteria, respectively.
   (b) To compared to histological evaluation, the CLE mucosal healing score for colonic Crohn's colitis and UC has higher sensitivity and specificity.
   (c) To targeted biopsy using iCLE using chromoendoscopy guidance (with methylene blue) that had a considerably greater diagnostic yield (4.75-fold) for neoplasia recognition in UC patients compared to conventional colonoscopy with non-targeted biopsies.

4.30 Optical Biopsy of Colonic Polyps

1. Utility of magnification endoscopy and chromoendoscopy—Kudo pit pattern.
2. Digital chromoendoscopy (NBI)—Sano classification and NICE classification.

1. **Utility of magnification endoscopy and chromoendoscopy—Kudo pit pattern**

(a) **Magnification endoscopy**

High-definition endoscopes display highly enhanced images (150× magnification) which enable detailed elucidation of the mucosal surface. These endoscopes are mainly from Asian countries since they are not available in Europe and the United States.

(b) **Chromoendoscopy**

- Chromoendoscopy (CE) is the topical instillation of dye through an endoscope to pinpoint mucosal details.
- Mucolyte and de-foaming agents containing solution are administered before endoscopy to eliminate mucus and residual faeces.
- The two most commonly used dye types are
  - **Vital or absorptive dyes** are taken up by intestinal cells, e.g. methylene blue (used as 0.1%) and crystal violet (0.05%). Absorptive contrasts are allowed to settle post-instillation for a minute before rinsing off. Compared to methylene blue, crystal violet is used at lower concentrations due to the risk of toxicity, and, therefore, minimal quantity is used with a special non-traumatic catheter.
  - **Contrast dyes** such as indigo carmine (0.1% large sections; –0.5% small sections) settle in the mucosal grooves and improve the subtle mucosal unevenness for better visualization. Dyes are transferred with a catheter for marking large mucosal areas or through the endoscope channel with a syringe.
- Chromoendoscopy using high-resolution endoscopes without magnification also permits the differentiation of neoplastic from non-neoplastic lesions. However, the precision is somewhat lower (87–92%) than those with magnification endoscopes.

4.30.1 Kudo Pit Pattern

Kudo pit pattern for colonic polyps was described with the use of magnification endoscopy (>100×) with concomitant use of chromoendoscopy.

Procedure and Kudo Pit Pattern (Figs. 4.8 and 4.9)

4.30.2 Digital Chromoendoscopy (NBI): Sano Classification and NICE Classification of Colorectal Polyps

- Narrow-band imaging (NBI) is a means of image enhancement which disallows white light (WL) and permits only two narrow bands of a light wavelength corresponding to the absorption wavelength of haemoglobin in the blood—415 nm (blue) and 540 nm (green), to pass through. NBI light is absorbed by haemoglo-

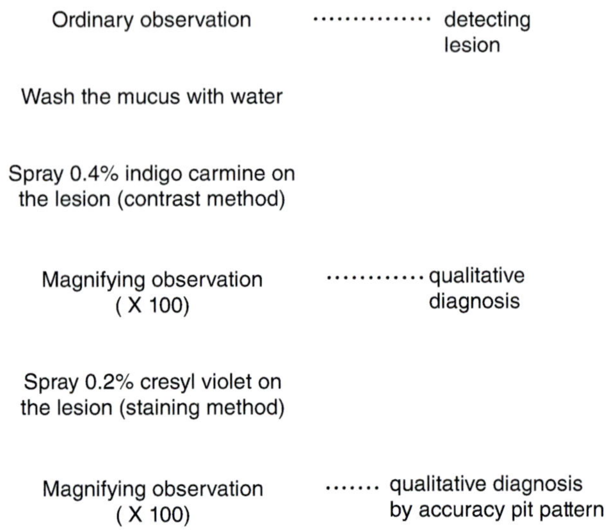

Fig. 4.8 Magnification endoscopy procedure

Type	Characteristics	Close up	Pit size (mm)	Interpretation
I	Round,normal		0.07+/-0.02	Normal
II	Asteroid		0.09 ±0.02	Hyperplastic
III_S	Tubular or round pit smaller than normal pit (type1)		0.003 ±0.01	Neoplastic Tubular Adenoma
III_L	Tubular or round pit larger than normal pit (type1)		0.22 ±0.08	Neoplastic Tubular Adenoma
IV	Dendritic/gyrus brain-like		0.93 +/-0.32	Neoplastic Tubulovillous or villous
Vi	Irregular arrangement (sizes of III_S, III_L, IV type pits)		NA	Neoplastic High grade or invasive
Vn	Loss or decrease of pits with amorphous structure (Non-structural)			

Fig. 4.9 Kudo pit pattern

bin in the vessel and is reflected by adjacent mucosa. This provides a good contrast between vessels and mucosa.

- The shorter NBI (blue) light is absorbed by superficial capillary vessels in the mucosa. This facilitates the detection of superficial tumours which have abnor-

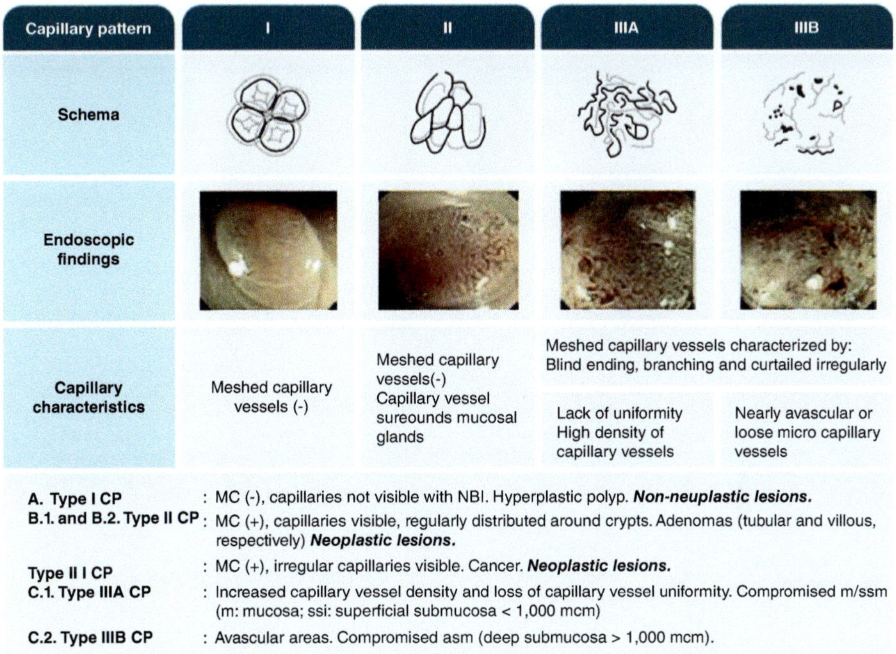

Fig. 4.10 Sano's capillary pattern classification for NBI in colorectal neoplasia

mal vascular patterns. The longer (green) light penetrates deeper and is absorbed by deeper vessels in the mucosal layer. Thus, NBI offers data regarding surface architecture and vascularisation, both of which are altered in neoplasia.

- The NBI with magnifying colonoscopy shows a detailed image of the colon, including its microarchitecture and microvasculature.

Sano NBI Classification

A classification based on capillary pattern (CP) has been prepared, which can be used with magnification endoscopes (Fig. 4.10).

Neoplastic lesions have peculiar vascular changes termed meshed capillary (MC) vessels which form the basis of CP classification known as Sanos Classification.

- Type 1—Normal capillary vessels not visible and hyperplastic polyps.
- Type 2—Capillary vessels regularly distributed around crypts and typical of adenoma.
- Type 3a—Meshed capillary vessel pattern present, high density, distributed irregularly, and cancers with superficial submucosal invasion <1000 μm.
- Type 3b—meshed capillary vessel pattern, avascular areas, and suggest deeper submucosal invasion >1000 μm.

	Type 1	Type 2	Type 3
	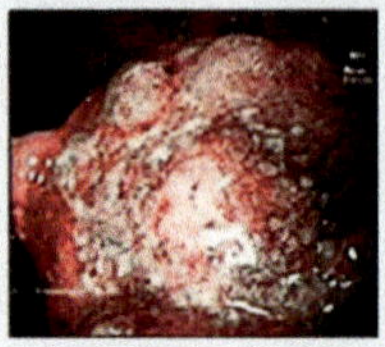	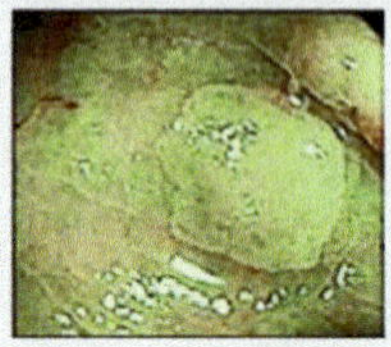	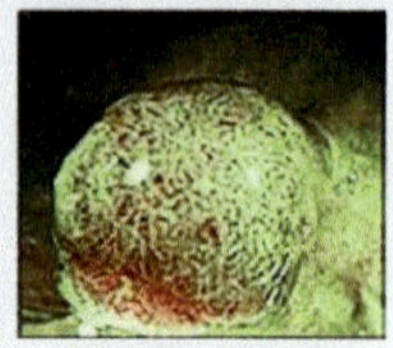
Color	Brown to dark brown compared to the background, occasionally white patches	Same or lighter than background	Slightly darker compared to the background
Surface pattern	Areas with distortion or absence of pattern	Dark or white spots of uniform size, or homogeneous absense of pattern	Oval, tubular, or branched white structures surrounded by brown vessels
Most likely pathology	Deep submucosally invasive cancer	Hyperplastic	Adenoma
Vessels	Some area(s) show considerably distorted or absence or vessels	None or isolated lacy vessels maneuvering across the lesion	Brown vessels surrounding white structures

Fig. 4.11 NICE classification of colonic lesions

- Sano's classification devoid of magnification gives satisfactory accuracy (91%), although less as compared to magnification endoscopy for differentiating non-neoplastic from neoplastic lesions.

Utility of Sano's Classification Based on NBI Versus Chromoendoscopy to Detect Deeper Invasion

- Sano's classification distinguishes between superficial and deep-seated invasions with more than 87% precision, but chromoendoscopy with magnification provides superior findings (98.8%). Hence, chromoendoscopy, along with magnification, is favoured in patients with a risk of invasive lesions.

4.30.3 NICE Classification of Colorectal Polyps

Magnification is not commonly used other than in Japan. Therefore, a joint classification with Asian and Western endoscopists has been recently devised. The NICE (NBI International Colorectal Endoscopic) classification is based on three findings: (1) colour, (2) vessels, and (3) surface pattern, and differentiates colonic lesions into three groups (Fig. 4.11).

Utility of NICE Classification

The NICE classification is a validated tool for separating non-neoplastic from neoplastic polyps submucosal-invasive from early colorectal carcinomas. These criteria have been validated in various studies.

- NBI can be used to distinguish between hyperplastic and adenomatous polyps by using the NICE classification (NICE1 and NICE2) with 89% accuracy.
- NICE2 and NICE3 patterns are confirmed based on the presence of any one of the three criteria (colour, vessels, and surface pattern) of deeply SM-invasive carcinoma. It has a high sensitivity (94.9%) and a high negative predictive value (95.9%).

4.31 Colonic Stents

4.31.1 Indications

1. **Palliation of locally advanced or metastatic malignant colonic obstruction**
   On comparing SEMSs with surgery for palliation of malignant colon obstruction, patients undergoing SEMS placement had a shorter duration of hospital stay, lower rate of early adverse events, and a lower stoma creation rate but a higher delayed adverse event rate and reintervention rate, with comparable overall survival rates.
2. **Acute malignant colonic obstruction**
   Colonic SEMSs may be used as a bridge in surgery in patients with acute left-sided obstruction who are considered surgical candidates. Preoperative stenting offers multiple advantages over emergent surgery, including
   (a) Avoidance of an ostomy, allowing a single-stage operation.
   (b) Appropriate staging of disease.
   (c) Correction of electrolyte and volume status and optimization of co-morbidities.
   (d) Conversion of emergency surgery to an elective operation with adequate bowel preparation and the possibility of laparoscopic resection.
   In patients with acute right-sided colon obstruction, preoperative SEMS placement does not alter the surgical course because a single-stage right hemicolectomy with an ileocolonic anastomosis can be performed emergently in the unprepared bowel.

4.31.2 Technique

- Colon SEMSs may be covered or uncovered through the scope or over the guidewire; smaller-diameter stents are generally used on the right side of the colon, and larger-diameter stents on the left side of the colon to prevent solid stool impaction.
- Ideally, a radiologic imaging study such as a barium enema or CT scan with rectal contrast should be obtained before stent placement to assess the degree of obstruction and the length and location of the stricture.
- A bowel preparation may not be possible; tap water enemas may be administered before the procedure in patients with a subtotal colon obstruction.
- Prophylactic antibiotics should be considered in patients with complete obstruction because air insufflation may lead to micro-perforation.

4.31.3 Procedure

- For left-sided obstruction, a therapeutic upper endoscope is ideal, whereas an adult colonoscope is often required for right-sided obstruction.
- The length of the stricture may be assessed endoscopically by traversing through it or by use of contrast material injection, with the use of fluoroscopy, through a balloon catheter, or a biliary cannula that may help delineate the strictured segment.
- A guidewire is advanced into the bowel proximal to the stricture, and the stent delivery system is then advanced over the guidewire, followed by deployment of the stent under fluoroscopic and endoscopic guidance.
- Ideally, a 2-cm segment of the stent should be spared beyond both the proximal and distal edges of the stricture to allow the formation of a waist in the middle and a flare at both ends.
- Rectal stents should be deployed at least 2 cm proximal to the anal verge to avoid pain and incontinence.
- Successful stent deployment is generally associated with the immediate passage of stool and flatus. Failure to achieve decompression could be a result of incomplete stenting of the entire length of the stricture, additional sites of intestinal obstruction, early stent migration, incomplete expansion of the stent, or faecal impaction.

4.31.4 *Adverse Events*

- Perforation: Bowel perforation may occur at the stent site because of tumour invasion or pressure from the stent or at a proximal site from colon distension (e.g. stent obstruction). A higher rate of colon SEMS-related perforation is noted in patients treated with the anti-VEGF bevacizumab.

Further Reading

Del Vecchio F, Mastroiaco V, Di Marco A, et al. Next-generation sequencing: recent applications to the analysis of colorectal cancer. J Transl Med. 2017;15:246. https://doi.org/10.1186/s12967-017-1353-y.

Dragomir MP, Kopetz S, Ajani JA, Calin GA. Non-coding RNAs in GI cancers: from cancer hallmarks to clinical utility. Gut. 2020;69(4):748–63. https://doi.org/10.1136/gutjnl-2019-38279.

Ghoshal UC. Pathogenesis of IBS. J Neurogastroenterol Motil. 2014;20(3):284–6.

Holtmann G. Pathophysiology of IBS. Lancet Gastroenterol Hepatol. 2016;1(2):133–46.

Kudo S, Rubio CA, Teixeira CR, Kashida H, Kogure E. Pit pattern in colorectal neoplasia: endoscopic magnifying view. Endoscopy. 2001;33:367–73.

Ruffle JK, et al. Artificial intelligence-assisted gastroenterology—promises and pitfalls. Am J Gastroenterol. 2019;114(3):422–8.

Sadler TW. Langman's—medical embryology. 14th ed. Wolters Kluwer; 2019.

Feldman M, et al. Irritable bowel syndrome. In: Sleisenger and Fordtran's Gastrointestinal and Liver Disease: Pathophysiology, Diagnosis, Management. 11th ed. Elsevier; 2021.

Tanaka S, Sano Y. Aim to unify the narrow band imaging (NBI) magnifying classification for colorectal tumours: current status in Japan from a summary of the consensus symposium in the 79th Annual Meeting of the Japan Gastroenterological Endoscopy Society. Dig Endosc. 2011;23(Suppl 1):131–9.

Uraoka T, Saito Y, Ikematsu H, Yamamoto K, Sano Y. Sano's capillary pattern classification for narrow-band imaging of early colorectal lesions. Dig Endosc. 2011;23(Suppl 1):112–5.

Chapter 5
Liver

A. C. Anand

5.1 Embryology of Liver

Around day 18, after fertilisation, the hepatic diverticulum arises from the ventral foregut (duodenum); the liver bud comprises a cluster of hepatoblasts. By week 6, hepatoblast migrates into the mesodermal mesenchymal cells that originates from the septum transversum of the diaphragm. The Interaction between hepatoblasts and cells of the septum transversum leads to the formation of hepatocellular tissue strand, and the capillary plexus of the septum transversum transforms into sinusoids. During the initial 8 weeks, the growing liver consists solely of hepatoblasts and bile ducts are not developed. These hepatoblasts can evolve into hepatocytes or cholangiocytes demonstrating their bipotential nature. The hepatoblast cell layer adjacent to the portal vein branch forms the single-layer ductal plate.

The ductal plate develops as a two-layered tube-shaped biliary epithelium structure surrounding the portal vein by the gestational age of 8 weeks. Around week 12 of gestation, remodelling of the ductal plate starts. Biliary channels are developed from the remodelling and fragmentation of these round and hollow ductal plates. Anomalous remodelling leads to DPM, as is shown in Figs. 5.1, 5.2, 5.3, 5.4, and 5.5.

The internal layer remains adjacent to the PV, while the external layer is adjacent to the primitive parenchyma. Sandwiched between these layers lies a marginal lumen where intrahepatic bile ducts develop. Fibro-polycystic liver illnesses, or ductal plate mutations (DPMs), are a range of anomalies seen in embryonic ductal plate development at different stages.

A. C. Anand (✉)
Department of Gastroenterology & Hepatology, Kalinga Institute of Medical Sciences, Bhubaneswar, India

G. Choudhuri et al. (eds.), *Gastroenterology and Hepatology*,
https://doi.org/10.1007/978-981-99-9287-4_5

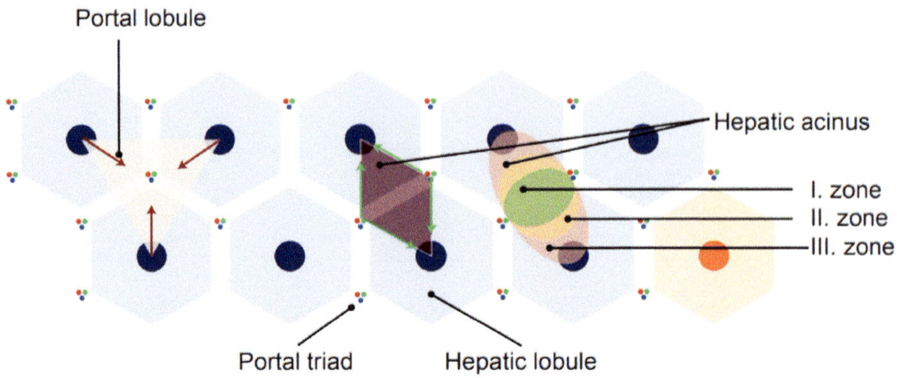

Fig. 5.1 Concept of hepatic lobule, portal lobule, and acinus

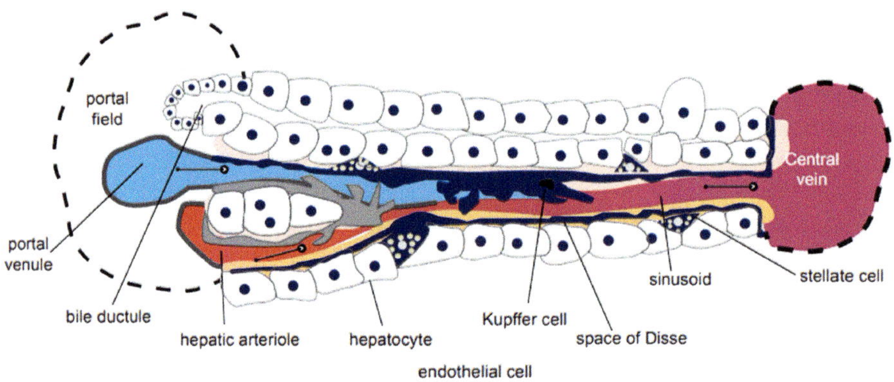

Fig. 5.2 Arrangement of hepatocytes in relation to sinusoids, endothelial cells portal venules, hepatic arteriole, and central venule

Ductal plate remodelling occurs in three phases. (a) Several focal dilatations, or peripheral tubules, arise between two ductal plate layers and emerge as intrahepatic bile ducts. (b) The hepatic artery branch appears in the periportal mesenchyma, and (c) peripheral tubules integrate into the periportal parenchyma and the residual nontubular segment of the ductal plate is eliminated via programmed cell death.

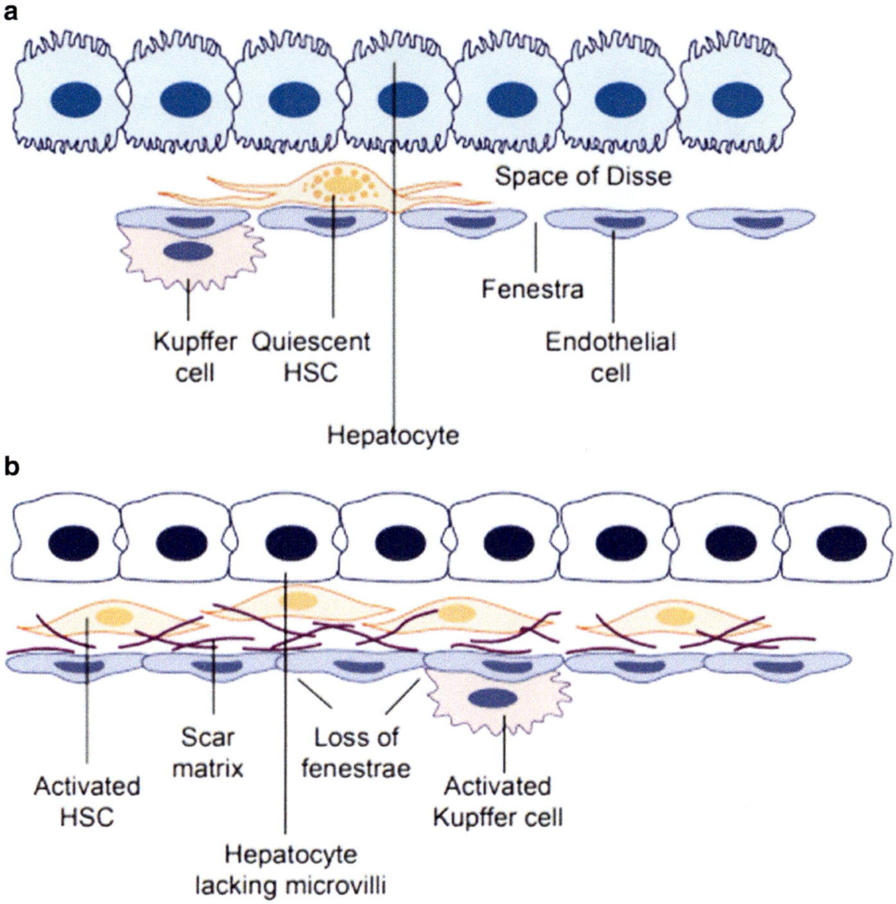

Fig. 5.3 Relationship of the space of Disse with HSC and Kupffer cells. (After Hui AY, Friedman)

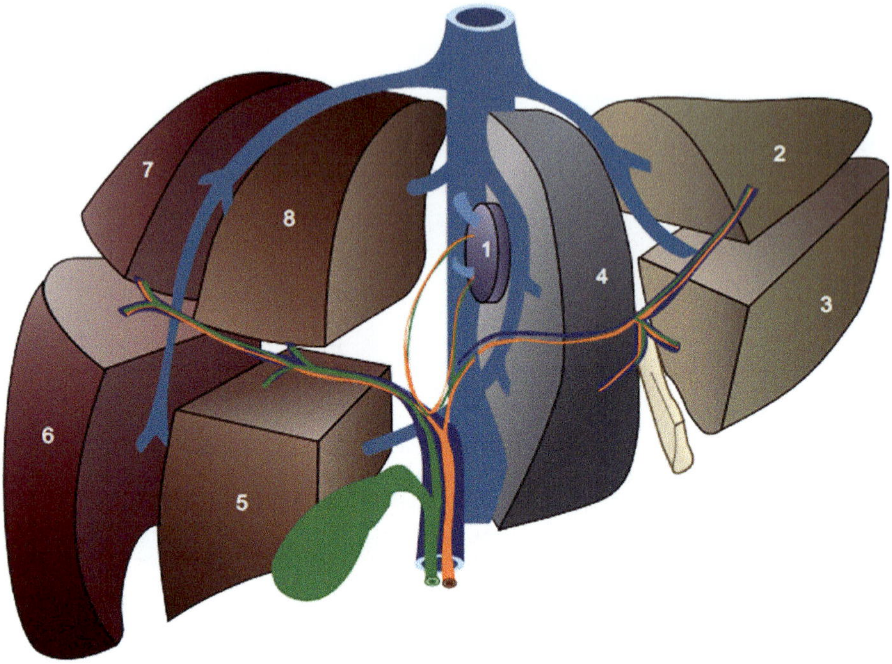

Fig. 5.4 Segmental anatomy of the liver showing how vascular channels divide the liver into eight relatively independent functional segments

Fig. 5.5 Selective Warren shunt

5.2 Ductal Plate Malformation/Congenital Hepatic Fibrosis

Several developmental abnormalities may appear in the form of fibro-cystic diseases. Abnormalities of small interlobular bile ducts may later appear as microhemartomas (von Meyerberg complexes) or as congenital hepatic fibrosis. Abnormalities

Table 5.1 Various forms of developmental cystic diseases of the liver

Duct size	Developmental cystic diseases
Small sized	Biliary hamartomas and congenital hepatic fibrosis
Medium sized	Adult polycystic liver disease
Large sized	Caroli's disease and choledochal cyst

from medium-sized bile ducts appear as autosomal dominant polycystic disease, and those of large segmental bile ducts may develop as Caroli's (intrahepatic) disease or choledochal cysts (extrahepatic) (Table 5.1).

Table 5.2 outlines the various ductal plate malformations and their comparison.

5.2.1 Congenital Hepatic Fibrosis

In intrahepatic fibrosis, the fibrotic tissue surrounds portal fibrosis and anomalous biliary channels/cysts. Patients with hepatic fibrosis commonly have irregularities liver on imaging (e.g. hypertrophic left lobe and atrophic right lobe), associated portal hypertension, renal anomalies (especially in polycystic kidney disease), and other related ductal plate aberrations, such as biliary hamartomas or Caroli's disease. MRI scans can help in making diagnoses in these cases.

Table 5.2 Summary of differences between various congenital abnormalities of bile duct development

Biliary harmartomas	Congenital hepatic fibrosis	ADPCLD	Caroli disease	Caroli's syndrome	Choledochal cyst
Pathology					
Disorganized clusters of dilated cystic ductal plate remnants unable to involute: lined by a single layer of cuboidal cells; surrounded by numerous fibrocollagenous stroma	Inconstant level of periportal fibrosis with non-uniformly shaped proliferating bile ducts	Malformation of the embryonic ductal plate, with the formation of von Meyerburg complexes that are lined with functional biliary epithelium	Incomplete ductal plate remodelling causes abnormal persistence of ductal plate remnants, inconstant degrees of destructive inflammation, and segmental dilatation	Defective remodelling of the large and small bile ducts	Aetiology not known; anomalous pancreaticobiliary ductal junction permits a mixture of pancreatic and biliary juices, which activates pancreatic enzymes resulting in inflammation and incapacitation of the duct wall may be the result of ductal plate anomaly
Imaging					
<10 mm lesions: predominantly subcapsular; hypoattenuating; comet-tail echoes	Dysmorphic liver (hypertrophic left segment, normal or hypertrophic medial segment, and atrophic right lobe with characteristics of portal hypertension; associated renal abnormalities)	>20 cysts; no communication with biliary tree; >4 cysts in patients with ADPKD	Diffuse or segmental type: saccular dilatation of intrahepatic bile ducts communicating with the biliary tree; central-dot-sign	Congenital hepatic fibrosis and Caroli's disease coexist	Todani's system classifies five types

5.3 Histology: Lobules Versus Acini

The liver lobules are anatomical subunits of a normal liver. The liver is enclosed in a thin connective tissue membrane (Glisson's capsule). It is thicker at the hilum (porta hepatis), where the portal vein and the hepatic artery enter the liver. Trabeculae that emerge from the capsule split the liver into lobules.

The structural unit of the liver has been conceptually viewed in many ways (Fig. 5.1).

1. *Classic Liver Lobule (Based on the Direction of the Blood Flow)*:

   Each lobule is a hexagonal tissue mass made of chords of hepatocytes that diverge from the central vein to portal tracts. The plates are partitioned by spaces called hepatic sinusoids. At the edges of the lobule, a branch of the portal vein, a branch of the hepatic artery, bile ductule, and lymphatic vessels form the 'portal triad' united by connective tissue. In the centre of this hexagonal structure are tributaries of hepatic veins. The blood circulates outward into each lobule (from the portal tract to central veins) while the bile diffuses into the portal regions.

   In the lobule, hepatocytes have somewhat different functions in different zones.

   (a) The centrilobular zone consists of hepatocytes encircling the central vein. This zone is farthest from the oxygenated blood vessel blood supply and, therefore, most susceptible to ischemic damage.
   (b) The periportal zone consists of hepatocytes that surround the portal tract. The external layer of periportal hepatocytes around the portal tract is known as the limiting or the restricting plate; it is part of the liver that gets blood supply from entry vessels and is damaged first by toxins that reach the liver through blood or portal veins.
   (c) The mid-zone of hepatocytes lies allying the centrolobular and periportal zones.

2. *Portal Lobule (Based Mainly on the Direction of Bile Flow):*

   The portal lobule has a central vein with one lobule each at the corners of its triagonal structure. It contains segments of triple adjacent liver lobules that evacuate bile into a single portal tract.
3. *Hepatic Acinus of Rappaport*

   It is the area of the liver which is vascularized by a terminal branch of the portal vein and hepatic artery and evacuated by a terminal branch of the bile channel. It contains two adjacent triangular areas of two nearby hepatic lobules. As in the classical lobule, cells in the hepatic acinus can be further split into three zones based on their contiguity with distributing vessels: (a) Zone I: cells in the zone I are closest to the disseminating vessels; (b) Zone II is middle of the road between zones I and III; and, finally, (c) Zone III, which is closer to the central vein. The blood in the zone I sinusoids has greater exposure to oxygen, nutrition, and toxins from incoming blood. As the blood flows towards the central vein, these constituents are, step by step, consumed by hepatocytes (zone II). Zone I hepatocytes have a greater metabolic rate and larger glycogen and lipid stores. They are likewise more vulnerable to harm by blood-borne poisons, and their vitality reserves are the earliest ones to be depleted amid fasting. This justifies the patterns of histopathologic necrosis seen in patients with hepatic damage due to various causes.

5.4 Cellular Composition of Liver

Hepatocytes from 60% to 80% of the liver. Nearly half of 20–40% of non-hepatocyte cells are liver endothelial cells that line the sinusoids. Approximately 5% are biliary cells (cholangiocytes), and the bulk of the remaining cells have primarily immunological functions. Nearly a fourth of non-hepatocytes are lymphocytes and about 20% of Kupffer's cells with resident macrophage function. Stellate cells and others constitute about 1% of total cells.

5.4.1 Hepatocyte

A **hepatocyte** is the primary parenchymal unit of the liver and makes up 60–80% of the hepatic cytoplasmic mass and has a usual lifespan of 5 months with a capacity to regenerate. They perform several metabolic, synthetic, endocrine, and secretory functions. These cells participate in protein synthesis, protein storage, conversion of carbohydrates, synthesis of cholesterol, bile salts, phospholipids, and formation and secretion of bile. They are responsible for detoxification, metabolism, and excretion/secretion of exogenous and endogenous substances.

Hepatocyte Histology

Hepatocytes have a round or polygonal shape and depict an eosinophilic cytoplasm, indicating multiple mitochondria. Basophilic stippling is noted microscopically due to the excess rough endoplasmic reticulum and free ribosomes. Brown lipofuscin granules and irregular unstained cytoplasm areas are also evident, indicating that cytoplasmic glycogen and lipid stores were eliminated during histological sample preparation (Fig. 5.2).

Hepatocyte nuclei are circular with scattered chromatin and conspicuous nucleoli. More than one nucleus may be present in a cell, and they show anisokaryosis.

Hepatocytes are organised into chords partitioned by sinusoids, a pattern reinforced by a reticulin (collagen type III) network. The hepatocyte plates are just a single layer of cells in mammals. Sinusoids exhibit an endothelial cell lining that has several tiny pores and is incomplete. The endothelial cells of the liver lack a basement membrane, and the space of Disse separates it from the hepatocytes, which empties lymph into the portal tract lymphatics.

Isolated hepatocytes can assess the biological and metabolic effects of xenobiotics. They can be isolated from the liver by a two-step process. (a) The liver is positioned in an isotonic solution containing calcium chelating agent. It removes the calcium and disrupts cell–cell tight junctions. (b) Subsequently, a collagenase solution is added to segregate the hepatocytes from the liver stroma, resulting in a suspension of hepatocytes. This can be used to develop a culture or plated on 96-well plates for instant use or even cryopreserved for later use. However, they do not proliferate adequately in culture. Hepatocytes are highly sensitive to destruction during the process of cryopreservation.

Protein Synthesis

The hepatocyte is the primary site for synthesising various proteins such as serum albumin, fibrinogen, and the prothrombin group of clotting factors (except for factors III and IV). Additionally, numerous essential molecules such as lipoproteins, ceruloplasmin, transferrin, complement, and glycoproteins are synthesised by the ER and secreted by both rough ER and smooth ER. The ER also conjugates the protein to both carbohydrates and lipids regardless of whether these are modified or synthesised inside the hepatocytes.

Carbohydrate Metabolism

The conversion of carbohydrates to fatty acids and, further, to triglycerides and glycerol occurs in the liver. Hepatocytes produce apoproteins, which club with lipoproteins (VLDL and HDL) for transportation. Gluconeogenesis also occurs in the liver.

Lipid Metabolism

The lipids collected by the systemic circulation are deposited in the liver for further metabolism. It also metabolises the chylomicron remnants from the GIT. It also synthesises cholesterol and bile salts.

Detoxification

Hepatocytes can metabolise and neutralise exogenous chemical compounds such as medicinal agents or toxins and endogenous ammonia to urea physiologically for excretion. The portal blood from the intestines brings several ingested extraneous substances/toxins to the liver for conversion to innocuous substances that can be utilised or excreted from the body.

5.4.2 *Cholangiocytes*

Epithelial cells lining the bile duct are cholangiocytes. In the smaller interlobular bile ducts, these cholangiocytes are cuboidal and gradually become columnar and mucus-secreting in the larger bile ducts in proximity to the porta hepatis and the extrahepatic ducts. Small cuboidal cholangiocytes are simple cells with a high nucleus/cytoplasm ratio, whereas large cholangiocytes are highly specialized with several organelles, including cilia and a small nucleus/cytoplasm ratio. The cilia imitate chemo- and mechanosensor within the bile duct lumen. In the healthy liver, cholangiocytes contribute to bile secretion. The bile flow in cholangiocytes is guided by the active transport of electrolytes, unlike hepatocytes, which are bile-acid dependent on bile flow. It is associated with canalicular secretion of bile acids via ATP-driven transporters and renders secretion of fluid and electrolytes via osmosis. Physiologically, large cAMP-dependent cholangiocytes are relatively predisposed to injury, while smaller ones are not. If damaged, large cholangiocytes are replaced by smaller ones.

Cholangiocyte-associated diseases, which are attributed to 18% of all liver transplantations, are 'cholangiopathies' and include primary biliary cirrhosis, primary sclerosing cholangitis, AIDS cholangiopathy, disappearing bile duct syndromes, Alagille's syndrome, cystic fibrosis, and biliary atresia. In a hepatic disease condition, reactive cholangiocytes encounter several complex alterations and attain a neuroendocrine-like phenotype that enables them to modulate complex molecular interactions.

5.4.3 Hepatic Stem Cells

The liver has enormous regenerative potential and contains both endogenous stem cells, i.e. oval cells and exogenous stem cells of bone marrow origin. Like most organs in an adult, the liver keeps up an ideal balance between cell formation and destruction. New liver cells can be produced from (a) mature hepatocytes, which multiply after localized necrosis of extreme liver cells; They are unipotent and react quickly to liver damage. (b) Oval cells multiply when the liver parenchymal damage is widespread; they exist in the canal of Hering (CoH); they are bipotent and react by multiplication in response to prolonged localized damage. (c) Exogenous liver stem cells are multipotent, obtained from circulating haematopoietic stem cells (HSCs). They can reestablish ordinary capacity in a murine model of innate tyrosinaemia.

Hepatocytes as Hepatic Stem Cells

Human hepatocytes are replaced at least once every year. The number of cells in an adult liver remains fairly constant throughout life. This replacement is done by division for existing hepatocytes in the normal course, which have been 'conferred' with limited potential to replicate. Such potential can become unlimited when there is cancerous growth.

Hepatic Stem Cells in CoH

Oval cells, like hepatoblasts, have the potential to grow into either hepatocytes or cholangiocytes. They come into play when there is widespread and incessant hepatic damage or if their replication is not possible, for example, due to viral infection. They proliferate and develop into hepatic or bile duct cells to reconstruct the liver.

5.4.4 Haematopoietic Stem Cells

Haematopoiesis and hepatic evolution occur around the same time in embryonic development. During fetal development, haematopoietic undifferentiated cells migrate from the yolk sac into the maturing hepatic bud. The fetus contains HSCs in the liver during the gestational period. The liver is a haematopoietic organ across gestation for roughly a few weeks after birth; however, the capacity reduces progressively towards delivery and post-partum. However, it surges again in the event of subsequent hypoplastic states. A few reports indicate that hepatic oval cells and HSCs have common CD34, Thy-1, and c-unit mRNA and protein. Oval cells also exhibit mRNA for the flt-3 receptor. Microenvironment and cell-to-cell signalling by neighbouring cells determine if HSCs will be quiescent to start differentiation.

Post-transplantation, bone marrow cells possibly course into the liver cell plates via the sinusoidal flow and intercalate arbitrarily into previous liver chords as hepatocytes. The oval cells from the CoH and the marrow cells may also translocate into the terminal branches of the biliary tree and develop into cytokeratin 19-positive small cholangiocyte-like cells. Expansion and evolution into hepatocytes depend on the extent and level of liver cell destruction. Thus, hepatic recovery via the marrow cells is also possible in certain circumstances.

Mesenchymal stem cells (MSCs) are pluripotent and can transdifferentiate into hepatoblasts and liver parenchymal cells. In contrast to liver transplantation, there are considerably more MSCs transplanted in hepatocyte transplantation. Patients' MSCs are an alternative potential choice to treat liver damage, as no immunosuppression is required.

5.4.5 Therapeutic Potential of Stem Cells

Hepatic stem cells, from whatever source, might be valuable for treating an assortment of maladies that influence the liver, including hereditary illnesses, for example, Wilson's infection (copper collection), Crigler Najjar disorder (absence of bilirubin conjugation), and tyrosinaemia, and situations where there is the extrahepatic cause of the ailment, e.g. Factor IX deficiency.

- Hepatic stem cells may be an option for conveying restorative qualities to the liver. HSCs can be promptly collected from a patient's blood and potentially help treat non-hereditary ailments dispensing with the requirement for immunosuppression.
- Hepatic stem cells may likewise be valuable for conveying anti-inflammatory cytokines (e.g. IL-10) for auto-immune liver illness, for example, primary biliary cholangitis (PBC).

- Hepatocyte transplantation is probably helpful for patients with acute hepatic failure, helping to recuperate their liver and avoid liver transplantation.
- Finally, roughly 33% of liver transplant patients can conceivably be weaned off long-haul immunosuppression and beneficiary bone marrow cell engraftment to the liver could be a huge factor in this toleration.

5.5 Hepatic Sinusoids

5.5.1 *Embryology*

As noted earlier, capillaries of the liver lobule are different from capillaries in other parts of the body. During the developmental phase, they are formed by the development of the columns of the hepatic cells into large blood spaces or sinuses, or 'sinusoids'.

5.5.2 *Anatomy*

Hepatocytes are arranged as plates in lobules, which can be seen to extend from a central vein to portal tracts. They are separated by complex interlacing spaces lined loosely by liver endothelial cells (LEC). Terminal branches of the portal vein and the hepatic artery discharge their blood into the sinusoids. Hepatocytes have at least three surfaces: (a) sinusoidal: facing the sinusoid; (b) cannalicular: facing the bile canaliculus; and (c) surface facing adjacent hepatocytes. Hepatic sinusoids have fenestrated endothelial cells unlike other capillaries (see Fig. 5.3a), and subendothelial stromal material is marginal, which permits a channel for large macromolecules like lipoproteins.

5.5.3 *Space of Disse*

The space between hepatocytes and sinusoidal endothelial cells is the space of Disse which consists of hepatic stellate cells (HSCs) or Ito cells apart from the fluid that flows out into lymphatics in portal zones. The elevation of sinusoidal pressure is accompanied by an increase in lymph production which causes ascites. Ito cells are activated in cirrhosis and contribute to the deposition of fibrous tissue (Fig. 5.3).

5.5.4 *Kupffer Cells*

Kupffer cells are macrophages in the periportal areas of liver sinusoids with pseudopodia attached to the sub-endothelial structures.

5.5.5 *Natural Killer (NK) Cells*

Liver-specific NK cells or pit cells also anchor to the sinusoidal surface of liver endothelium. It has cytotoxic activity against tumor and virus-infected hepatocytes.

5.5.6 *Sinusoidal Pathology*

During the process of cirrhosis development, the sinusoids undergo several changes: (a) the space of Disse becomes widened with the proliferation of new collagen; (b) endothelial fenestrations become smaller and less numerous; and (c) hepatocellular microvilli are effaced. These changes reduce transport across the sinusoidal walls. Weakening of reticulin fibres may predispose to sinusoidal wall rupture and blood-filled cysts, a condition known as peliosis hepatis.

5.6 Hepatic Stellate (Ito) Cells

5.6.1 *Introduction*

Previously known as lipocytes or fat-storing cells, they are currently called HSCs or Ito cells (named after the Japanese physician Toshio Ito). These are cells present along the perisinusoidal space or the space of Disse and also in the pancreas, lung, kidney, and intestine, forming the stellate cell system.

5.6.2 *Ultrastructure and Morphology*

In a healthy liver, stellate cells are spindle-shaped structures with elliptical nuclei. Their perikaryons lie in the gaps between adjacent parenchymal cells. These cells have several cell organelles, some moderately developed like rough endoplasmic reticulum (rER), Golgi complex adjoining the nuclei, and conspicuous dendritic processes. The subendothelial projections with microscopic spine enclose the sinusoids allying the endothelial cells and hepatocytes. A recent study showed that these protrusions are the cell's leading organelle to 'sense' chemotactic signals which are subsequently transferred to the cell's mechanical apparatus, which invokes a contractile force.

Two or more sinusoids are generally bordered by a single stellate cell. Several projections from hepatocytes are spread across the space of Disse and to adjacent hepatocytes which allows the movement of soluble mediators and cytokines between the cells. A few nerve endings are also identified adjacent to stellate cells, confirming this cell's neurohumoral responsiveness.

5.6.3 Staining

Cytoplasmic lipid droplets on the Ito cells can be stained with gold chloride so that Ito cells can be easily recognized. Besides, Reelin demonstrated that these cells also make them distinct from other myofibroblasts, which are known to be elevated post-hepatic damage.

5.6.4 Function

Although the physiology of stellate cells in a healthy individual is unknown, it has been suggested to be a hepatic APC presenting lipid antigens (CD1, MHC-I, and II) to NKT cells and triggering their multiplications.

In the event of a hepatic injury, activation of stellate cells demonstrates increased multiplication, contractility, and chemotaxis. This also utilises reserve vitamin A to produce enormous extracellular matrix (ECM) constituents, including collagen, proteoglycan, and adhesive glycoproteins. Besides, experimental studies have found that in the event of hepatic fibrosis, these cells undergo senescence indicated by elevated SA-beta-galactosidase staining, p53 accumulation, and activation of Rb. This probably attenuates hepatic fibrosis by triggering the immune processes and stimulating NK cells.

Stellate cells present antigenic peptides to $CD8^+$ and $CD4^+$ T cells and facilitate cross-priming of $CD8^+$ T cells. Experimental studies have also shown that these cells activate peptide-specific T cells when presented with a mutational antigen. In the case of infectious disease, these cells trigger antigen-specific T cell-mediated protection, which is associated with a shape change similar to fibroblasts.

5.7 Kupffer Cells

Named after Von Kupffer, a pathologist, Kupffer cells (KC) are hepatic macrophages with self-renewal potential accounting for 80–90% of tissue macrophages in the body, together with characteristic NK cells, dendritic cells, and several macromolecules secreted by liver form innate immunity component of the liver.

5.7.1 Structure of Kupffer Cells

Kupffer cells represent approximately 5–15% of the cellular component of the liver and reside adjacent to liver endothelial cells in sinusoids. Kupffer cells are ameboid and hold fast to the wall of fenestrated sinusoidal endothelial cells. The surface of

Kupffer cells bears microvilli and lamellipodia, with which they sense particulate material in sinusoids and communicate with neighbouring hepatocytes through endothelial fenestrae. Kupffer cell cytoplasm contains thick granules, electron-lucent vacuoles, lysozymes, Golgi contraption, covered vesicles, pinocytic vesicles, ribosomes, centrioles, microfilaments, and microtubules. The nuclei of Kupffer cells are ovoid or indented and, at times, lobulated. The presence of CD163 and CD68 helps to recognise KCs, which are eliminators, cell-surface receptors, or scavenger receptors. Macrophages M1 or classically activated and M2 alternatively activated have different functions; the former secretes inflammation-enhancing mediators like cytokines while the latter are inhibitory and participate in the repair of KCs. These eliminate particulate antigenic strictures from sinusoids, for example, infected erythrocytes and microscopic organisms by phagocytosis and ingest smaller particles/molecules through pinocytic vesicles.

Kupffer cells situated in the periportal regions of the liver lobule differ from those in the centrilobular areas. Periportal KCs bigger have more noteworthy lysosomal granules and are more phagocytic.

5.7.2 *Functions*

1. *Host defence*

   Several antigens in the form of non-self-proteins belonging to the dietary components or gut microbiota are presented, which would produce an immune response to trigger inflammation. The aim is to suppress the replication of the pathogens and encourage the participation of other immune system components, which would release cytokines and other mediators. Other immune cells include dendritic cells (DCs) triggered by hepatocyte growth factor and macrophage colony-stimulating factor (M-CSF) via activation of the signal of STAT3 and SMAD, then suppressing NF-κB, with the production of anti-inflammatory cytokines. Parallelly, the secretion of various factors [Fms-related tyrosine kinase 3 ligands (Flt3L), granulocyte colony-stimulating factor (G-CSF), or granulocyte-macrophage colony-stimulating factor (GM-CSF)] triggers DCs to overturn endotoxin-related immunoparalysis.

   Kupffer cells are vigilant endogenous guards trapping antigens and pathogens and, therefore, important components of the host immune system against several pathogenic bacteria (*L. monocytogenes* is a facultative intracellular model microorganism trapping which elicits huge recruitment of monocytes forming Lm-containing microabscesses 2–3 days after inoculation). These microabscesses are rich in M1 macrophages, TNF/iNOS-producing dendritic cells, and neutrophils. Exposure to infective disease stimulates the secretion of inflammatory mediators (IL-1β and IL-4) by KCs that suppress microbial proliferation with the simultaneous release of cytokines [MIP-1α (CCL3), MIP-1β (CCL4), MCP-1 (CCL2), and MIP-2 (CXCL2/-3)] that enhance M1 recruitment for suppression of infection. On apoptosis or necrosis of KCs, hepatocytes release cytokine, facilitating switching from M1 to M2 phenotype. This involves the release of alarmin interleukin-33 (IL-33), which stimulates basophil IL-4 production and enhances the recruitment and proliferation of monocyte-derived macrophages.

Similar to bacterial invasion KCs are also protective against viral ones, including Hepatitis B and C virus infection. Trapping the virus by KCs, followed by the production of inhibitory interferons, prevents viral replication and dissemination. Besides KCs responses also involve the enhanced expression of MHC II that activates virus-specific $CD4^+$ T cells in the liver, producing interferons. The antiviral status also enhances the KCs phagocytic potential, which may also improve control of viral replication.

In patients with chronic HBV, phagocytosis of the virus activates KCs, followed by activation of transcription-3 (STAT-3) binding to the TGF-β promoter, enhancing TGF-β synthesis. Thus, KCs release the profibrogenic/anti-inflammatory cytokine TGF-β, not pro-inflammatory cytokines.

HCV infection also could make KCs and liver-infiltrating lymphocytes the major sources of TGF-protein, leading to liver fibrosis. TGF plays a pivotal role in generating Treg cells from precursor cells, such that a GRP94 inhibitor would improve immunity against HCV infection by blocking the induction of Treg cells.

2. *Alcoholic and Non-alcoholic Liver Disease*

Ethanol administration to rodent models alters the gut microbiota, and the use of antibiotics results in the reduction of intestinal microbial populations and liver injury is also attenuated. Nonetheless, pre-and-probiotics have not shown consistent results in alcoholic liver disease. Such associations have also been observed in models of obesity and fatty liver wherein altered gut permeability triggers the release of bacterial products in portal circulation.

Multiple pathogen-recognition receptors (PRRs) classified as Toll-like receptors (TLRs) and nucleotide-binding oligomerization domain (NOD)-like receptors (NLRs) recognise both pathogen- and danger-associated molecular patterns (PAMPs and DAMPs). PAMPs indicate external stimuli, whereas DAMPs are endogenous and denote stress. Lipopolysaccharide (LPS), present in gram-negative bacterial cell walls, acts as a TLR4 ligand. Similar roles of TLRs, specifically TLR2 and TLR4, have been established in ALD and NAFLD/NASH.

Chronic ethanol administration to rodent models stimulates mRNA expression for TLR-2, 4, 6, 7, 8, and 9 and increases the sensitivity of these models to several bacterial ligands. Similar changes have been observed in liver biopsies of patients with ALD. Thus, it is essential to decipher the role of PRRs and elucidate probable complex interactions between KCs and PRRs in the perpetuation of hepatic injury.

3. *Kupffer Cells in Ischemia–Reperfusion Injury*

Liver ischemia–reperfusion (I/R) injury occurs in liver transplantation, trauma, hepatic resection, or hypovolemic shock. As mentioned earlier, it involves aggravation of ischemia due to the return of blood flow and oxygen delivery post the procedures. The activated KCs generate reactive oxygen species (ROS) and proinflammatory cytokines such as TNF-α, IL-1β, nitric oxide, and chemokines. These increase hepatocyte death, endothelial dysfunction, and leukocyte activation and recruitment. Increased CCL2 secretion by KCs increases CCR2-expressing neutrophil recruitment from the bone marrow, followed by hepatic infiltration during I/R, and secrete matrix metalloproteinases (MMPs) to worsen graft dysfunction. Additionally, thrombocytes too stick to the KCs, resulting in leukocyte accumulation diminishing sinusoidal perfusion, culminating in hepatic failure.

In later stages of liver grafts, KCs become protective against liver-transplant–induced I/R injury via the release of anti-inflammatory IL-10 and the production of antioxidant haeme oxygenase. Prior treatment with IL-10 guards the steatotic livers against I/R, which suggests that active KCs retain their hepatoprotective role in the steatotic milieu.

4. *Liver Regeneration*

   To reach the optimal regenerative potential of the liver, the activation of KC is essential since it releases TNF-α and IL-6. Subsequently, at least, partial hepatocyte expansion occurs via NFkB and STAT-3 translocation. This activity is enhanced by leukocyte–KC connection by ICAM-1 binding.

5. *Role in Haemoglobin Degradation*

   Senescent or distorted red platelets are eliminated by KC and tissue macrophages in the spleen in IgG-dependent manner by recognising senescence-associated clustering of certain antigens on the RBC surface. Iron in the human body is primarily recovered and reused from the turnover of haemoglobin in RBCs. The KCs and tissue macrophages of the spleen process RBCs liberating haemoglobin which is lysed to produce biliverdin and subsequently reduced to unconjugated bilirubin released into the plasma. It is combined with albumin and moved to the hepatocytes. The released iron in Ferrous form binds to apo transferrin, forming transferrin which is facilitated by the action of the copper-dependent enzyme ceruloplasmin. The iron can also be stored as apoferritin. Over some time, oxidation and degradation of ferritin lead to hemosiderin formation. It appears inside macrophages on staining with Prussian blue. Iron may also be combined with proteins such as cytochrome P450.

6. *KCs in Transplant Immunology*

   The KCs are capable of increased expression of MHC and other such substances that can stimulate naïve T cells and simultaneously can be activated to release T1 cytokines IL-2, IL-1, IL-6, TNF-α, and IFN-γ. These interact with the recipient T cells that migrate into the graft and are vital in eliciting immune response. Some donor liver KCs would be substituted by the macrophages/KCs of the recipient marrow. Analysis of mRNA expression levels in liver grafts demonstrates a move in Th1/Th2 proportions favouring reduced rejection. There is a positive correlation between cytokines TNF-α expression in KCs and acute rejection, so it can be targeted to reverse the rejection. Depletion of KCs by GdCl3 resulting in bile duct cell apoptosis reduced the expression of ALT, ALP, TBIL, and TNF-α, and inhibition of Fas-FasL-Caspase signal transduction reduces the risk of rejection. KCs also recruit and aid in the maturation of DCs to elicit the release of inflammatory cytokines and chemokines. KCs can also trigger T lymphocyte death which may be important in modulating transplant tolerance. The KCs may also protect the graft by synthesising high levels of Th2 cytokines, IL-10, and TGF-β and low levels of IL-12.

7. *KCs in Liver Cancer*

   The KCs have an opposing role in oncogenesis. In some rodent models, such as diethylnitrosamine-induced hepatocellular carcinoma (HCC), pro-inflammatory activation of KCs increases tumour development. While activated KCs subdue tumour cells via the ADCC pathway via FcγRIII (CD16) and cytokines. The activated KCs use lysosomal enzymes, NO, and peroxidase to eliminate target cells. The release of various cytokines by KCs may also affect tumorigenesis, e.g. IL-6 is released in large amounts by KCs, which causes tumour progression and angiogenesis and is overproduced in HCC. KCs are an important component of the host tumoral surveillance system due to their strategic position in the liver. It enables distinguishing and eliminating neoplastic cells in the liver. In the event of metastatic cell migration to the liver and subsequent entrapment in sinusoids, KCs release numerous growth factors such as PDGF-β, vascular endothelial growth factor, TGF-β, and EGFR ligands, cytokines, chemokines, and soluble factors in various circumstances and have a crucial role in metastasis.

5.8 Histopathological Patterns in Chronic Hepatitis

Inflammation of the portal vessels, parenchyma, and the plate between the portal tract and liver parenchyma, along with necrosis and cirrhosis, are peculiar manifestations of chronic hepatitis.

5.8.1 Portal Inflammation

Portal inflammation, mild, prominent, or diffuse, is highly evident in chronic hepatitis. It includes majorly lymphocytes and variable numbers of plasma cells. A smaller proportion of scattered macrophages, neutrophils, and eosinophils are also present. In certain places, one may encounter lymphoid follicles, especially in HCV.

5.8.2 Interface Hepatitis

Interface hepatitis or piecemeal necrosis involves mononuclear inflammatory infiltration with hepatocytes damaging the limiting plate and necrosis of periportal hepatocytes.

5.8.3 Lobular Necroinflammatory Activity

The extent and severity of hepatocyte necrosis varies, and a patchy distribution exists throughout the lobule. Apoptotic hepatocytes are centrally located on periportal fields, and mononuclear cells are inclined to accumulate, encircling the worn-out hepatocytes.

5.8.4 Fibrosis

Gradually, fibrosis replaces the areas of inflammation. Worsening fibrosis causes enlargement of portal tracts and stellate periportal fibrous extension. Subsequently, it leads to the development of portal–portal or portal–central fibrous bridges, eventually causing cirrhosis. The cirrhosis may be either micronodular, macronodular, or mixed type.

5.9 Anatomy and Ligaments of the Liver

The liver is the body's biggest organ, weighing between 2% and 3% of body weight. It is wedge-shaped and positioned in the right upper quadrant of the abdomen and lower thorax beneath the diaphragm. More often than not, it expands from the fifth intercostal space in the mid-clavicular line to the right costal margin. The base of the wedge rests on the right arch of the diaphragm and the right thoracic cage. The liver's anterior, superior, and right lateral surfaces are positioned against the diaphragm. The posterior surface has troughs from the intestine, the right kidney on the right lobe, and the stomach on the left lobe.

The capsule of Glisson encloses the liver. The capsular peritoneum reflects onto the diaphragm on the superior surface of the liver and proceeds as the parietal peritoneum. At porta hepatis, the connective tissue of the capsule lies in continuity with a fibrous sheath of portal vessels. The liver is upheld by the peritoneal reflections that shape the coronary ligaments, the right and left triangular ligaments, and the falciform ligament.

The lower unanchored margin of the falciform ligament contains the circular ligament, which is composed primarily of the annihilated umbilical vein. Anteriorly the falciform ligament links the liver to the diaphragm, and superiorly the peritoneal reflections to the left of the vena cava.

The fibrous capsule on the rear aspect of the liver reflects onto the diaphragm and posterior abdominal wall, leaving an uncovered area where the liver lies in a continuum with the retroperitoneum.

The hepatoduodenal ligament links the liver to the superior part of the duodenum, which contains the portal vasculature, bile duct, innervations, and lymphatic vessels in the transverse portal fissure. With reference to this fissure, the tail end or caudate lobe of the liver is posterior, and the quadrate lobe is anterior. The quadrate lobe is additionally separated by the gallbladder and the umbilical fissure on the right and left aspects.

5.10 Morphological Anatomy

Classically, the lobes of the liver are formed due to the presence of *falciform* ligament. The ligament stretches from the liver to the anterior abdominal wall and the diaphragm to the umbilicus. The inferior edge of the falciform ligament encloses the circular ligament (ligament teres), which traverses the longitudinal fissure on the lower surface of the liver. The fissure proceeds toward the posterior surface, creating fossa for the ligamentum venosum. A characteristic H-shaped pattern is present on the inferior surface of the liver: (a) the right limb formed by ligament teres and ligamentum venosus; (b) the left limb formed by the gallbladder bed and IVC fossa. (c) The horizontal limb of porta hepatis contains the hepatic artery, common hepatic duct, portal vein, and lymphatics. (d) The quadrate lobe lies between ligament teres and gallbladder fossa, and (e) the Caudate lobe lies between ligamentum venosum and IVC fossa.

5.11 Functional Anatomy

Couinaud (1957) divided the liver into eight self-sufficient units with independent own vascular inflow-outflow (hepatic vein and hepatic artery) and biliary drainage (bile duct) systems. Right and left hepatic veins separate the liver into anterior and posterior segments and medial and lateral parts, respectively. The middle hepatic vein splits it into the right and left lobes. This axis extends from the inferior vena cava to the gallbladder fossa. The left and right portal veins split superiorly and inferiorly to venture into the middle of each segment (Fig. 5.4).

Due to the structural aspects of the liver, certain segments can be resected with no injury to the remaining ones. However, resections should parallel the hepatic veins without encroaching into the central vascular and drainage system. Because the branching of the left portal vein is irregular, Strasberg has recommended that the divisions of the arteries or ducts define the segments. This can be done without modifying the segments defined by the Couinaud.

5.12 Portal Circulation

The portal vein, ~8 cm in length, lies at the same level as the second lumbar vertebra by the joining superior mesenteric and splenic veins. These veins merge to the anterior of the inferior vena cava and posterior to the neck of the pancreas. It moves superiorly on the posterior side of the intestine and subsequently in the lesser omentum's right border to the porta hepatic's right end. At the porta, it bifurcates into a right and a left branch, along with the corresponding branches of the hepatic artery into the liver. The portal vein returns the venous blood to the liver from the spleen and the intestines. In the liver, the portal vein branches like an artery forming sinusoids, from which the blood is passed to the inferior vena cava via the hepatic veins. The cystic vein joins the right branch of the portal vein, which vascularizes the right lobe, while the left branch, which is longer with lesser calibre, vascularises the caudate lobe, followed by the left lobe.

5.12.1 *Tributaries of Portal Vein*

The tributaries of the portal vein are:

1. The splenic vein: the union of 5–6 branches from the spleen merges to form a single vessel, which passes from left to right, grooving the anterior and posterior sections of the pancreas. The short gastric, the left gastroepiploic, the pancreatic, and the inferior mesenteric veins reach the splenic vein. The inferior mesenteric vein returns blood from the intestine and terminal parts of the gastro intestinal tract (GIT). It originates in the haemorrhoidal plexus in the rectum as the superior hemorrhoidal vein, via which it also remains in contact with the central and inferior haemorrhoidal veins.
2. The superior mesenteric vein returns the blood from the small intestine, cecum, and ascending and transverse colon. It originates in the right iliac fossa by merging the veins from the ileum, cecum, and appendix.

More than 60% of the hepatic blood is supplied by portal venous blood. The high-pressure, well-oxygenated arterial blood blends completely with the low-pressure, low-oxygen-containing, nutrient-rich portal venous blood perfusing the sinusoids. Subsequently, it flows into the hepatic venules, veins, and inferior vena cava.

In hepatic circulation, blood flow in the portal vein and the hepatic artery are closely linked; therefore, when portal blood flow increases, hepatic arterial flow decreases and vice versa, termed the hepatic arterial buffer response. It is regulated by an adenosine-mediated vascular reflex which secures the sustenance of comparatively uniform sinusoidal perfusion in the event of variation in portal inflow, such as after food consumption.

Table 5.3 Locations of porto-systemic anastomosis

Site	Portal	Systemic
Anal canal	Superior rectal (portal)	Middle and inferior rectal veins (systemic)
Bare area	Small veins	Phrenic veins
Periumbilical region	Small paraumbilical veins drain into the left vein	Superficial veins of the anterior abdominal wall
The terminal part of the esophagus	Tributaries of the left gastric vein	Esophageal veins (azygous and hemiazygous)
Gastric (extrahepatic portal vein obstruction)		Esophageal sub-mucosal vein ↔ short gastric vein ↔ splenic vein

5.13 Sites of Portosystemic Collaterals

The portal and systemic blood flow are linked in several ways. In cirrhosis, when the direct pathway via the liver is congested, which leads to a surge in portal venous pressure, portosystemic anastomoses develop, substituting for the congested route. Porto-systemic anastomosis loci are detailed in Table 5.3.

5.14 Venous Drainage of the Liver: Importance in Budd–Chiari Syndrome

The portal venous system directs the GIT blood to the liver. The absorbed components from the intestine are always transported to the liver for screening before release to other parts. The hepatic artery brings oxygenated blood. The superior (right, left, and middle) and inferior (6–8 smaller veins draining caudate section) hepatic veins that drain hepatic blood are created by merging the central veins. The superior hepatic veins drain into the inferior vena cava, inferior to the diaphragm. The inferior group consists of 6–18 small veins, which drain the blood from the caudate lobe, including part of the right lobe. In the majority of individuals, the middle and left hepatic veins join together before draining into the vena cava. The caudate lobe of the liver and adjoining liver parenchyma usually drains directly into the vena cava through one or more small veins caudal to hepatic veins.

5.14.1 The Clinical Relevance of Budd–Chiari Syndrome

Budd–Chiari syndrome is a malady due to an impediment to hepatic venous outflow at any point across the small hepatic veins to the intersection of the inferior vena cava with the right atrium. The aetiology of blockage of the venous system is usually the development of a thrombus in hepatic veins, with or without vena cava thrombosis, but it can also be due to other causes, such as tumours or abscesses. Small veins draining the caudate lobe are often separate from the main hepatic veins and are usually not involved, allowing sparing of the caudate lobe and consequent compensatory hypertrophy in Budd–Chiari syndrome. This hypertrophy can cause inferior vena cava compression, further jeopardising the situation. This anatomical variation is important from a therapeutic point of view.

Clinical presentation often involves pain, rapid ascites accumulation, and, later, all features of cirrhosis. Diagnosis is based on Doppler ultrasound or CT angio findings. The usual approach to treatment includes (a) anticoagulants (in early stages), (b) angioplasty or stenting of hepatic veins with or without IVC stenting by an interventional radiologist, (c) transjuglar intrahepatic portosystemic shunt (TIPSS), or (d) surgical shunting.

Angioplasty and/or HV stenting/IVC stenting is usually required for stenosis of the short segment of the hepatic veins (<3 cm) or membranous blockages of the inferior vena cava. In comparison, a TIPSS or shunt surgery is the preferred option for long-segment HV block (>3 cm), failed angioplasty or stenting, and refractory variceal bleeding.

5.15 Surgical Splenoportal Shunts

Surgical shunts aim to reduce portal pressure below a threshold by various operating techniques. Surgical shunts are classified as (a) total, (b) selective, and (c) partial shunts.

5.15.1 Total Shunts

These shunts can completely redirect the portal bloodstream to the vena cava. Total shunts can reduce portal pressure significantly and prevent variceal bleeding in nearly 95% of cases. There is an associated adverse outcome in the form of hepatic encephalopathy in about 40% of patients. These shunts may be life saving in an intractable variceal bleeding emergency. Although shunting reduces death from variceal bleeding, death from liver failure cannot be prevented.

End-to-Side Portocaval Shunts

The portal vein is separated at the hilum of the liver, and its stump is ligated. The terminal portion of the portal vein is connected to the side of the infrahepatic vena cava. The portal flow is thus diverted to the systemic circulation, decompressing the system while the hepatic artery maintains the inflow. This shunt is rarely if ever, done now.

Side-to-Side Portocaval Shunt

The portal and the infrahepatic vena cava are connected and anastomosed side to side. The portal vein thus decompresses into the inferior vena cava and supports all the hepatofugal bloodstream. Any portocaval shunt with a diameter greater than 12 mm is expected to lead to complete portal blood shunting. A smaller size side to side can be created by interposition graft. Besides, patients with a history of the portocaval shunt are predisposed to intraoperative morbidity and need a transfusion when undergoing liver transplantation. Thus, such shunts may be avoided if the candidate is likely to undergo liver transplantation.

Interposition Mesocaval Shunt

The subhepatic vena cava and mesenteric vein are anastomosed end-to-side to either end-to-side or via a prosthetic graft. This shunt is sufficiently far from the porta hepatitis and, therefore, does not cause any serious trouble to ensuing liver transplantation. The graft appears like the horizontal limb of the English letter 'H'; its length makes it susceptible to kinking or thrombosis. Results are similar to meso caval shunts.

5.15.2 Selective Shunts

Selective shunts were evolved to lower the chances of hepatic encephalopathy. Anastomosing the distal splenic vein to one side renal vein specifically decompresses the gastric and splenic veins while keeping up a generally high flow in the mesenteric and portal veins. The Warren distal splenorenal shunt is particularly suitable for patients with additional portal vein thrombosis, of whom around 80% may have a patent splenic vein and gastric varices (Fig. 5.5). A variation of this surgery is splenectomy with anastomosis of the proximal end of the splenic vein to the renal vein. The mesenteric–left portal venous bypass, or Rex shunt, can be performed in patients with preserved patency of the portal vein but with extrahepatic portal vein thrombosis. Since the hepatic portal blood flow is reinstated, the risk of hepatic encephalopathy or long-term learning disability in pediatric patients is attenuated. A jugular vein graft may connect the superior mesenteric vein and intrahepatic section of the left portal vein in the embryonic umbilical vein or Rex recess remnants. This surgery is preferred, particularly in pediatric patients with extrahepatic portal vein thrombosis with complications associated with portal hypertension and adults with delayed post-transplant portal vein thrombosis. The benefits of this method are that only varices at the gastroesophageal intersection and spleen are decompressed, and portal hypertension is sustained in the superior mesenteric vein and portal vein limiting variceal bleeding; however, the risk of ascites remains.

5.15.3 *Partial Shunts*

Partial portocaval shunts are designed to control variceal haemorrhage while preserving portal perfusion of the liver. It is performed using 8–10-mm-diameter polytetrafluoroethylene (PTFE) portocaval H-grafts with ligation of collaterals which offers a higher blood flow rate. The principle of a partial shunt is maintaining the pressure lower than the cut-off of the pressure gradient (<12 mmHg) between the portal vein and vena cava above which variceal bleeds occur. This is achieved by incompletely decompressing the complete portal system to attain a pressure lower than the threshold but adequate to maintain antegrade splanchnic nutrient blood flow to the liver. Expanded PTFE is considered less prone to thrombosis compared to Dacron. The reinforcing rings avert compression on the exterior of the graft. The portal vein conducts more blood flow than the superior mesenteric vein, thus increasing inflow. Certain features of these methods confer long-lasting patency and significantly lower the risk of perioperative thromboses. These features include long anastomotic slants, inverted suture lines, and heparin-infused prosthetic material. The frequency of averting variceal rebleeding and encephalopathy is comparable between partial and distal splenorenal shunt. Due to its lack of effect on sinusoidal pressure, ascites may occur in approximately 20% of patients with a partial portosystemic shunt.

5.16 Transjugular Intrahepatic Portocaval Shunt

The transjugular intrahepatic portosystemic shunts (TIPS) are side-to-side portocaval ones that are widely used instead of surgical shunts. It can lower high portal pressure by forming a conduit between the hepatic vein and an intrahepatic tributary of the portal vein. A TIPS is useful in managing complications of portal hypertension, primarily variceal bleeding and unyielding ascites, hepatic hydrothorax, hepatorenal syndrome, and Budd–Chiari syndrome.

5.16.1 *Methodology*

A TIPS can be positioned under sedation by an interventional radiologist and has a negligible mortality rate (<1–2%). The criteria include but are not mandatory in an emergency: a platelet count higher than 60,000/mm^3 and an INR less than 1.5.

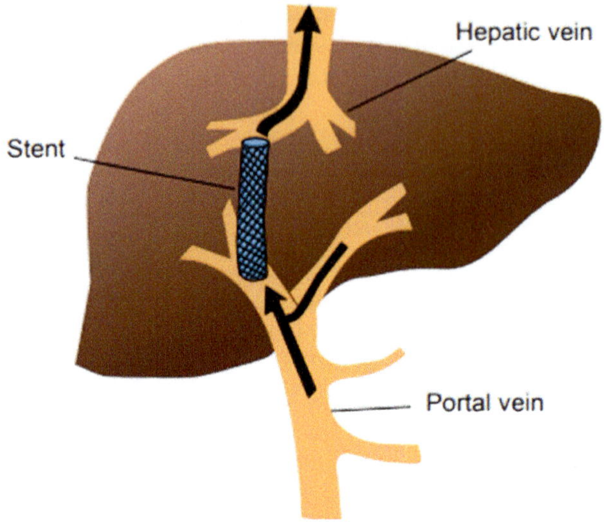

Fig. 5.6 Transjugular intrahepatic portosystemic shunt (TIPS)

Broad-spectrum antibiotics are recommended in patients with PSC and operated on due to emergency. The procedure involves penetration of a guidewire via a cut over the inner jugular vein under fluoroscopy and ultrasonography up to the suitable hepatic vein (Fig. 5.6). Subsequently, a needle is lead aligned along the guidewire into a close-by branch of the portal vein. The subsequent tract is enlarged with the inflatable balloon. An expandable stent of 8–10 mm in width is placed to maintain the communication between hepatic and portal veins. A patent portal vein is essential for the creation of TIPS.

5.16.2 *Efficacy*

The main advantage of TIPS is (a) immediate portal decompression, (b) avoidance of general anesthesia, and (c) lack of need for dissection in the portal hepatis. However, there are some disadvantages too. (a) There may be a technical failure in a small percentage of cases. (b) About 30–50% of cases may have stent thrombosis on follow-up, requiring additional intervention. (c) Rarely stent migration and intra-bdominal bleeding may occur. (d) Emergency TIPS has a 30% risk of mortality and hepatic encephalopathy. A meta-analysis that compared TIPS with endoscopy found a lower risk of rebleeding with TIPS. Nevertheless, the frequency of encephalopathy was higher, while there was a lack of effect on survival. Hence, in the absence of significant advantages over existing options, TIPS is not recommended over endoscopy or medical therapy.

5.17 Bilirubin Metabolism and Congenital Disorders

Bilirubin is a tetrapyrrole formed by the degradation of haemoglobin in the reticuloendothelial system. In adults, the daily bilirubin production averages approximately 250–300 mg/day. The majority (70–80%) of daily bilirubin production is obtained from the breakdown of ageing erythrocytes and a small part from ineffective erythropoiesis. The balance is obtained from the degradation of hemoproteins in the liver (catalase and cytochrome oxidases). The conversion of haeme to bilirubin occurs in two steps:

1. Microsomal enzyme *heme oxygenase* converts haeme to *biliverdin.*
2. The cytosolic enzyme *biliverdin reductase* converts biliverdin to bilirubin.

5.17.1 Conjugation of Bilirubin

Bilirubin is a toxic and lipophilic substance which requires conjugation in the liver before elimination. Bilirubin uptake across the sinusoidal membrane of hepatocytes is mediated by transporters (sinusoidal organic anion transport protein-2, OATP-2, and SLC21A6). After uptake, binding proteins (e.g. *ligandins* and *fatty acid binding protein*) direct bilirubin to the endoplasmic reticulum, where it is conjugated with uridine diphosphate (UDP)-glucuronic acid in the presence of bilirubin UDP-glucuronyl transferase (B-UGT).

5.17.2 Excretion of Conjugated Bilirubin

The ATP-dependent export pump (multidrug resistance–associated protein-2 or **MRP2/ABCC2**) transports the conjugated bilirubin into the bile canaliculus. Some bilirubin glucuronides are also released into plasma beyond the sinusoidal membrane (multispecific organic ion export pump **MRP3/ABCC3**). Resorption of conjugated bilirubin in the biliary system and intestine is negligible. In the terminal parts of the small intestine and the colon, conjugated bilirubin is converted to colourless urobilinogen, of which ~20% are resorbed and eventually eliminated in bile and urine.

5.18 Congenital Disorder of Bilirubin Metabolism and Excretion

Bilirubin biotransformation and excretion involve multiple steps, and abnormalities associated with these are usually hereditary.

5.18.1 Defects in Bilirubin Uptake and Conjugation

(a) Gilbert syndrome
(b) Crirglar Najjar syndrome types 1 and 2

5.18.2 Defects in Hepatic Excretory Function

(c) Dubin Johnson syndrome
(d) Rotor syndrome
(e) Benign recurrent intrahepatic cholestasis
(f) Progressive familial intrahepatic cholestasis

5.19 Gilbert's Syndrome

Gilbert's syndrome is marked by excessive levels of unconjugated bilirubinemia, but individuals are asymptomatic. Bilirubin levels are usually <3 mg/dl (<51 mol/l). It fluctuates substantially, increasing during intercurrent illnesses, fasting, and stress. Population prevalence as high as up to 8% has been described. Males predominate over females by 1.5:1 to >7:1. It is usually diagnosed shortly after puberty. Apart from hyperbilirubinemia, other liver function tests are normal. A histologically modest increase in lipofuscin pigment in hepatocytes may be noted. UGT1A1 activity is characteristically lowered to 10–35% than usual. Phenobarbital usually settles the serum bilirubin concentration and hepatic bilirubin elimination to normal levels. But refractory patients raise a high suspicion of concomitant haemolytic defect. The diagnosis is usually confirmed with an elevation of S. bilirubin on being stimulated by 48 h of fasting or systemic nicotinic acid administration. This condition has an excellent prognosis.

5.20 Crigler–Najjar Syndrome, Type I

Crigler–Najjar Syndrome, Type I, is an autosomal recessive disease characterised by striking unconjugated hyperbilirubinemia of about 20–45 mg/dl and undetectable expression of UGT1A1 activity in hepatic tissue. It is initially present in the neonates and stays all along. No bilirubin is found in the urine. Other classical liver function tests are within normal limits, as is histological analysis with the absence of hemolysis. Infrequently, bile plugs are observed inside canaliculi. There is an improvement with phenobarbital or other enzyme inducer administration. Years ago, due to a lack of phototherapy, infant or childhood mortality was attributed to bilirubin encephalopathy (kernicterus). Liver transplantation is the most critical management option to avert the risk of imminent brain injury and subsequent fatality.

5.21 Crigler–Najjar Syndrome, Type II (CN-II)

This illness is in between the above two illnesses in severity. It is associated with reduced levels of UGT1A1 (typically 10% of normal). Like Type I, elevated unconjugated hyperbilirubinemia does not impact hepatic enzymes or histology. Haemolysis is also not evident in the absence of abnormalities of other conventional hepatic biochemical tests, hepatic histology, or hemolysis. CN-II can be differentiated from CN-I based on serum bilirubin concentrations decreased by >25% on administration of phenobarbital. Generally, average bilirubin concentrations are lower than CN-I and only infrequently associated with kernicterus.

5.22 Dubin–Johnson Syndrome

Dubin–Johnson syndrome (DJS) is an autosomal recessive asymptomatic disorder involving defective bilirubin conjugate excretion and some organic compounds, but bile acids remain unaffected. Defective expression of MRP2, an ATP-dependent canalicular membrane transporter, impairs secretion of conjugated bilirubin into bile canaliculus. Some patients may develop non-specific constitutional symptoms. The disease may surface during pregnancy or when oral contraceptives are used since both increase hyperbilirubinemia, transforming from subclinical to clinical jaundice. Histological features include the presence of multiple lysosomes of centrilobular hepatocytes of dark, roughly granular pigment, which gives the liver a grossly black appearance.

5.23 Rotor Syndrome

Rotor syndrome is largely similar to DJS, except that hyperpigmentation is absent. However, total serum bilirubin is increased, attributed to increased conjugated bilirubin.

5.24 Benign Recurrent Intrahepatic Cholestasis (BRIC)

BRIC is a familial autosomal recessive condition with ATP8B1 (earlier called familial intrahepatic cholestasis gene 1 or *FIC1*) mutation, normally responsible for canalicular bile excretion. The typical features of BRIC include repetitive bouts of pruritus and jaundice. Typically, it starts with slight malaise and a progressive rise in liver enzymes and itching. Therefore, it is at a high risk of being misdiagnosed as acute viral hepatitis. Such episodes of varying duration manifest in pediatric or adult patients and resolve entirely with liver function tests and physical parameters completely normal in between them. BRIC does not increase the risk of cirrhosis.

5.25 Progressive Familial Intrahepatic Cholestasis (PFIC)

This condition includes three phenotypes. All three phenotypes manifest as cholestasis worsening to malnutrition and end-stage liver disease. PFIC type 1 (Byler disease) occurs in early infancy. The progressive FIC type 2 occurs due to a mutation in *bile salt excretory protein* (BSEP), while type 3 involves mutation of MDR3, required to eliminate phospholipids across the bile canaliculus. The three phenotypes can be distinguished biochemically since only PFIC-3 shows elevated serum gamma-glutamyltransferase activity, while its levels are normal or negligibly elevated in FIC types 1 and 2.

5.26 Enterohepatic Circulation of Bile Salts

Bile salts are important constituents of bile and are formed by conjugating bile acids (synthesized from cholesterol) with taurine or glycine residues. These salts are required for the digestion of fat components of meals and the absorption of fat-soluble vitamins. Primary bile acids are secreted into bile and deposited in the gallbladder. Secondary bile acids (deoxycholic acid and lithocholic acids) are synthesised in the colon by bacterial action. They are the main driving force of bile flow. Nearly 600 mg/day of bile salts are produced to replenish the ones eliminated in stools; however, the quantity of bile salts secreted, reabsorbed, and recycled due to enterohepatic circulation is considerably greater.

5.26.1 Components of Enterohepatic Circulation

Anatomic components of the enterohepatic circulation include the liver, biliary tract, small intestine, portal venous circulation, and, to some extent, the colon, systemic circulation, and kidneys. This system includes sequence repositories (gallbladder and small intestine), valves (SOD, ileocecal valve), mechanical pumps (small intestine, bile canaliculi, and biliary tract), and chemical pumps (cholangiocyte, hepatocyte, and ileocyte).

5.26.2 Enterohepatic Pathway

Fasting moves the bile acids in the lower biliary tract, concentrating ~10-fold in the gallbladder. On meal consumption, the small intestine releases CCK, the gallbladder contracts, and SOD undergoes relaxation. The concentrated gallbladder content is drained into the small intestine to improve the fat absorption by stimulating the action of pancreatic lipase on triglyceride, solubilizing the hydrolytic products, and delivering lipids to the mucosal surface since it contains micelles. Micelle formation necessitates an intraluminal bile acid concentration higher than 1.5 mmol/l. The bile acid concentration across various compartments varies, with the highest in the biliary tract (20–50 mmol/l) and gallbladder (50–200 mmol/l). The bile acids are directly transferred into the duodenum when large meals are consumed, bypassing the contracted gallbladder. Post-meal, the SOD contracts, and the gallbladder relaxes, allowing the excess bile acids to return to the bladder (Fig. 5.7). Thus, this shows that the enterohepatic cycling of bile acids depends on meals that accelerate digestion and decelerate during gaps and fasting periods. Post-cholecystectomy, bile acids are accumulated in the proximal small intestine and pitched into the distal ileum due to contractions of the small intestine after a meal.

Being an extremely organized endogenous system, the enterohepatic circulation of bile acids allows a loss of less than 10%, which dodges reabsorption. Absorption of bile acids in the small intestine largely occurs by transporters (active) in the terminal ileum and, to a negligible extent, by a passive mechanism. Actively absorbed bile acids are the only conjugated anions reabsorbed in the small intestine and undergo an enterohepatic circulation. In healthy adults, the enterohepatic circulation sustains a 2–4 g bile acid (50–60 μmol/kg body weight). Usually, with every

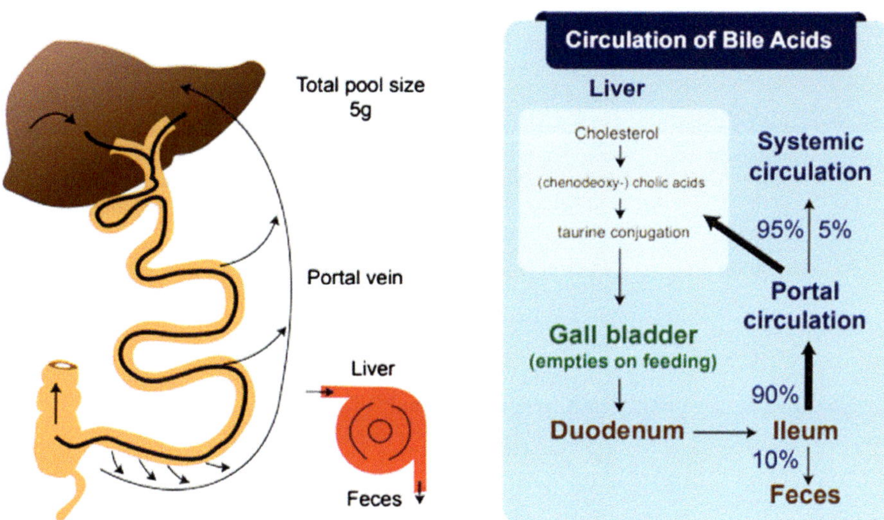

Fig. 5.7 Enterohepatic circulation of bile acids

meal, the bile acid pool cycles and 6–10 cycles/day, the intestine may reabsorb 10–30 g of bile acid with only 0.2–0.6 g escaping recirculation and being excreted in faeces. Due to efficient enterohepatic circulation, less than 5% of the bile acids are synthesized de novo from cholesterol. Thus, it also is essential for removing cholesterol from the system.

5.27 Liver as an Immunological Organ

5.27.1 Introduction

The liver is strategically placed in the human body such that antigen-containing fluids from the GIT enter it, and the sinusoids filter out APCs and lymphocytes. The hepatic lymphocyte populace is specifically advanced in innate and adaptive immune response, protecting systemic circulation against attacking pathogens and modulating systemic immune functions (Table 5.4).

Table 5.4 Common immunological cells in the liver and their functions

S. No.	Cells in liver	Remarks
1.	NK and iNKT cells (innate natural killer and natural killer T cells)	Liver-resident, patrolling lymphocytes; produce both pro- and anti-inflammatory cytokines; recognise infected or abnormal cells by the absence of MHC antigens. Modulate local and systemic immunity
2.	Kupffer cells	Intravascular liver-resident macrophages express several immune receptors such as (Fc, CR), PRRs (TLRs), and scavenger receptors; Have a key role in pathogen detection, capture, and clearance; present antigen via MHC-I and -II and CD1d; maintain immune hyporesponsiveness/immunotolerance; produce a wide variety of cytokines
3.	Liver sinusoid endothelial cells (LSEC)	Scavenger receptors such as Hyluronan and CD44; can act as antigen-presenting cells; trigger and produce immunological tolerance to $CD4^+$ and $CD8^+$ T cells to blood- and gut-derived antigens; also, express PD-L1 and help develop tolerance to gut-derived bacterial antigens
4.	Hepatocytes	Present antigen (express MHC-I and can be induced to present MHC-II and CD1d); maintain immunotolerance; produce acute-phase and complement proteins
5.	Dendritic cells	Classical antigen-presenting cells; present in the space of Disse, exposed to liver lymph. Two types (a) $CD123^+$ plasmacytoid (pDC) associated with PD-L1 and (b) CD11c high myeloid (mDC) that produce IL-10. A subpopulation CD11c high $CD8^+$ proinflammatory

Table 5.4 (continued)

S. No.	Cells in liver	Remarks
6.	HSCs	In space of Disse, contain vitamin A and lipid storage cause fibrogenesis; can also function as immune Sentinel with APC: expresses MHC-I and II, CD80/86, can endocytose Ag process and present with MHCs, communicate with lymphocytes (CD4 and CD8) and NK/NKT cells through LSEC sieve; can also constrict sinusoids
7.	Monocytes	Multiple subtypes (CCR2+ proinflammatory, CX3CR1+ reparative); produce a variety of cytokines; involved in damage mitigation and tissue repair; differentiate into macrophages
8.	Neutrophils	Few resident cells; but rapidly recruited during liver infection/injury; one of the principal phagocyte types responsible for pathogen clearance; release neutrophil extracellular traps; cause collateral damage to the liver and involved in tissue repair
9.	$CD4^+$ T cells	• Multiple subtypes (Th1, Th2, and Th17); immunomodulatory; few resident cells; rapidly recruited during liver infection/injury
10.	$CD8^+$ T cells	• Cytotoxic; release inflammatory cytokines; clear viral infections; few resident cells; rapidly recruited during liver infection/injury

5.28 Innate Immune Responses in the Liver

The liver is rich in KCs, NK cells, and NKT cells which are principal components of the innate immune framework. KCs are initiated by different pathological insults, including lipopolysaccharides (LPS) and bacterial superantigens, and released cytokines play a vital role in balancing the maturation and expansion of different clones of cells. Constant exposure to KCs leads to the secretion of initially inflammatory cytokines and later cytokines that dampen the inflammatory response through Tregs. IL-10 and 23. This leads to upregulation of MHC-II expression in sinusoidal endothelial cells and dendritic cells and down-regulates T cell activation. Which confers tolerance to antigens in the portal vein, at least partially. KCs also release IL-12 and IL-18 to control NK-cell activation and expansion of cytotoxic NK cell subpopulations that express a lot of antiviral IFN-gamma. Newer antigens induce KCs to secrete proinflammatory cytokines such as IL-1, IL-6, TNF-alfa', and chemokines to attract neutrophils. Neutrophils destroy extracellular microscopic organisms attached to KCs and hepatocytes via surface phagocytosis, activating the innate immune system by releasing cytokines and chemokines to recruit CD4 and CD8 T cells.

The maximum number of natural killer cells in the liver accounts for half of the lymphocytes but are rarely found in other parts. These immune cells act vigilant towards entering pathogens and abnormalities in the tissues by recognising the absence of MHC-1 instead of the presence of PAMPs or particular antigens. The marked absence of MHC-1 occurs in an attempt to escape the immune mechanism when oncogenic or infectious abnormalities occur in the body. Interestingly, NK-1 cells use this absence to eliminate abnormal cells, enhance cell activation, and stimulate cytokine secretion. NK cells display comparable excitatory and inhibitory receptors for self-molecules in various cells. The absence of either can lead to the switch in its favour and subsequent activating or inhibiting effects. NK-cell memory provides faster and better responses to subsequent encounters with pathogens.

The majority of human immune system responses can be attributed to iNKT, which is produced in the thymus. These usually counter infringements of lipid antigens from the host or pathogen via CD1d recognition. The frequency of iNKT cells is dynamic and can be with the help of flow cytometry. iNKT cells initially are $CD4^{-}CD8^{-}$ double-negative precursors, which evolve into $CD4^{+}CD8^{+}$ double-positive cells, and arbitrarily reposition their TCRs.

The presence of a high number of CCR5 and CXCR3 receptors on hepatic iNKT guides them during inflammatory events. iNKT cells have effector capacity and are activated on stimulation by the TCR via lipid antigens even without co-stimulations. The monocytes, neutrophils, and iNKT cells constantly move along the sinusoid walls scrutinising for cognate ligands regardless of blood flow direction under basal conditions. The KC cells arrest the pathogens in the circulation, prepare for glycolipid manifestation, and release chemokines to activate the iNKT cells and facilitate their migration to the required loci. Subsequently, iNKT cells release IFN-γ to eliminate pathogens. While in the case of dead cells, iNKT releases IL-4 and cytokines while anchored to the vasculature. On the other hand, α-galactosylceramide (α-GalCer) is a ligand that inhibits iNKT activity by binding to CD1d on endothelial cells and macrophages in the systemic circulation.

5.29 Adaptive Immune Responses in the Liver

Not much information about B cells exists in the liver because their numbers are modest, and they are difficult to study in the liver. Traditionally, B-cell activation, transformation, and clonal expansion happen in the lymphoid follicles of auxiliary lymphoid organs or apparent 'ectopic' germinal foci found in some non-lymphoid organs. The HCV-infected liver may contain some lymphoid follicles, which allow B cells to proliferate. These intraportal lymphoid follicles display a germinal central structure in which activated B cells are found in association with follicular dendritic cells. A T-cell zone consisting of CD4- and CD8-T cells is identified at the fringe of such centres. B cells in these intrahepatic follicles are clonally controlled to react to local antigens.

Adaptive response to many pathogens includes robust CD4 and CD8 T-cell reactions. Peptides of intracellular pathogens are displayed through MHC class I. CD8 T cells can recognise peptides of 8–11 amino acids linked with MHC class I molecules and destroy the infected cells. These effector components consist of the generation of cytokines, for example, IFN-gamma and TNF-alfa, and cytolytic enzymes such as perforin and granzyme and activation of Fas-interceded apoptosis. The functions of various subsets of T cells are shown in Table 5.5.

Table 5.5 Various T-cell subsets in the liver and their main functions

Cell group	Subset	Function
CD4 T cells	Th1	Secretion of IFN-γ, IL-2, TNF-α; activation of CD8 T cells; induction of Ig class shifting to complement-fixing antibodies; cell-mediated and delayed-type hypersensitivity responses
	Th2	Secretion of IL-4, IL-5, IL-9, IL-10, IL-13 and IL-25; induction of Ig class shifting to IgG1 and IgE; facilitating in antibody-dependent cell-mediated cytotoxicity; linked with allergic responses; inhibition of Th1 cells and Th17 cells
	Th17	Secretion of cytokines IL-17, IL-21, IL-22, IFN-γ; suppression of Treg cell function
	Tfh	B-cell proliferation, differentiation, and maturation in lymphatic tissue; antibody synthesis; immune reaction; triggering CD8 T cells
	Treg	Secretion of IL-10 and TGF-β; inhibition of Th17 proliferation and secretion of IL-17; inhibition of proliferation of CD4 CD25(−)T, CD8 T cells, and their secretion of IFN-γ; attenuation of innate immunity
CD8 T cells	Tc	Secretion of granzymes, perforin, IFN-γ, TNF-α, IL-4, IL-5, IL-9, IL-10, IL-13, IL-17, and IL-21; cytotoxicity against tumors and intracellular pathogens; promotion of Th2-mediated allergy; dissemination of autoimmunity
	CD8 Treg	Secretion of granzymes, perforin, TGF-β, and IL-10; suppression of proliferation of Th cells; inhibition of immunity in response to infection
γδ-T cell	γδ-T cells	Secretion of IL-10, IL-17, IL-22, TNF-α, and IFN-γ; inhibition of CD8 T cells; phagocytosis of bacteria; cytotoxicity against hepatic tumors; stimulating of hepatocyte apoptosis; attenuate hepatic inflammation and fibrosis

Modified from Shuai Z, Leung MWY, He X, Zhang W, Yang G, Leung PSC, Gershwin ME. Adaptive immunity in the liver. Cell Mol Immunol. 2016;13(3):354-68

IFN-γ interferon γ, *Ig* immunoglobulin, *IL* interleukin, *Tc* cytotoxic T cell, *Tfh* follicular helper T cell, *TGF-β* transforming growth factor-β, *Th* helper T cell, *TNF-α* tumour necrosis factor α, *Treg* regulatory T cell

5.30 Ammonia Handling in the Liver, Kidney, Brain, and Muscles

In mammals, ammonia is synthesised by the degradation of nitrogen-containing compounds (purines and pyrimidines). Ammonia is water soluble physiologically (~98%) found as ammonium radicle ($NH4^+$). Intestinal bacterial flora also synthesises ammonia via urea using urease.

Transaminase is unique for a set of substrates but not for all the remaining. L-Alanine acts as a substrate for glutamate transaminase and its amino nitrogen can be transferred to α-ketoglutarate to form L-glutamate and pyruvate. L-Glutamate can undergo oxidative deamination in mammalian tissue to ammonia and α-ketoglutarate glutamate dehydrogenase. Thus, glutamate functions as an ammonia buffer or carrier.

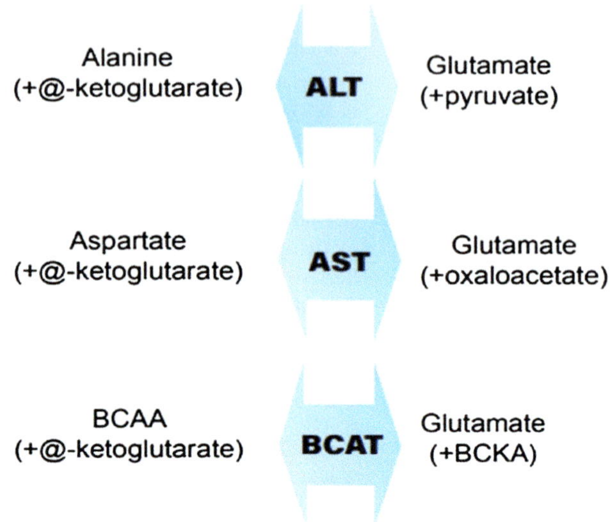

Fig. 5.8 Reactions catalysed by aminotransferases. *ALT* alanine aminotransferase, *AST* aspartate aminotransferase, *BCAA* branched-chain amino acids, *BCAT* branched-chain amino acids aminotransferase, *BCKA* branched-chain keto acids

Glutamate can take up another ammonium radicle to become glutamine (catalysed by glutamine synthetase), while the glutaminase enzyme requires reverse action. Similarly, aspartate and branch-chain amino acids can be converted to glutamate, as shown in Fig. 5.8.

Glutamine, the most abundant amino acid in the blood, forms nearly half the free amino acids. It is a crucial source of energy for enterocytes of the small intestine. The intestine consumes glutamine in large quantities from either blood or intestinal lumen and utilises it through glutaminase enzyme mainly in the jejunum and, to a small extent, in the colon to synthesise glutamate and ammonia from glutamine. The large bowel consumes less glutamine but more glucose, short-chain fatty acid, and ketones. But the large intestine considerably adds to portal venous ammonia concentration by bacterial lysis of urea and amino acids. Nearly half the ammonia in circulation comes from the intestine.

5.30.1 Liver

The liver synthesises and metabolises glutamine due to both glutaminase and glutamine synthetase. The liver's main function is to convert ammonia to urea through the urea cycle (Fig. 5.9). Enzymes necessary for urea synthesis and glutaminase exist in periportal and glutamine synthetase in pericentral hepatocytes. The portal blood first comes into contact with hepatocytes equipped for urea synthesis, where glutaminase releases ammonia to be used to form urea. Glutamine synthetase in pericentral areas soaks up any ammonia that has escaped the periportal urea cycle. The net effect is that the liver almost completely removes the ammonia in the portal or systemic blood.

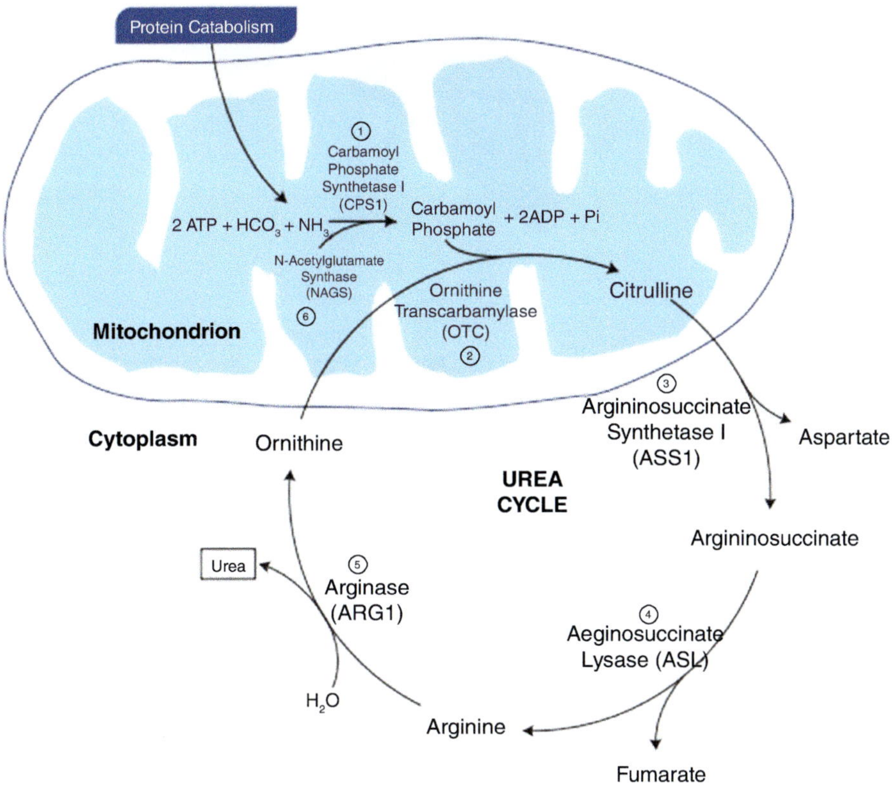

Fig. 5.9 The urea cycle produces urea from ammonia. The six enzymes of the pathway are numbered 1–6, with their associated gene in brackets

In case of liver failure or cirrhosis, ammonia is released into circulation, and additional mechanisms are required to keep ammonia levels low (see Fig. 5.9). Besides, ammonium (45–240 μmol/l; average 110 μmol/l) in the hepatic bile in otherwise healthy individuals undergoing cholecystectomy for gallstones. It possibly adds to ammonia being absorbed from the intestines. Nearly 1000 mmol/day of ammonia is formed in a day. Some of it is utilised in biosynthesis and for the metabolism of enterocytes, while the remainder must be disposed of quickly as it is a neurotoxin.

The blood inflow into the liver is about 900–1350 ml/min and ~300 ml/min from the portal vein and hepatic artery, respectively. Thus, the portal vein brings in about 137–150 mmol/24 h. The liver removed most of it, and hepatic veins contain little ammonia. The ammonia is around 40μmol/L at the inferior vena cava level. Thus, the liver excretes the majority of ammonia by either the urea cycle or glutamine synthesis. Ammonia is finally excreted in urine as urea and as ammonia buffer. Since the liver is the only organ that can convert ammonia into an excretable form, ammonia levels rise significantly in liver failure (Fig. 5.10). Ammonia is a

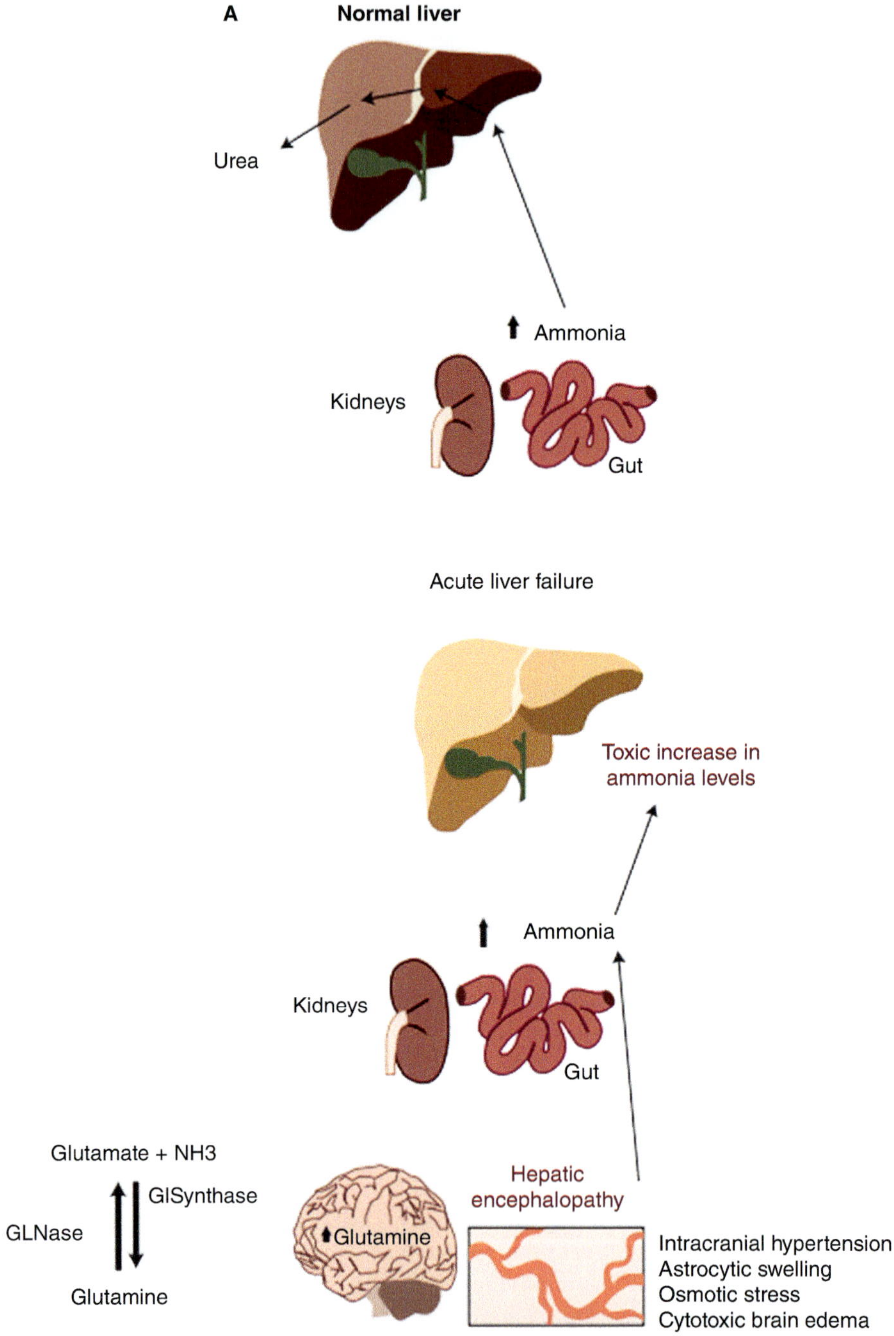

Fig. 5.10 Metabolism of ammonia in normal liver and liver failure

neurotoxin and also affects neutrophilic function and liver regeneration. Therefore, excess ammonia in hepatic failure can result in cerebral dysfunction.

5.30.2 *Brain*

In physiologic conditions, the arterial ammonia inflow and the jugular vein outflow ammonia concentrations are similar. In the brain, it is 200–300 μmol/kg, which to some extent undergoes biotransformation in the brain per se. Ammonia readily traverses BBB with positive arterial–venous gradients indicating net brain ammonia uptake, which occurs at <5 mmol/24 h and releases glutamine. Astrocytes contain most of the brain glutamine synthetase, while neurons contain virtually all brain glutaminase. Astrocyte glutamine synthetase favours ammonia uptake for glutamine synthesis, which is deaminated to form GABA and glutamate. When brain ammonia concentrations spike due to increased arterial ammonia, the inflow of ammonia surpasses the outflow. Among patients with intra- or extrahepatic collateral, the brain eliminates ammonia at a rate of 27 mmol/24 h (brain blood flow = 750 ml/min) which is attributed to the difference of 25 μmol/l between the inflow and the outflow in the portal-systemic circulation. The association between tissue ammonia concentration and the extent of encephalopathy is established; however, a similar association between plasma ammonia and encephalopathy has not been proven. Thus, factors in addition to ammonia also regulate the neurological status.

5.30.3 *Muscle*

Ammonia can be utilized or discharged from the skeletal muscle. At rest, when arterial ammonia is not elevated, arterial and venous blood ammonia concentrations are similar. However, when it is high, a definite arterio-venous difference is demonstrated, but venous ammonia concentration usually stays normal. In individuals with comparatively greater ammonia levels, half of the arterial ammonia is metabolised by skeletal muscle. Although glutamine synthetase activity in striated skeletal muscle is not very high, its large mass, compared to other glutamine synthetase-containing organs, makes it a major site for glutamine synthesis. The skeletal muscle at both the resting stage and in the disease takes up ammonia at a rate of <5 mmol/24 h with progressive increase with increased severity. Muscle ammonia uptake followed by glutamine synthesis is detoxification pathways prevalent in severe hepatic disease. Strenuous exercise also elevates ammonia (120–130 mmol/24 h to more than 150 mmol/24 h) in the veins draining the exercising muscles in persons with normal liver and cirrhotic liver derived primarily from the purine nucleotide cycle activity. Also, in patients with hepatic disease, marginal elevation is noted peri- and post-hyperventilation due to the higher workload of the respiratory muscles. Post-exercise, the half-life is 20–30 min for ammonia in the blood.

5.30.4 *Kidney*

Glutaminase and glutamate dehydrogenase are present in kidneys, suggesting that glutamine may be a substrate for ammonia genesis. Utilising venous catheterisation and arterial sampling, renal venous and blood vessel ammonia levels uncovered a positive contrast of 8 μmol/l showing that the renal discharge of ammonia occurs at a rate of around 13 mmol/24 h in the renal veins (renal bloodflow: 1100 ml/min). Around 33% of the ammonia is eliminated into the urine, while 66% returns to the systemic circulation. About 9 mmol/24 h is discharged into the urine. In the event of metabolic acidosis, the situation changes and aggregate renal ammonia genesis is increased, and approximately three-fourths of this additional quantity are excreted in the urine. Ammonia excretion is one of the mechanisms for buffering acid–base balance by kidneys. Thus, blood pH and ammonia levels may direct urinary ammonia excretion. In physiological conditions, the kidneys are major regulators of blood ammonia levels. In liver failure and cirrhosis, the kidney excreted ammonia into the urine.

Thus, the liver plays an important role in nitrogen metabolism and disposal of ammonia. Normally ammonia is present at nontoxic levels in the blood. When ammonia levels are raised, it becomes toxic and significantly contributes to hepatic encephalopathy in liver failure. Hyperammonemia in chronic liver disease is regularised to some extent by increased uptake in the muscle. In acute liver failure, this mechanism fails, leading to a rapid increase in ammonia and subsequent cerebral edema. Ammonia-lowering therapy can help with liver failure (Fig. 5.15).

5.31 Ethanol Metabolism

All the alcoholic beverages used are derived from plants by a process of fermentation in which glucose is converted to ethanol. It is absorbed by passive diffusion through the upper gastrointestinal tract; absorption predominantly occurs in the duodenum and proximal jejunum (80%), followed by the stomach (20%). The absorption depends on the concentration of alcohol, ingestion rate, gastric emptying, and dietary components in the stomach. After absorption, most alcohol consumed (>90%) is metabolised in the liver by oxidation into its metabolites such as acetaldehyde and acetate. The rest of the amount is excreted in breath, sweat, and urine. In the liver, ethanol is oxidised to acetaldehyde by three different enzymes such as alcohol dehydrogenase (ADH), the microsomal ethanol-oxidizing system (MEOS) consisting of cytochrome enzymes (mostly CYP2E1), and catalase

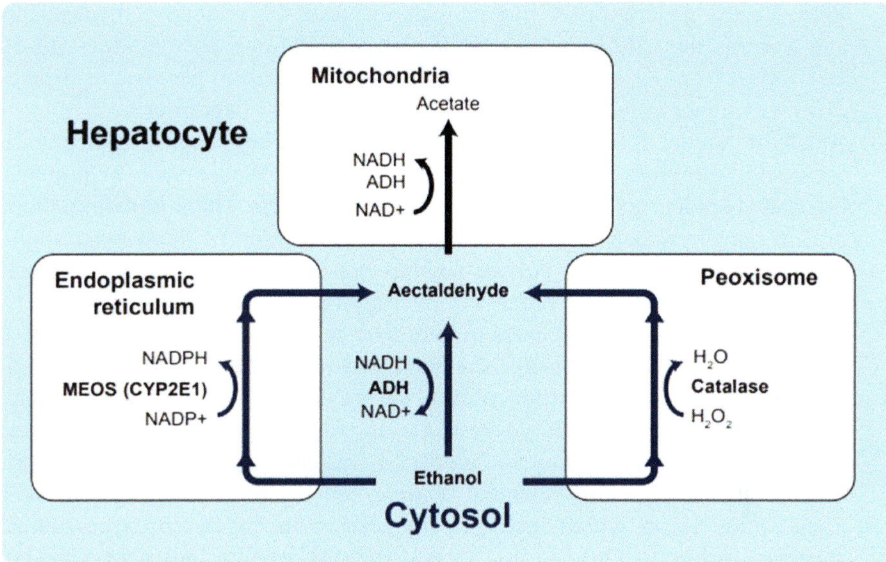

Fig. 5.11 Oxidizing pathways of ethanol in the liver

(Fig. 5.11). In the hepatocyte, ADH is present in the cytosol (cytoplasm), MEOS in the microsomes, and catalase in the peroxisomes. ADH is usually saturated at low concentrations of alcohol; thus, MEOS plays a major role in alcohol oxidation in excessive ethanol intake. MEOS activity is also induced by alcohol. Acetaldehyde generated from ADH and MEOS in hepatocytes is further oxidised in the mitochondria by the enzyme acetaldehyde dehydrogenase (ALDH) to acetate. Acetate passes into the blood and becomes part of various metabolic pathways. ALDH competes with cytosolic ADH for NAD+. A higher NADH/NAD+ ratio in hepatocytes leads to acetaldehyde accumulation, a toxic chemical unlike acetate. Mitochondrial impairment due to acetaldehyde results in a further decrease in the activity of ALDH and eventually initiates a vicious circle leading to the accumulation of toxic acetaldehyde.

5.31.1 Mechanisms of Liver Injury

The liver is damaged by several mechanisms.

(a) Oxidative stress: Chronic alcohol intake induces hepatic MEOS and CYP2E1 levels, enhancing reactive oxygen species (ROS) production. ROS, along with accumulated acetaldehyde, cause the destruction of hepatocytes. The KCs also generate ROS after activation by intestinal bacterial translocation (discussed later). Moreover, along with HSCs, KCs produce pro-inflammatory cytokines leading to inflammation and neutrophilic infiltration, contributing to

ROS. Excess generation of free radicals exhausts the protective mechanisms and worsens the oxidant burden. ROS and proinflammatory cytokines (IL-1β and TNFα) from KCs enhance programmed cell death and necrosis of hepatocytes which incites local inflammation and, eventually, liver injury.

(b) Aldehyde adduct formation: Being a toxic and highly reactive molecule, it covalently combines with various cellular macromolecules such as proteins, lipids, and nucleic acids giving rise to aldehyde adducts. These transformations lead to the disruption of normal structure and function of these macromolecules. Furthermore, these bulkier adducts can induce an immune response, which adds to the ongoing process of liver injury.

(c) Lipid metabolism: Alcohol leads to fatty liver in more than 90% of patients. It is related to enhanced lipogenesis due to alteration in the cellular redox potential. Ethanol downregulates the activity of the enzyme adenosine monophosphate-activated kinase (AMPK). A decrease in AMPK activity stimulates the mammalian target of rapamycin complex 1 (mTORC1), leading to the activation of sterol regulatory element-binding protein-1c (SREBP-1c). It is a transcription factor which increases lipogenesis via increased expression of necessary enzymes. There is also decreased β-oxidation of fatty acid related to altered hepatocyte redox. Second, acetaldehyde impedes the peroxisome proliferator-activated receptor alpha (PPAR-α) pathway transforming the liver from a lipid-metabolising organ to a lipid-storing organ.

(d) Arrested lipophagy: Chronic ethanol exposure inhibits the process of autophagy by decreasing the activity of AMPK and activating mTORC1. Furthermore, alcohol also reduces the activity of proteins (such as Rab7 and Dyn2) which are critical for the fusion of autophagosomes and lysosomes, leading to faulty lysosomes. Reduced autophagy enhances lipid accumulation.

(e) Reduced adiponectin: Chronic alcohol abuse usually increases pro-inflammatory and pro-fibrogenic leptin in the circulation, whereas reduced adiponectin expression and secretion are associated with ALD. The reasons for decreased adiponectin production are ethanol-induced oxidative stress due to the activity of MEOS and altered methionine metabolism leading to increased homocysteine levels. Leptin promotes hepatic steatosis (by suppression of fatty acid biosynthesis), fibrogenesis (because of up-regulation of TGF-β1), and inflammation (through stimulation of secretion of proinflammatory chemokines). On the contrary, adiponectin inhibits steatosis (by suppressing fatty acid production and stimulating oxidation of fatty acids), TGF-β1-induced fibrogenesis, and inflammation (through inhibition of proinflammatory cytokines and chemokines), leading to alcoholic fatty liver and hepatitis.

Impaired very low-density lipoprotein (VLDL) secretion and increased fatty acid influx to liver.

(f) Gut liver axis: Excessive alcohol consumption causes bacterial overgrowth (due to reduced gastro-intestinal motility) and alters the normal composition of the intestinal microbiome, which involves diminished beneficial bacteria such as bacteroidetes, lactobacillus, and increased harmful bacteria like proteobacteria and bacilli. Moreover, alcohol-induced inhibition of prostaglandin synthesis injures normal mucosa causing intestinal barrier dysfunction. The resultant increased gut permeability leads to intestinal bacterial translocation. These microbes are delivered to the liver by the portal venous system and stimulate KCs and HSCs by lipopolysaccharides (LPS) mediated toll-like receptor 4 (TLR4) activation. Subsequently, KCs and HSCs produce pro-inflammatory cytokines and mediators (IL-1, IL-6, TNFα, and ROS), promoting hepatic necro-inflammation and fibrogenesis.

(g) Impact on the immune system: Alcohol affects the immune system profoundly. The pro-inflammatory state is created by the mechanisms given above. LPS-mediated TLR4 activation to external antigens (PAMPs) stimulates the nuclear factor κB (NF-κB), triggering the gene transcription for several pro-inflammatory cytokines such as TNFα and IL-17. In addition, ROS and acetaldehyde-mediated liver damage also activates innate immune cells, triggering an inflammatory response even in the absence of invading pathogens. This sterile inflammation results from the activation of nucleotide-binding oligomerisation domain-like receptors, or NOD-like receptors (NLRs) in KCs by the molecules released by breakdown hepatocyte necrosis, also known as alarmins or DAMPs. These immune pathways are usually necessary for clearing the damaged cells and cellular debris but can also cause liver damage. With their persistent activation by ethanol, a vicious circle ensues, leading to unrestricted liver damage. Even adaptive immunity is affected. Chronic alcohol use reduces the peripheral T cells pool by enhancing T cell apoptosis due to increased expression of pro-apoptotic molecules like BCL2 associated X, apoptosis regulator (bax), and diminished expression of anti-apoptotic ones such as B-cell lymphoma 2 (bcl-2) protein. Ethanol use also causes chronic activation of the T cells resulting in impaired T-cell functioning with reduced migration and diminished ability to respond to pathogens. This state of anergy contributes to an increased susceptibility to infections. On the other hand, there is a rise in B-1 cells which secrete increased immunoglobulins, including antibodies directed against liver autoantigens liberated during hepatocyte death.

(h) Ubiquitin proteasome pathway: Liver cells' function is regulated by a specialised mechanism of protein degradation known as the ubiquitin–proteasome pathway. Chronic ethanol ingestion results in impairment of proteolytic activity of the proteasome. This leads to the accumulation of abnormal proteins in the form of Mallory–Denk bodies. Hepatocytes dying due to inhibition of the proteasome pathway release inflammatory cytokines, leading to neutrophilic infiltration and hepatic necro-inflammation.

(i) Fibrogenesis: In ALD, KCs and HSCs are stimulated by gut microbiota translocation-mediated TLR4 activation and damaged hepatocytes-mediated NOD-like receptors activation. Besides, KCs secrete prostaglandin D2 which further stimulates the HSCs. HSCs transform into myofibroblasts leading to irregularly TGF-induced extracellular matrix deposition, especially collagen type 1. Moreover, activated HSCs contribute to the inflammatory response by recruiting and stimulating circulating leucocytes in the liver. Hence, the perpetuation of ongoing inflammation sustains the process of fibrosis.

5.32 Coagulation Pathway

On the injury to a blood vessel, hemostasis sets in via vasoconstriction, platelet plug formation or clot formation, and the development of fibrous tissue to seal the injury. The liver is almost entirely responsible for coagulation. Coagulation is a dynamic procedure that maintains a delicate balance between clotting and clot dissolution mechanisms.

After trauma to the vessel wall, the clotting is initiated within 15–20 s. Activator substances are released by the platelets that aggregate at the site and from the proteinic substances anchored to the damaged vessel wall. When the damage is not too large, the ruptured vessel is clogged with a clot within 3–6 min. Subsequently, 20 min to an hour later, the clot withdraws from the site, further sealing the vessel. This clot can now be invaded by fibroblasts, which may repair the damage or dissolve if clot formation is excessive. When excess blood seeps into the tissues forming clots, certain activated enzymes in the clot dissolve it.

5.32.1 Coagulation Pathways

Multiple substances directly or indirectly involved in blood coagulation have been documented, grouped as pro-coagulants, anticoagulants, or coagulation factors. A fine balance between these substances dictates the outcome in terms of whether coagulation is supported or anticoagulation. Normally, the anticoagulants surpass the coagulants in concentrations to maintain blood flow across the vasculature. However, when an injury occurs, procoagulants have site-specific activation to activate the clot formation to prevent excessive bleeding, thus exceeding anticoagulant levels.

Clotting occurs in three main steps: (a) an intricate sequence of chemical events with the participation of more than a dozen factors leads to the generation formation of a complex of activated substances known as prothrombin activator. (b) The latter converts prothrombin into thrombin. (c) Thrombin converts soluble fibrinogen into insoluble fibrin that entangles thrombocytes and RBCs to generate the clot. Main coagulation factors are named by Roman numerals and include Factor I: fibrinogen factor; Factor II: prothrombin factor; Factor III: tissue factor (also called tissue thromboplastin); Factor IV: calcium; Factor V: proaccelerin or labile factor [Ac-globulin (Ac-G)]; Factor VII: serum prothrombin conversion accelerator (SPCA) or proconvertin 9; Factor VIII: Antihemophilic factor (AHF) or antihemophilic globulin (AHG) or antihemophilic factor A; Factor IX: plasma thromboplastin component or (PTC) or Christmas factor or antihemophilic factor B; Factor X: Stuart factor or Stuart-Prower factor; Factor XI: plasma thromboplastin antecedent (PTA) or antihemophilic factor C; Factor XII: Hageman factor; and Factor XIII: fibrin-stabilising factor. Also, there are additional compounds such as Prekallikrein or Fletcher factor; high-molecular-weight kininogen or Fitzgerald factor or HMWK and then there are platelets. Numeral VI remains unassigned.

5.32.2 Initiation of Coagulation

Either extrinsic or intrinsic pathways can initiate step 1 of the coagulation cascade.

The extrinsic pathway is triggered when blood contacts damaged non-vascular tissue. (a) The damaged cells secrete tissue thromboplastin (tissue factor) consisting of membranous phospholipids and a lipoprotein complex acting as a proteolytic enzyme. (b) In the presence of calcium ions, the tissue factor complexes with Factor VII activate Factor X (Xa), rapidly interacting with available phospholipids released and Factor V to form a prothrombin activator complex. (c) Prothrombin cleaves into thrombin facilitated by calcium ions and Factor V in seconds.

Another mechanism is the intrinsic pathway, which begins with blood exposure to collagen from an injury to the vasculature. (a) When Factor XII is exposed to collagen, it transforms into a proteolytic enzyme called 'activated Factor XII'. (b) The activated Factor XII activates Factor XI in the presence of HMW (high-molecular weight) kininogen and catalysed by prekallikrein. (c) The activated Factor XI activates Factor IX enzymatically on Factor IX to also activate this factor. (d) The activated Factor IX, acting in concert with which activated Factor VIII, platelet phospholipids, and platelet factor III from the damaged thrombocytes, activates Factor X. (e) The subsequent last step is similar to the intrinsic pathway. These pathways do not occur simultaneously. Instead, thrombin generated by extrinsic pathways is further amplified by intrinsic pathways.

In step 2 of coagulation, prothrombin is converted to thrombin by the prothrombin activator catalysed by calcium ions. The generation of prothrombin activator is the rate-limiting factor in blood coagulation. Platelets have prothrombin receptors; therefore, prothrombin first interacts with these receptors in the event of tissue damage. Hence, the platelet also contributes to the formation of thrombin. Prothrombin is an alpha2-globulin present in the plasma with a molecular weight of 68,700 at a concentration of ~15 mg/dl. Prothrombin is chemically unstable and, therefore, is easily lysed into compounds, one of which is thrombin, with half its molecular weight of 33,700. Prothrombin is synthesised in the liver continuously in the presence of vitamin K. In the case of liver disease or deficiency of vitamin K, the risk of bleeding increases due to low prothrombin levels.

In step 3, soluble fibrinogen is converted to insoluble fibrin. Fibrinogen is a high-molecular-weight protein (MW = 340,000) formed in the liver, maintaining plasma levels between 100 and 700 mg/dl. Normally, due to the large size, small amounts of fibrinogen escape into the interstitial fluids, but no coagulation occurs. However, when the integrity of capillary walls is weak in a disease state, larger amounts may seep and clot the interstitial fluids like blood. Thrombin converts fibrinogen to fibrin by eliminating four low-molecular-weight peptides, and the resultant multiple fibrin monomers polymerize to form fibrin fibres.

In the initial polymerisation phase, the non-covalent hydrogen bonds that bind the fibrin are weak, and the clot is prone to lysis. However, the fibrin-stabilising factor from clot platelets forms covalent bonds increasing cross-links with enhanced strength to the three-dimensional strength of the fibrin meshwork. It also stimulates contractile proteins (thrombosthenin, actin, and myosin), which reduce the clot size and retract. As the clot withdraws, the margins of the ruptured vasculature are brought closer, and bleeding ceases. The levels of fibrinogen are affected by liver disease.

Factor VIII and von Willebrand factor (VWF) are two different but closely linked glycoproteins circulated in a tightly bound complex (FVIII/VWF). Deficiency of these factors or mutations affecting their composition leads to certain hereditary bleeding disorders, namely hemophilia A (HA) and von Willebrand's disease (VWD). The VWF has two main functions: (a) it promotes platelet adhesion to thrombogenic surfaces and other platelets, and (b) it is the transporter for FVIII in plasma. VWF is important for FVIII synthesis, stabilisation, conformation, and immunogenicity. The duo plays a vital role in replacement therapy for patients with HA.

5.32.3 Anticoagulant Mechanisms

The clotting or contact activation that triggers clotting is averted in normal conditions and is attributed to the smoothness of the endothelial cell surface and a layer of glycocalyx lining the endothelium, which wards off platelets and clotting factors. Besides, thrombomodulin on the endothelial surface decelerates the clotting process by eliminating thrombin and activating Protein C and Protein S. (a) Protein C is a serine protease with significant anticoagulant, profibrinolytic, and mitigating characteristics. It inactivates activated Factors V and VIII. (b) (Protein S is a vitamin K-dependent plasma glycoprotein produced in the liver and exists in a free form (cofactor) and a complex form bound to complement protein C4b-binding protein (C4BP). It acts as a cofactor to Protein C in the deactivating Factors Va and VIIIa. (c) Antithrombin III and fibrin are among the most important anticoagulants in the blood and lead to the removal of thrombin from the blood. Fibrin is formed due to the coagulation process, while antithrombin III (also known as antithrombin-heparin cofactor) is an alpha globulin in the blood. In the clotting process, most of the thrombin formed from the prothrombin is absorbed in the fibrin fibres, preventing its escape in circulation. The remaining unabsorbed thrombin interacts with antithrombin III and is inactivated in 12–20 min. (d) Heparin is a negatively charged conjugated polysaccharide anticoagulant, produced by many but primarily by basophilic mast cells and is present in low physiological concentrations. Its combination with antithrombin III increases the thrombin-eliminating potential of antithrombin III by 100–1000 times. This complex also eliminates various other (activated Factors XII, XI, X, and IX) activated coagulation factors acting as effective anticoagulants. (e) Protein Z is a gla domain vitamin K-dependent protein. Therefore, warfarin therapy negatively affects its function. It primarily accelerates (1000 times) the breakdown of Factor Xa performed by Protein Z-related protease inhibitor (ZPI).

5.32.4 Lysis of Blood Clots: Plasmin

The plasma proteins also contain plasminogen (or profibrinolysin) which is activated to plasmin (or fibrinolysin). It is a proteolytic enzyme that can cause fibrinolysis and degrade other protein coagulants (fibrinogen, Factors V, VIII, XII, and prothrombin), leading to clot lysis. This causes hypocoagulability of the blood since the clot formed has a large amount of plasminogen trapped in the clot along with other plasma proteins. After the bleeding ceases, the traumatised tissues and vascular endothelium gradually release tissue plasminogen activator (tPA). Finally, plasminogen is converted to plasmin, which dissolves any balance clots leading to the return of patency in the smaller blood vessels.

5.32.5 Coagulation System in Liver Disease

The liver is the site where most clotting and anticoagulant factors are synthesised; therefore, liver disease like cirrhosis can increase or decrease coagulation via modulation of various factors or intermediate steps. Many hemostatic processes may be abnormal, such as platelet adhesion and activation, generation, and crosslinking of fibrin, and even clot dissolution.

The hepatocyte produces nearly all coagulation factors except Factor VIII, which is synthesised in the endothelial cells, and Factor XIII A-subunit, produced in megakaryocytes. In addition to synthesising coagulation proteins, hepatocytes also lead to post-translational modifications, including glycosylation and gamma-carboxylation of some factors. These functions can be impaired in liver disease, affecting the coagulation cascade. Patients with cirrhosis also have thrombocytopenia and diminished platelet function due to concomitant uremia, infection, and/or endothelial abnormalities. Vitamin K deficiency may coexist, especially in alcoholic patients, and may exacerbate deficiencies of vitamin K-dependent factors (II, VII, IX, and X).

Nearly half the patients with cirrhosis have evidence of systemic fibrinolysis although clinically evident hyperfibrinolysis is seen in 5–10% of patients with decompensated cirrhosis leading to premature clot dissolution. Hypofibrinolysis has also been reported in decompensated cirrhosis, causing spontaneous thrombosis. Hepatocytes and KCs also have the task of clearing coagulation factors, and fibrinolysis remains from the systemic circulation. There are reports to show several coagulation abnormalities in Cirrhosis patients, such as (a) elevated levels of tPA; (b) reduced alpha 2 antiplasmin, coagulation Factor XIII, and thrombin-activatable fibrinolysis inhibitor (TAFI) levels; (c) increased fibrin degradation products such as D-dimer levels; and (d) fibrinolytic activity of ascitic fluid that may be passed into the circulation through the thoracic duct.

Since both procoagulant and anticoagulant factors are equally diminished, it leads to a relatively balanced steady state in most patients. The risk of bleeding and thrombosis is both increased. The risks of bleeding and thrombosis are not reflected in conventional coagulation tests. VWF is often elevated, and the ratio of Factor VIII to protein C carries independent prognostic importance in patients with cirrhosis.

5.33 Urea Cycle and Congenital Disabilities

5.33.1 Introduction

The urea-cycle disorders (UCD) are attributed to the defective metabolism of the excess nitrogen formed by proteolysis and other nitrogen-containing compounds. The excess nitrogen is normally transformed into ammonia (NH_4) and processed via the urea cycle in the liver. This process involves five enzymes in the direct pathway and a single NAGS (Table 5.6), a necessary cofactor.

5.33.2 Clinical Manifestations

Severe deficiency of even one of the initial four enzymes in the pathway (CPS1, OTC, ASS1, and ASL) or the cofactor maker (NAGS) leads to a collection of ammonia and a few antecedent metabolites amid the initial couple of days after birth. As no viable alternative mechanism for the removal of ammonia exists, disturbance of this pathway brings about the quick collection of alkali in the body and the evolution of related adverse effects. Infants with defective urea cycle seem healthy during

Table 5.6 Urea-cycle disorder

Enzyme deficiency	Plasma ammonia	Serum citrullin	Urine–serum argino-succinate	Urine orotic acid	Serum arginine/orthinine
(a) Carbamyl phosphate synthase I (CPSI)	Increased	Reduced	Reduced	Reduced	Reduced
(b) Ornithine transcarbamylase (OTC)	Increased	Reduced	Reduced	Increased	Reduced
(c) Argininosuccinic acid synthetase (ASS)	Increased	Increased	Reduced	Normal or increased	Reduced
(d) Argininosuccinic acid lyase (ASL)	Increased	Increased	Increased	Normal or increased	Reduced
(e) Arginase (ARG) 1	Increased	Increased	Increased	Normal or increased	Increased

childbirth but quickly suffer cerebral edema, unconsciousness, anorexia, hyper- or hypoventilation, hypothermia, seizures, neurologic acting, and trance-like state. Usually, babies are discharged home from the hospital within a couple of days after birth; these manifestations frequently occur at home and may not be perceived by the family domiciliary healthcare worker. The usual initial symptoms are the inability to thrive, maintain thermoregulation with a low central temperature, and sleepiness or sluggishness.

Symptoms advance from sluggishness to unconsciousness. Decerebrate posturing and coma are related to cerebral edema and possibly coning. Seizures may be the only presentation in nearly half the patients. The presentation may be more acute in babies with fused cranial sutures. Hyperventilation due to hyperammonaemia's impact on the brainstem, a typical early finding in hyperammonemia assaults brings about respiratory alkalosis. Subsequently, hypoventilation and respiratory arrest may occur later.

In partial deficiencies of the urea-cycle enzyme, symptoms are usually milder than the neonatal presentation of a UCD. Hyperammonemia, however, can be triggered unexpectedly by other illnesses or certain stressors like prolonged fasting, surgery, or peripartum. Hyperammonemic episodes manifest as anorexia, emesis, fatigue, and behavioural disturbances. One may see encephalopathic (slow-wave) EEG during hyperammonemia, and an MRI scan displays nonspecific brain atrophy.

Defects in the ARG1 enzyme in the pathway result in hyperargininemia, which manifests as an indistinct disorder with neurologic symptoms and neonatal hyperammonemia. Defects in the amino acid transporters (ORNT1 and citrin deficiency) lead to hyperammonemia, and the former is also seldom associated with chronic hepatic dysfunction. Citrin deficiency commonly manifests as hyperammonemia in adolescents or adults; however, even neonates can be affected and present with intrahepatic cholestasis, and slightly older children fail to thrive. Excess ammonia can cause brain damage by causing cerebral edema by elevating glutamine in astrocytes. The neurological manifestations of acute hyperammonemia in infancy are identical to those of hypoxic–ischemic events or stroke. The insular cortex appears to be the most vulnerable area. Long-standing hyperammonemia may also impact the other regions, such as the parietal, frontal, and occipital. T_2-weighted MRI sequences or diffusion tensor imaging can display these abnormalities. Early disease is not reflected in the MRI findings, while clinical abnormalities are observed. Chronic hyperammonemia also causes abnormalities in the alpha-ketoglutarate/glutamate/glutamine ratio, ion gradients, movement of metabolites, neurotransmitters, and mitochondrial function. Acute exacerbations of hyperammonaemia may commonly cause subclinical seizures in neonates and alter cerebral metabolism in an otherwise compromised state.

Previously, neonatal hyperammonemia was associated with a bleak prognosis; however, the prognosis has significantly improved with early diagnosis and newer ammonia-lowering strategies.

5.34 Specific Urea-Cycle Disorders

Carbamoylphosphate synthetase I deficiency (CPS1 deficiency) is an extremely severe form of the urea-cycle disorder that quickly develops hyperammonemia in the neonatal phase. Although children recover with treatment, they are highly susceptible to recurrent episodes of hyperammonemia.

Ornithine transcarbamylase deficiency (OTC deficiency) manifests the same way as in CPS1 deficiency in men, while women only develop mild cognitive deficits being carriers. It is as severe as CPS1 deficiency in males, while females are carriers with minor cognitive defects; however, ~15% are at risk of experiencing hyperammonemia, necessitating therapy.

Citrullinemia type I (ASS1 deficiency): The disorder causes severe hyperammonemia.

Argininosuccinic aciduria (ASL deficiency), **like CPS1**, frequently manifests as rapid-onset hyperammonemia in the neonatal phase, along with hepatic enlargement and abnormalities of hepatic enzymes. The liver biopsy reveals augmented hepatocytes and later fibrosis. Other signs may be trichorrhexis nodosa (fragile hair that appears like a node), but it responds to arginine supplementation. Less severe cases present with developmental defects.

Arginase deficiency (hyperargininemia and ARG1 deficiency). Severe symptoms are less common, while many experience exacerbating spasticity, tremor, ataxia, and choreoathetosis. Retardation of growth is common.

Ornithine translocase deficiency (ORNT1 deficiency, hyperornithinemia-hyperammonemia-homocitrullinuria syndrome) presents at a variable age of onset ranging from infancy to adulthood chronic neurocognitive defects, hyperammonemic crisis, or chronic liver dysfunction.

Citrin deficiency causes neonatal intrahepatic cholestasis (NICCD). Older children as failure to thrive and dyslipidemia (FTTDCD), and adults suffer from recurrent hyperammonaemia with neuropsychiatric symptoms in citrullinemia type II (CTLN2).

N-Acetyl glutamate synthase deficiency (NAGS deficiency) frequently occurs with a clinical presentation similar to CPS1 deficiency, as *N*-acetyl glutamate is a cofactor for CPS1.

5.34.1 Diagnosis

Although the diagnosis is based on clinical evaluation, biochemical investigations, and molecular genetic data, it requires a high index of suspicion.

A three-generation relevant family medical history should be carefully recorded, focusing on neurologic signs and symptoms in pediatric patients. Direct examination of affected family members and other investigations similar to the index patient or autopsy examination may also be required if death has occurred. A familial history comparable with X-linked inheritance is suggestive of OTC deficiency. Physical evaluation usually shows no neurological abnormalities, which may be similar in all defects though trichorrhexis nodosa, when present, suggests a deficiency of ASL, and worsening spasticity of the legs indicates arginase deficiency.

Biochemical tests that suggest the diagnosis of a specific disorder are summarised in the table above (Table 5.6). These are the main methods for coming to a definitive diagnosis.

5.34.2 *Molecular Genetic Testing*

Molecular testing has significantly improved the estimation of enzyme activity as the definitive diagnostic test. These include single-gene testing, multigene panels with the eight genes involved, or more comprehensive genomic testing.

5.34.3 *Newborn Screening*

Tandem mass spectrometry can determine abnormal concentrations of analytes related to ASS1 and ASL deficiencies. CPS1, OTC, arginase, and citrin deficiency screening tests are available.

5.34.4 *Differential Diagnosis of Urea-Cycle Disorders*

One must consider alternative diagnosis in terms of other diseases of the liver and biliary tract (e.g. Herpes simplex virus infection, vascular bypass of the liver, biliary atresia, or acute liver failure) or use of certain pharmacological agents such as valproic acid, cyclophosphamide, or 5-pentanoic acid, which can lead to somewhat similar presentation.

5.34.5 *Treatment*

Once the diagnosis is made, all protein supplementation should be withdrawn in neonates and infants. Ammonia-lowering strategies should be started.

The use of oral lactulose to reduce excess nitrogen is yet to be evaluated in this condition. Since ammonia levels are extremely high, constant arteriovenous haemodialysis or hemofiltration is frequently necessary, but transfusions and peritoneal dialysis are not beneficial. Sodium benzoate and sodium phenylacetate are administered systemically. Oral phenylbutyrate is a more palatable alternative than phenylacetate. Certain amino acid deficiencies may occur, so supplementation is necessary. Low-dose arginine (100 mg/kg/day) rather than 500 mg/kg/day, as recommended in some guidelines, along with an ammonia scavenger, is effective in repleting arginine stores, maintaining low ammonia levels, and minimising liver enzyme elevations compared with high-dose arginine. After the patient improves, protein (0.5–1 g/kg) containing diet with small increments should be introduced if tolerated.

Hyperammonemic coma and a delayed diagnosis are associated with poor prognosis.

Patients with a UCD who deteriorate or fail to improve despite therapy may need orthotopic or auxiliary liver transplantation.

Liver transplantation is preferred prior to the onset of neurological deterioration. Usually, liver transplantation can normalise the metabolic abnormality and stabilise the neurologic status. Hepatocyte transplantation enhances metabolic function and serves as a bridge to auxiliary liver transplantation. Detecting the detrimental mutations in a patient with a UCD cannot be overstated. This reveals data due to which the carrier can be investigated and prenatal diagnosis is feasible, which may guide treatment strategy.

5.35 Branched Chain Aminoacids in Liver Disease

The essential branched-chain amino acids (BCAAs) leucine, isoleucine, and valine are among humans' nine essential amino acids. BCAA levels are found to be low in chronic liver disease. BCAAs participate in the metabolism of all three macronutrients. They stimulate the synthesis of glycogen and albumin by activating the mammalian target of rapamycin (mTOR) signalling, which via the phosphoinositide-3-kinase-protein kinase B (PI3K-Akt) signalling pathway, improves cell growth and proliferation, and modulates insulin physiology.

5.35.1 Immunity

BCAAs, particularly valine, are essential for mitogen-induced lymphocyte multiplication. BCAA supplementation elevates the intrahepatic lymphocytes and activates natural killer (NK) cells and hepatic lectin-dependent cytotoxic activities. Valine also has a critical role in the maturation of DCs and may attenuate carcinogenesis in the hepatic tissue of cirrhotic livers by reinstating immune functions. BCAA supplementation in patients with chronic hepatitis C has been shown to resolve abnormalities of interferon signalling that occur due to malnutrition through the mTOR and FoxO pathways. BCAAs enhance IgA secretion, which may improve the mucosal surface defence. In cirrhosis, one finds decreased serum concentrations of BCAA and elevated concentrations of aromatic amino acids (AAAs phenylalanine and tyrosine), decreasing the Fischer ratio (BCAAs: AAAs). Fischer ratio decreases with cirrhosis and hepatic encephalopathy exacerbation and may have prognostic significance.

5.36 Apoptosis and Regeneration of Hepatocytes

BCAA supplementation delays of CCl4-initiated incessant liver damage in a rodent demonstrated by diminishing hepatic apoptosis. It has been shown to enhance hepatocyte recovery in a rodent hepatectomy model. BCAAs also promote hepatocyte growth factors. Evidence indicates that BCAA supplementation may lessen hepatocyte apoptosis and promote liver recovery and regeneration after liver damage.

5.36.1 Albumin Synthesis

BCAAs, (especially leucine) via the mTOR pathway, improve protein synthesis. Phosphorylation of p70S6 kinase 1 (S6K1) and 4E-binding protein 1 (4EBP1) with an assembly of eukaryotic initiation factor 4E (eIF4E) in mTOR signalling occurs for albumin synthesis. It also triggers the translation of albumin mRNA by importing polypyrimidine-tract-binding protein in the nucleus. BCAA metabolites such as branched-chain α-keto acids (BCKA), β-hydroxy-β-methyl butyrate (HMB), and glutamate regulate protein synthesis. At the same time, HMB participates in muscle protein synthesis.

5.36.2 Insulin Resistance

The activation of mTORC1 performs serine phosphorylation of insulin receptor substrate (IRS)-1 and IRS-2, resulting in insulin resistance. Experimental studies with fat-diet-induced IR demonstrated that the administration of BCAA enhances insulin sensitivity. The mechanism includes stimulation of hepatic glucokinase and GLUT-4 and increased expression of glucose-6-phosphate, which lead to enhanced insulin sensitivity, insulin levels, and improvement in HOMA-IR in patients with chronic liver ailments.

5.36.3 Gene Expression and Mitochondrial Biogenesis

BCAAs attenuate oxidative stress by stimulating the expression of PGC1α or SIRT-1 or by triggering the genes that regulate antioxidant effects; consequently, hepatocyte proliferation increases. It also upregulates the expression of peroxisome proliferator-initiated receptor (PPAR) γ coactivator-1α (PGC-1α), which regulates mitochondrial ROS production and sirtuin-1, protecting the mitochondria. Experimental models have shown improved survival and a lesser risk of hepatic carcinogenesis.

5.37 Therapeutic Implications of BCAA

5.37.1 BCAAs for Liver Cirrhosis

Patients with cirrhosis have hypoalbuminemia, which leads to ascites, and energy deficiency, causing loss of fat and muscle mass, reducing the overall quality of life. Clinical evidence has demonstrated considerable improvement in overall quality of life, reduced weakness, fatigue, and sleep disturbances with BCAA supplementation. In patients with cirrhosis, both compensated and decompensated, it improved glycemic control and maintained albumin levels. It has also shown a reduction in the frequency of various hepatic complications and associated mortality with greater benefits if initiated in the early stages. The effect of BCAA is possibly more marked in the early stages of cirrhosis than in the advanced stage. BCAAs, thus, are a preferred nutrient for cirrhotic patients.

5.37.2 BCAAs for Hepatic Encephalopathy

Hepatic encephalopathy (HE) is a complication in advanced cirrhosis associated with reduced QOL in these patients. Pathogenesis of HE involves increased blood ammonia levels. BCAAs benefit its treatment since they lower the blood ammonia levels and correct the poor Fischer ratio in patients with HE. Gastrointestinal bleeding, a precipitating event, adds Leucine to the gut, but isoleucine is absent in the haemoglobin molecules. A systematic review has shown that BCAAs are beneficial to HE and well-tolerated. A recent study has demonstrated that the supplementation of BCAAs may not attenuate the repeat episode of HE, but HE could be prevented marginally. It could also help in the recovery of muscle mass. Clinical evidence suggests that only oral supplementation improves him, unlike intravenous which does not show a significant difference. It should be considered a second-line treatment of HE after non-absorbable disaccharides, which are more cost-effective than BCAAs but show comparable efficacy. A combination of two molecules may also be effective.

5.37.3 Sarcopenia

Sarcopenia is loss of muscle bulk, strength, and function usually related to old age. In cirrhosis, one finds accelerated development of sarcopenia, indicating an increased frequency of adverse outcomes, including mortality. One of the causes of sarcopenia could be a deficiency of BCAAs. It has been shown that leucine can stimulate mTOR signalling and increase ribosomal protein translation. Clinical evidence suggests that supplementation with leucine-enriched BCAA can reverse the impaired mTOR signalling and increased autophagy in cirrhotic patients. Exercise also improves the benefits of BCAA supplementation since it upregulates amino acid transporters in the skeletal muscle, and, therefore, a combination of both may be more beneficial.

5.37.4 Insulin Resistance

Insulin resistance is a common finding in patients with chronic hepatitis C virus (HCV) infection, which is related to steatosis, abnormal glucose metabolism, and carcinogenesis in the liver. BCAAs improve glycemic status by acting on insulin-related organs such as adipose tissue, skeletal muscles, and the liver. Long-term BCAA administration can improve glycemic control in patients with nonalcoholic steatohepatitis (NASH)-associated cirrhosis. Clinical evidence also demonstrates that BCAA treatment in patients with HCV reduces HbA1c levels and improves insulin resistance.

5.37.5 BCAAs for HCC

Chronic inflammation and fibrosis increase the risk of HCC, as do chronic hepatitis B virus (HBV) or HCV infection. Viral inhibition significantly reduces the risk of HCC but does not eliminate the risk. Increased Fisher ratio suppresses HCC cell proliferation in vitro. BCAAs reduce the expression of insulin-like growth factor-1 (IGF-1) receptors on HepG2 cells, resulting in insulin-mediated multiplication of HepG2 cells. With elevated insulin levels, BCAAs reduced the expression of vascular endothelial growth factor (VEGF) by HepG2 cells by trimming VEGF mRNA. BCAAs increase the sensitivity of the tumours to 5-fluorouracil by decreasing the expression of a cancer stem cell marker, epithelial cell adhesion molecule, via the mTOR pathway. It also enhances the antitumor benefits of acyclic retinoid or sorafenib. Thus, these data suggest that adjunctive BCAAs with antineoplastic agents produce synergistic effects.

5.38 Albumin, Prealbumin, and Other Biochemical Measures of Nutritional Status

Traditionally, serum proteins which are negative acute-phase proteins such as albumin and prealbumin (i.e. transthyretin) indicate the nutritional status of patients. Other nutritional status markers that are investigated include retinol-binding protein (RBP), transferrin, and total cholesterol. C-Reactive protein (CRP) and total lymphocyte count (TLC) are indicators of systemic inflammation (Table 5.7).

Albumin is a protein that circulates in the serum; its half-life is 14–20 days. It is an important carrier of various minerals, hormones, and fatty acids. It also preserves the capillary oncotic pressure. Greater than half of its total quantity lies in the extravascular compartment, with a negligible 5% being synthesized daily by the liver. Therefore, daily protein intake rarely affects albumin levels. Thus, a patient's protein consumption in a day hardly affects the patient's albumin level. However, in several inflammatory pathological states, certain medicinal agents metabolised by the liver or those conditions (hepatic failure, burns, sepsis, and trauma) affecting hepatic physiology influence its levels. Albumin levels can diagnose/predict malnutrition better than BMI, particularly in patients undergoing surgeries (cardiac transplant and orthopaedic), as corroborated in a study in cardiac transplant patients.

Prealbumin (PAB) is also synthesised in the liver; therefore, its levels are impacted in inflammatory states such as infections involving the hepatic tissues. It has a shorter half-life than albumin and is present in quite lower quantities than the latter. It also is an indicator of nutritional status, but since it is also modified and degraded by the renal tissue, impairment of kidneys can also alter its levels similar to the liver.

Transferrin can also indicate nutritional status. However, false positives or negatives can occur due to its role in iron metabolism. Its levels are elevated in iron deficiency and reduced in overload. In iron-deficiency states (including chronic blood loss anemia), the transferrin levels are elevated due to increased iron absorption. Consequently, the levels are decreased in iron-overload states. Besides a study that compared its level in individuals with varying body weight,

Table 5.7 Some characteristics of plasma proteins

Protein	Half-life	Conditions
Albumin	20 days	Low in malnutrition, also in infection, burns, fluid overload, hepatic failure, cancer, and nephrotic syndrome
Transferrin	10 days	Low in protein energy malnutrition, but also affected by iron status
Prealbumin	2–3 days	Low in malnutrition, also in infections. Liver failure and increased in renal failure
C-reactive protein	Positive acute phase reactant. Helos determine whether above proteins are reduced because of inflammatory process or due to inadequate substrate, as in malnutrition	

no correlation was observed between fat-free mass and transferrin levels. Hence, its utility as a malnutrition marker needs to be investigated.

Retinol-binding protein (RBP) is primarily part of the retinol-circulating complex. Since its appropriate physiological actions need vitamin A and zinc, deficiency of these negatively affects the levels and effects of RBP. Complete degradation of RBP occurs only in the kidneys, so any renal impairment elevates serum levels of RBP. Very low-calorie diets (<500 kcal/day) over 20 days significantly decrease PAB and RBP while albumin is unaffected. Therefore, given the quick turnover, these two could be better markers of nutritional status.

TLC could be a useful marker for nutritional status since it varies with the severity of malnutrition. The TLC levels below 1500/mm^3 indicate malnutrition, while lower levels <900/mm^3 suggest its severe status.

5.39 Necroinflammatory and Fibrosis Scores

A liver biopsy confirms the diagnosis and determines the extent of disease severity. The latter is described in two different terms:

(a) Stage: the extent of disease dissemination.
(b) Grade: reflects the rapidity of disease spread.

Histopathologic features of liver biopsies considered for grading and staging by looking at

(a) Hepatocellular changes
(b) Inflammation in the periportal area
(c) Biliary changes
(d) Fibrosis and architectural changes

There are several systems of classifying the severity of liver disease. Common staging systems are

(a) Metavir staging (Fig. 5.12)
(b) Ludwig's score
(c) Modified Ishak's score—complex numerical score
(d) NASH activity score and staging system—complex numerical score

Grading of activity is based on features of the activity, namely (a) degree of interface hepatitis (interface hepatitis)—minimal, mild, moderate, and marked; and (b) parenchymal injury—mild, moderate, and marked.

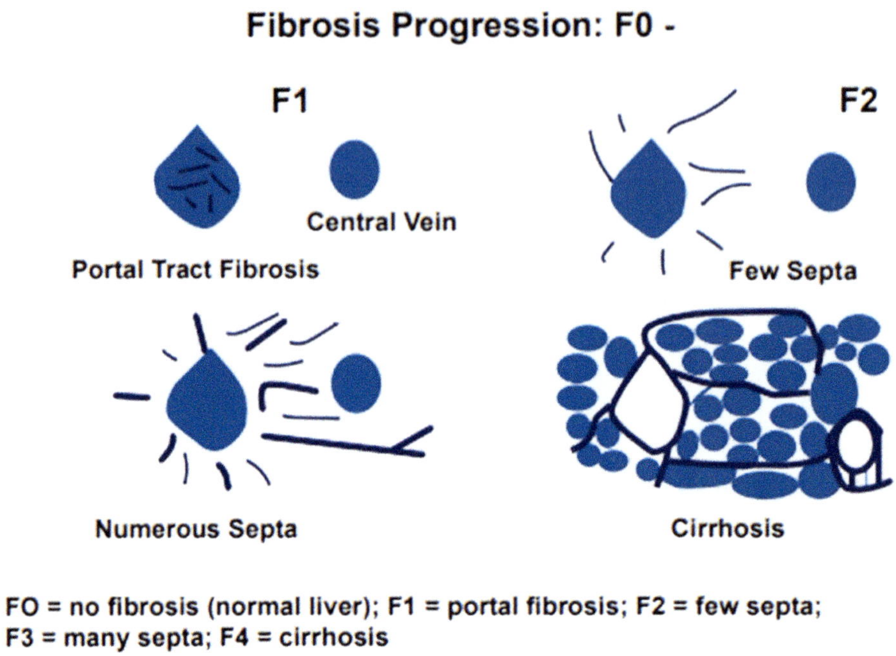

Fig. 5.12 Concept of staging fibrosis

Table 5.8 Ludwig's grading of interface hepatitis

Terminology Semiquantitative	Terminology Descriptive	Lymphocytic Piecemeal Necrosis	Lobular inflammation and necrosis
0	Portal inflammation only, no activity	None	None
1	Minimal	Minimal, patchy	Minimal, occasional spotty necrosis
2	Mild	Mild, involving some or all portal tracts	Mild, little hepatocellular damage
3	Moderate	Moderate, involving all portal tracts	Moderate, with noticeable hepatocellular change
4	Severe	Severe, may have bridging fibrosis	Severe, with prominent diffuse hepatocellular damage

Numerical scores are useful for statistical analysis of liver biopsies obtained in clinical trials and are accurately reproducible to be relevant for managing individual patients. Numbers are ordered categories rather than values. The algorithm for grading hepatitis activity-(Ludwig) is presented in Table 5.8.

Similarly, fibrosis staging is also presented in Table 5.9. A guide to fibrosis assessment is presented in Table 5.9.

Table 5.9 Comparison of Ludwig and Metavir scoring for fibrosis assessment

Definition	Ludwig	Metavir
No fibrosis	Stage 0	F0
Fibrous portal expansion	Stage 1	F1
Few bridges or septa	Stage 2	F2
Numerous bridges	Stage 3	F3
Cirrhosis	Stage 4	F4

5.40 Pathogenesis of NAFLD

The nomenclature of non-alcoholic fatty liver disease (NAFLD) is in flux. There have been attempts to change it to metabolic dysfunction associated fatty liver diease (MAFLD) in 2020 and to metabolic dysfunction associated steatotic liver disease (MASLD). Till the dust settles down, we will use the conventional nomenclature i.e. NAFLD. This disease i.e. NAFLD involves a multitude of abnormalities, including mild steatosis to steatohepatitis and cirrhosis. While NAFLD is called nonalcoholic, a small amount of alcohol intake is considered acceptable as non-significant for this diagnosis. It has varied between various studies. AGA technical report on NAFLD uses 20–30 g/day. The estimated prevalence of NAFLD is 15–30% in various countries across the globe. In a large community-based study from Mumbai, the prevalence of NAFLD based on the above ultrasound was 18.9% in adults over 20 years.

NAFLD is usually associated with metabolic syndrome, which includes diagnosis of any three of the listed abnormalities:

1. Hypertension [>130/85 mmHg]
2. Impaired glucose tolerance [>110 mg/dl]
3. Hypertriglyceridaemia [S. triglycerides >200 mg/dl]
4. Low HDL [HDL <40 mg/dl in males; <35 mg/dl in females]
5. Waist circumference [>90 cm in males; >80 cm in females]

Among patients with NAFLD, the prevalence of these risk factors is high. Obesity may be present in 60%, Hypertriglyceridaemia in 50% and diabetes mellitus in over 20%

5.40.1 Pathogenesis of NAFLD

An earlier two-hit hypothesis was used to explain NAFLD. Steatosis indicates the 'first hit', which predisposes the liver to injury intervened by 'second hits', which include inflammatory cytokines. However, even deposits of FFA alone can cause liver damage in the absence of or need of a second hit. However, we have realized that liver damage may be multifactorial. Lipid deposition is associated with insulin resistance, mitochondrial dysfunction, and oxidative stress resulting in hepatocyte lipotoxicity. Also, there exists immune-mediated inflammation. These factors activate the HSCs due to signals from damaged hepatocytes and macrophages. This forms the basis of worsening NASH. There are significantly diverse components of NAFLD. We may need to segregate it depending on the predominant characteristics:

ER stress, apoptosis, macrophage, or B-cell-predominant NASH. Additionally, each type could have varying activation of HSCs. It also suggests that a combination therapy rather than a single approach may be needed to treat it.

Pathogenic factors that need to be addressed while planning therapy include

(a) **Intake determinants**: This includes hunger/satiety mechanism, eating behaviour, genetic/epigenetic factors, food availability, and food desirability.
(b) **Energy consumption**: The above factors determine the intake of sugars, fats, and proteins.
(c) **Energy absorption**: The mechanisms that allow energy extraction by the digestive system for absorption.
(d) **Extrahepatic diversion**: The handling of fat by the body is an important determinant. Extrahepatic consumption of fatty acids by exercising muscles, oxidation in brown adipose tissue, and factors determining its storage in white adipose tissue.
(e) **Liver uptake and processing**: Sugars and fatty acids are taken up from the blood. Excess energy availability over and above consumption initiates insulin resistance. Fatty acid undergoes beta-oxidation and conversion to triglycerides and the production of lipotoxic products such as diacylglycerols (DAGs), lysophosphatidylcholine (LPC), ceramides, lysophosphatidic acid (LPA), and omega oxidation products.
(f) **Liver stress response**: The above mechanisms result in ER stress, tissue restoration, programmed cell death followed by cytokine and chemokine release, and stellate cell activation.
(g) **Liver outcomes**: Persistent Inflammation, hepatocyte ballooning and damage, and fibrogenesis response lead to cirrhosis and associated genomic instability leads to HCC.

5.41 Wilson's Disease

5.41.1 Introduction

Wilson's disease is a rare autosomal recessive hereditary copper metabolism disorder characterized by the liver and neurological abnormalities. The manifestations usually surface between the age of 20 and 30 years since the un-excreted excess copper accumulates in large quantities.

5.41.2 Hepatic Pathology

In the initial disease stage, copper is visible in the diffused form in the hepatocyte cytoplasm with immunohistochemical staining, manifesting as macrosteatosis, microsteatosis, and glycogenated nuclei. There is expansion and segregation of mitochondrial internal and external membranes, with increased intercristal spaces, elevation in the density and granularity of the matrix, or the formation of bigger vacuoles. These signs suggest Wilson's disease, especially when no cholestasis is observed.

Periportal inflammation, mononuclear cellular infiltrates, rupture of the limiting plate, necrosis of hepatic lobes, and bridging fibrosis are some features of the intermediate stage common to autoimmune hepatitis. Half of the biopsy samples also display Mallory's body. These features are warning signs of impending cirrhosis with either a micronodular or a mixed macronodular–micronodular histological pattern. In patients with fulminant hepatic failure, cirrhosis is accompanied by programmed cell death of parenchymal cells, cell necrosis, and death. The histochemical comfy ratio of excess copper may aid in the diagnosis. However, its absence does not suggest the absence of Wilson's disease.

5.41.3 Molecular Pathogenesis

The gene attributed to Wilson's disease (on chromosome 13) (ATP7B) is widely present in hepatic, renal, and placental tissues. This gene regulates a transmembrane protein, a copper-dependent P-type ATPase. The ATP7B transporter transfers the copper into the trans-Golgi compartment, to be integrated with ceruloplasmin, and into the bile to eliminate excess stores. Therefore, defects in ATP7B cause the accumulation of copper in the liver resulting in various features of Wilson's disease.

5.41.4 Clinical Applications of Wilson's Disease Genetics

As per the Human Genome Organisation (HUGO) database, the gene ATP7B demonstrates ~300 mutations at different loci, with most patients diagnosed with the heterozygous disease.

5.41.5 Clinical Manifestations and Range of Disease

The clinical range of Wilson's disease is wide, and knowledge of the various disease presentations is important. In broad terms, patients can present acutely with liver failure, haemolysis, or both, or more chronically with liver disease, neurological disease, or both. Manifestations can be hepatic, neurological, ophthalmic, psychiatric, or other systems.

Hepatic manifestations include:

(a) Constant elevation of serum transaminases
(b) Chronic hepatitis
(c) Fulminant hepatic failure (with or without haemolytic anemia)
(d) Decompensated or compensated cirrhosis

Neurological manifestations will include:

(a) Tremor
(b) Choreiform movements
(c) Parkinsonism or akinetic rigid syndrome—i.e. partial
(d) Parkinsonism
(e) Gait disturbances
(f) Dysarthria
(g) Pseudobulbar palsy
(h) Rigid dystonia
(i) Seizures
(j) Migraine headaches
(k) Insomnia

Ophthalmic findings may be Kayser–Fleischer (K–F) rings or Sun flower cataracts

Psychiatric syndromes are:

(a) Depression
(b) Neuroses
(c) Personality changes
(d) Psychosis

Renal abnormalities such as aminoaciduria and nephrolithiasis could manifest rarely.

5.41.6 Diagnosis

Currently, there is no single test for the diagnosis of Wilson's disease. The modified Leipzig Scoring System is generally used, as shown in Table 5.10. Liver abnormalities accompanied by movement issues new in onset should raise suspicion for Wilson's disease. Along with routine assessments, serum copper, ceruloplasmin, and 24-h urinary copper excretion should also be evaluated. However, quantitative estimation using liver biopsy is the best alternative but is not commonly used. Another method is K–F rings using slit-lamp examination. Genetic diagnosis should be considered when clinical and biochemical investigations are inconsistent, along with screening remaining family members.

5.41.7 Treatment

The pharmacological treatment includes copper chelators to increase the excretion of copper or the use of zinc to limit absorption, promote copper excretion from the body, or zinc to reduce copper absorption or both (Table 5.11).

Table 5.10 Modified Leipzig Scoring System is used for the diagnosis of Wilson's disease

Features	Evaluation	Points allotted
K–F rings	Present	2
	Absent	0
Serum ceruloplasmin normal (>20 mg/dl)	0–5 mg/dl	3
	6–11 mg/dl	2
	11–20 mg/dl	1
24-h urinary copper (in the absence of acute hepatitis)	>100 mcg	2
	40–100	1
Coomb's-negative hemolytic anemia with liver disease	Present	1
	Absent	0
Mutational analysis	On both chromosomes detected	4
	On one chromosome detected	1
	No mutation detected/test not done	0
Liver biopsy for histology S/O WD	Orcein- or rhodanine-positive granules	1
Neurobehavioral symptoms		2
Typical features on MRI brain		1
History of Wilson's disease in a family member	Sibling death from liver disease/neurological disease suggestive of WD	1
The total score of points allowed	*Evaluation*	
≥4 or more	Diagnosis established	
3	Diagnosis is possible, but more tests are needed	
≤2	Diagnosis very unlikely	

K–F Kayser–Fleischer, *MRI* magnetic resonance imaging, *S/O* suggestive of, *WD* Wilson's disease

Penicillamine

Penicillamine has immunosuppressant effects, chelating action, and also suppresses collagen cross-linking. Although the neurologic status of patients, especially those with neurologic symptoms, deteriorates during the initial phase, most recover with prolonged use. Skin rashes, pemphigus, and elastosis perforans serpiginosa have been reported along with mild loss of taste, gastrointestinal symptoms, and arthralgias to severe proteinuria, leukopenia, and thrombocytopenia. Aplastic anemia may occur rarely and has a low probability of reversal on penicillamine withdrawal. Some other conditions reported include nephrotic syndrome, Goodpasture's syndrome, and myasthenia.

Trientine

Trientine is an alternative to penicillamine. Trientine elevates urinary excretion, reduces intestinal absorption of copper, and has fewer tolerability issues except for occasional gastritis and iron deficiency. Unlike penicillamine, bone marrow suppression occurs rarely.

Table 5.11 Treatment for Wilson's disease

Penicillamine (+ pyridoxine, 25 mg daily)		Trientine		Zinc	
Dose					
Initial: 1–1.5 g/day (adults) or 20 mg/kg/day (children) divided into two or three doses	Maintenance: 0.75–1 g/day as needed to maintain cupriuresis	Initial: 1–1.5 g/day divided into two or three doses	Maintenance: same	Initial: 50 mg elemental zinc three times daily (adults)	Maintenance: titrate dose against efficacy monitoring data
Tests for monitoring efficacy					
24-h urinary copper: 200–500μg (3–8 mmol) per day as the target; estimated non-ceruloplasmin	Bound copper <200 mg/l	24-h urinary copper: 200–500 μg (3–8 mmol) per day as the target; estimated non-ceruloplasmin	Bound copper <200 mg/l	24-h urinary copper: <75 μg (1.2 mmol) per day as the target	Estimated non-ceruloplasmin-bound copper <100 mg/l
Tests for monitoring side effects					
Complete blood count; urinalysis; skin examination		Complete blood count; iron studies		Serum zinc level	

Zinc

Zinc restricts GI absorption, increases copper excretion, and may cause gastritis.

Ammonium tetrathiomolybdate: The tetrathiomolybdate obstructs copper absorption and chelates it with high affinity. It is reserved for treating advanced neurologic Wilson's disease and does not lead to neurologic deterioration even in the early phase.

5.41.8 Approach to Treatment in Wilson's Disease

Diet

Copper-containing dietary components should be avoided in the initial year of treatment. Such food items include shellfish, mushrooms, meats, and chocolates. If high levels of copper are suspected in the water, appropriate measures to eliminate it are necessary.

Pregnancy

Treatment should be maintained throughout for pregnant patients with Wilson's disease due to the risk of hepatic decompensation postpartum if treatment is withdrawn.

Family Screening

Family screening is an important preventive measure. Screening the patient's siblings is mandatory when Wilson's disease is diagnosed because the best outcome is associated with the treatment that begins in the presymptomatic period.

Liver Transplantation for Wilson's Disease

In patients with Wilson's disease associated with acute fulminant hepatic failure and those intolerant or refractory to medical therapy, liver transplantation is the last recourse. Individuals with liver failure, severe cerebral oedema, long-term reduction of cerebral perfusion pressure, active infections, malignancies, or those with the severe psychiatric disease are not candidates for transplant. Similarly, patients with chronic neurological impairment are improbable to gain from transplantation.

Prognosis

Early treatment initiation helps in living a normal age span and improved quality of life, while the absence of treatment may be fatal by the fourth decade of life.

5.42 Hemochromatosis

5.42.1 Definition

Excess iron deposition in various iron-storing organs, including the liver, occurs in hemochromatosis.

5.42.2 Types of Hemochromatosis

Primary Hemochromatosis [Hereditary Hemochromatosis (HH)]

HH includes various inherited disorders of iron homeostasis, which involves elevated levels of iron absorption leading to an excess in tissue iron deposition, the most common being HFE-related HH. The liver is affected in symptomatic HH. Non-HFE-related HH are listed in Table 5.12.

Table 5.12 Classification of hemochromatosis

Hereditary hemochromatosis	Secondary iron overload
HFE-related HH	Iron-loading anemias
• C282Y homozygosity • C282Y/H63D compound heterozygosity • Other HFE mutations	• Thalassemia major • Sideroblastic anemia • Chronic hemolytic anemia • Aplastic anemia • Pyruvate kinase deficiency • Pyridoxine-responsive anemia
Non-HFE-related hereditary hemochromatosis	Chronic liver disease
• Hemojuvelin (HJV) mutations • Hepcidin (HAMP) mutations • Transferrin receptor 2 (TfR2) mutations • Ferroportin 1 (SLC40A1) mutations • African iron overload	• Hepatitis C • Hepatitis B • Alcoholic liver disease • Nonalcoholic steatohepatitis • Porphyria cutanea tarda • Portacaval shunt
	Parenteral iron overload
	• Red blood cell transfusions • Iron-dextran injections • Long-term haemodialysis
	Dietary iron overload
	Dysmetabolic iron overload syndrome

Secondary Hemochromatosis

When the cause of excess iron deposition is other than genetic, these are grouped under secondary hemochromatosis as listed in Table 5.12.

5.42.3 Two Additional Forms of the Disease

Juvenile Hemochromatosis

It is associated with extreme iron overload and hepatic and cardiological abnormalities between 15 and 30 years. The HJV gene on chromosome 1q that codes hemojuvelin and the hepcidin gene (HAMP) that down-regulates iron absorption are mutated.

Neonatal Hemochromatosis

It causes rapid iron accumulation in the hepatic tissue of the neonate, which may be fatal.

5.42.4 Pathophysiology

HH may occur via three mechanisms:

1. Dysfunctional HFE protein
2. Elevated absorption of dietary iron in the intestine
3. Tissue injury and fibrogenesis due to iron

The primary detrimental effects of HH occur in the liver since it may induce iron-associated lipid peroxidation, causing hepatocyte damage or death. The KCs may be activated by profibrogenic cytokine release by damaged hepatocytes which subsequently stimulate HSCs to produce collagen, further worsening the fibrosis.

5.42.5 The Population at Risk

The disease is rare in Indians and uncommon in African Americans, Hispanics, Asians, and American Indians, but it is the commonest genetic disease in the United States and Caucasians of Northern European descent. Younger children seldom develop hemochromatosis. Usually, it manifests in older people, and women manifest signs and symptoms post-menopause. Those with alcoholism, familial history of cardiological ailments, liver disease, diabetes, arthritis, and erectile dysfunction (impotence) are at risk.

5.42.6 Symptoms

The predominant symptoms are joint pain, fatigue, lack of energy, abdominal pain, loss of libido, and cardiological issues. Undetected disease increases the risk of:

- Arthritis
- Liver disease, including hepatomegaly, cirrhosis, malignancy, and hepatic failure
- Pancreatic damage increases the risk of diabetes
- Arrhythmia or congestive heart failure
- Impotence
- Early menopause
- Hyperpigmentation of skin which makes it appear grey or bronze
- Thyroid deficiency
- Adrenal gland damage

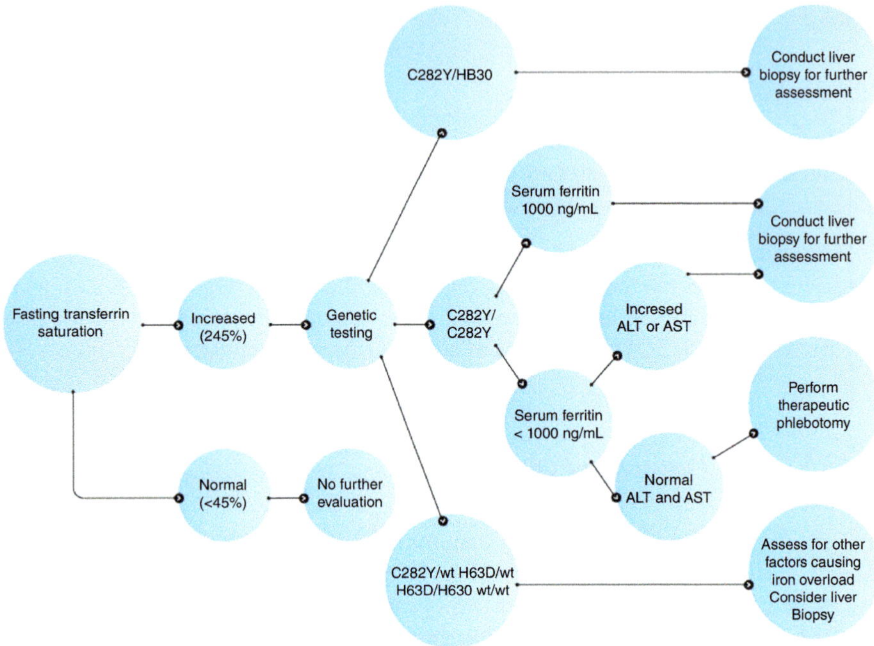

Fig. 5.13 Algorithm for evaluating possible HFE-related hereditary hemochromatosis in a person with a negative family history. *ALT* alanine aminotransferase, *AST* aspartate aminotransferase, *wt* wild-type

5.42.7 Tests and Diagnosis

Blood Tests

Serum transferrin saturation values >45% are indicative of HH. Serum ferritin is elevated significantly. However, it is a non-specific marker since several other non-hereditary (infectious and inflammatory) conditions also lead to elevations (Fig. 5.13).

Additional Testing

Testing for gene mutations in the HFE gene can aid in confirming an HH diagnosis.

A liver biopsy reveals the presence of iron, scarring or cirrhosis but has a risk of bruising, bleeding, and infection.

Screening healthy people for hemochromatosis: People at risk should undergo screening for early diagnosis that can avert complications.

5.42.8 Treatment

Treatments for hemochromatosis include

- Dietary modifications
- Therapeutic phlebotomy
- Iron chelation therapy
- Managing complications

Goals of Treatment

- Control the body's iron content and restore it to normal levels for life.
- Iron overload should be prevented so that organ damage is delayed.

Therapeutic Phlebotomy

- Perform phlebotomy of 500 ml of whole blood weekly to decrease hematocrit below 37%.
- Monitoring transferrin saturation and ferritin levels every 2- to 3-month.
- Change schedule from once weekly to once in 2–3 months when iron stores are depleted (ferritin <50 ng/ml; transferrin saturation <50%) and continue to maintain these levels.

Iron Chelation Therapy

Dietary Changes

- Avoid taking iron supplementation in any form and vitamin C that enhances the absorption of dietary iron.
- Avoid uncooked fish and shellfish.
- Limit alcohol intake.

Treatment for Complications

- Complications due to HH should be treated appropriately.

5.42.9 Prevention

- People with hemochromatosis or family history and prospective parents should opt for genetic testing and counselling, which may reveal any mutation in HFE genes.

5.43 Glycogen Storage Disorders

Glycogen storage disorders are inherited metabolic diseases with varied manifestations from infancy to childhood. They have inborn errors of metabolism (IEM) of carbohydrate metabolism, usually attributed to single gene mutations encoding particular proteins. IEMs should be contemplated as a differential diagnosis in a sick neonate. These errors affect the metabolism of carbohydrates, fats, and proteins, which accumulate in the tissue to toxic levels, as in glycogen storage disorders (Fig. 5.14). Generally, all GSDs cannot store or utilise glycogen. Excess glycogen is stored in the cytoplasmic glycogen vesicles in the liver (10%) and the muscle (2%) after the metabolism of dietary glucose. Since the total muscle mass is greater than the liver mass, glycogen storage is double that of the liver. When required, the glycogen polymer is metabolised into glucose monomers and utilised. Several GSDs have been detected although some are extremely rare. Earlier named after the discovering physician, GSDs now follow a different nomenclature, as depicted in Table 5.13.

5.43.1 Etiology

The GSDs are attributed to gene mutations encoding enzymes and transporters necessary for glycogenesis or glycogenolysis (see Fig. 5.14). Table 5.13 shows various common types of GSD, although many more exist. The reported frequency is 1 in 20,000 and 1 in 40,000, primarily from Western regions. Considering the probability of mild disease manifestations or asymptomatic nature and sudden neonatal death, many cases could be undetected and unreported; the reported frequency is possibly underestimating.

5.43.2 Diagnosis

A high index of suspicion is necessary due to non-specific and sometimes mild manifestations. A liver or the muscle biopsy is necessary to measure glycogen content, suitable enzymatic assays, and blood and urine investigations. In older individuals, urine should be assessed for aminoaciduria, proteinuria, microalbuminuria, uric acid, and calcium. Blood tests should include glycaemic and lipid parameters, anion gap, serum lactate, serum uric acid, liver function tests, calcium, phosphorus, urea, and creatinine levels, along with a complete blood count and an MRI/CT scan for adults and ultrasonography for children.

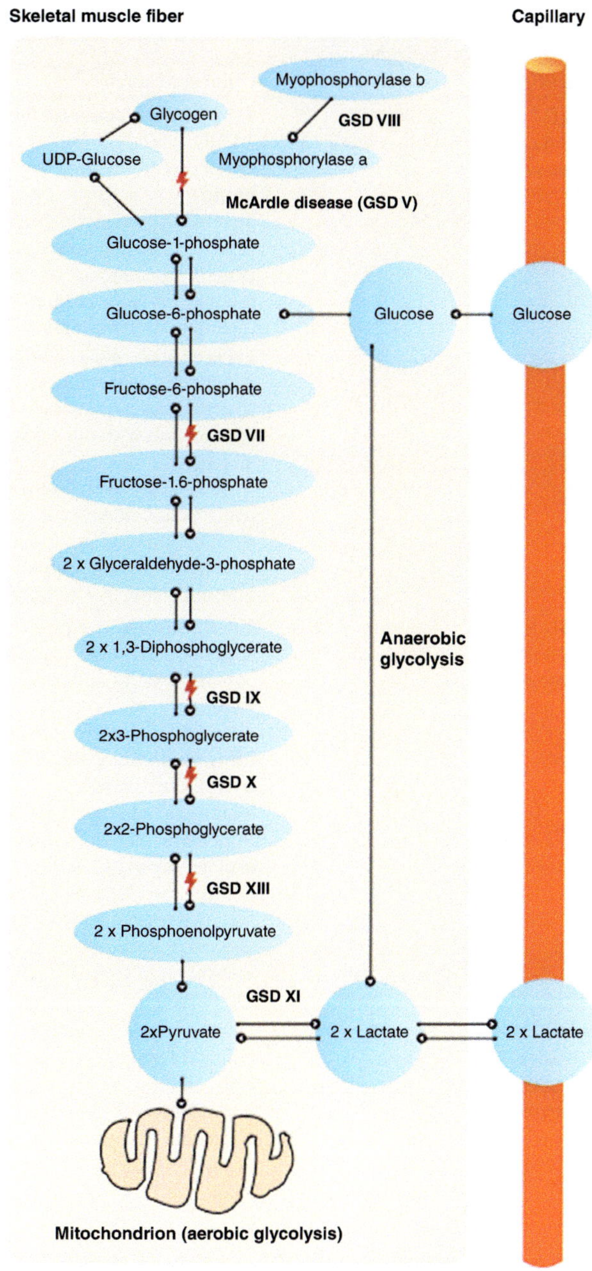

Fig. 5.14 Common glycogen metabolism and abnormalities in GSD

Table 5.13 Common types of glycogen storage disorders

Disease	Enzyme deficiency	Clinical features	Treatment
GSD I (Von Gierke's disease)	Glucose-6-phosphatase	Type Ia: most common type 1 GSD (>80%) Onset: before the first birthday Severe hypoglycaemia, lactic acidosis, and hepatomegaly later, hepatic adenomas, renomegaly with progressive renal insufficiency and hypertension, short stature, hypertriglyceridemia, hyperuricemia, platelet dysfunction with epistaxis, and anemia Type Ib: less severe but with neutropenia, neutrophil dysfunction with recurrent infections, and inflammatory bowel disease	Uncooked cornstarch 1.5–2.5 g/kg po q 4–6 h or lactose-free formula with maltodextrin to maintain normoglycaemia; nocturnal feedings important; fructose and galactose restriction; for lactic acidosis, bicarbonate 0.25–0.5 mmol/kg qid; allopurinol to keep uric acid to <6.4 mg/dl; liver and kidney transplantation; G-CSF for type Ib patients with neutropenia
GSD II (Pompe's disease)	Lysosomal acid α-glucosidase	Onset: infancy, childhood, or adulthood; residual enzyme activity in child and adult forms *Clinical features*: in infantile form, cardiomyopathy with heart failure, severe hypotonia, and macroglossia In juvenile and adult forms, skeletal myopathy with delayed motor development and progressive peripheral and respiratory muscle weakness; in type IIb, mental retardation	Treatment: none known; for cardiomyopathy, heart transplantation
GSD III (Forbes' disease, Cori's disease)	Debrancher enzyme	Onset: infancy or childhood Clinical features: in type IIIa, liver, and muscle involvement with features of types Ia and II; in type IIIb, involvement of the liver only with features of Ia	Treatment: uncooked cornstarch and continuous feeding to maintain normoglycaemia, high-protein diet to stimulate gluconeogenesis

Table 5.13 (continued)

Disease	Enzyme deficiency	Clinical features	Treatment
GSD IV (Andersen's disease)	Branching enzyme	Onset: early infancy Clinical features: hepatomegaly with progressive cirrhosis; hypoglycaemia, esophageal varices, and ascites; splenomegaly; failure to thrive; in neuromuscular forms, hypotonia, and muscle atrophy	Treatment: none known; for cirrhosis, liver transplantation
GSD V, VII	McArdle disease, Tarui's disease respectively	Onset: adolescence or early adulthood Clinical features: exercise intolerance due to muscle cramps, rhabdomyolysis	Carbohydrate administration before exercise, high-protein diet
GSD VI	Hers disease Hepatic phosphorylase deficiency or related enzymes deficiency	Like type-1, although milder	
GSD VIII	Phosphorylase kinase deficiency	Mild liver enlargement and hypoglycaemia	

5.44 HCC-Molecular Pathway

Cirrhosis may predispose 80% of the patients to HCC, attributed to various environmental risk factors. These include genetic and epigenetic modifications, chromosomal abnormalities, mutations, and modified molecular pathways. Hepatic carcinogenesis is a multi-step process and involves phenotypes such as uncontrolled expression of growth factors, refractoriness to growth inhibition, and resistance to apoptosis. Limitless replication, sustained angiogenesis, and propensity for infiltration and metastasis. Persistent inflammation may be the main driver of carcinogenesis as inflammatory cytokines such as TNF-α activate NFκB and induce c-Jun and c-jun amino-terminal kinase (JNK). In turn, NFκB activates the anti-apoptotic genes *A1/Bfl1*, c-*IAP1*, *GADD45*β, and Jun and JNK cooperate to inhibit p53-induced apoptosis. Around 30–40% of HCCs have mutations in Wnt/Catenin signalling, which is required for para and autocrine growth regulation.

Similarly, sites of hepatitis B DNA integration often lead to the suppression of tumour suppressor genes such as FHIT, WWOX, and Parkin. Tumours are known to secrete several angiogenic factors. Different pathways are activated in different stages of carcinogenesis. Different aetiologies take different routes to the same end. Knowledge of these molecular pathways has enabled man to develop drugs to inhibit these molecular pathways to treat HCC. Several multikinase, angiogenesis, and checkpoint inhibitors are already available for clinical treatment, while several others are in the pipeline.

5.45 Hepatic Fibrogenesis

Significant advances in treating chronic liver disease have renewed interest in unravelling the pathogenesis of hepatic fibrosis. Activating the resident HSCs to transform into a proliferative, contractile, and fibrogenic mode during hepatic injury still drives the research in this area.

Studies defining the cellular sources of extracellular matrix (ECM) and the roles of cytokines and paracrine interactions among resident liver cells and inflammatory cells have yielded a more nuanced understanding of how the liver responds to injury. Fibrogenesis is an evolving event with different parts of the liver moving forward at different speeds towards a common endpoint, i.e. cirrhosis.

5.45.1 Initiating Events

Hepatocyte damage and initiation of inflammation are the usual initiating events. Quiescent Stellate cells are activated via growth factors successively released by inflammatory cells. Activation involves swift stimulation of the β-PDGF receptor and forming a contractile and fibrogenic phenotype. The initial stimulus involves paracrine signals, such as reactive oxygen species from injured hepatocytes and cholangiocytes. The release of chemoattractant PDGF, monocyte chemoattractant protein-1, and CXCR3 ligands by the injured tissue regulates the migration of stellate cells. This PDGF-stimulated chemotaxis triggers cell spread at the edge, followed by the cell body and the rest of the projections associated with temporary myosin phosphorylation. Adenosine can retard chemotaxis and, therefore, is a counter-regulatory mechanism. Paradoxically, enhanced adenosine signalling stimulates stellate cell fibrosis leading to alcoholic fibrosis. This also substantiates the beneficial effect of caffeine, which suppresses adenosine production.

5.45.2 Changes in ECM Composition

Activation of stellate cells in the space of disse (SOD) is associated with increasing alterations in the ECM environment. ECM normally comprises loose type IV collagen, heparan sulphate proteoglycan, and laminin transforms to fibril-forming collagens, especially types I and III. Deposing cellular fibronectin from sinusoidal endothelial cells activates stellate cells in the presence of TGF-β. Such modifications in the content of ECM aggravate fibrosis via positive feedback. (a) The integrin in the membrane detects abnormal matric signals and interacts with RHO and Rac, which are proteins attached to the membrane. These proteins signal the actin cytoskeleton to enhance migration and contraction. (b) The activated cellular matrix metalloproteases release cytokines that trigger cellular proliferation and fibrosis. (c)

The so-formed denser ECM has increased stiffness, which activates the stellate cells. An increase follows oedema and inflammation that increases stiffness in the density of the ECM. Thus, these observations justify the role of Fibroscan and magnetic resonance elastography in the non-invasive evaluation of hepatic stiffness as a marker of ECM composition.

5.45.3 Fibrogenesis

TGFβ1 is the classic fibrogenic cytokine. Smad proteins carry out downstream signals in the stellate cell. There are two types of Smad protein, of which 2 and 3 are stimulatory, and 7 is inhibitory and opposed by Id1. TGFβ1 stimulates the collagen transcription via hydrogen peroxide and C/EBPβ-dependent. Glycaemic abnormalities like hyperglycaemia and excess insulin levels also affect connective tissue growth factor (CTGF/CCN2) levels, known as inducers of stellate cell fibrogenesis. The production and regulation of CTGF are TGFβ-dependent. Cannabinoids have also been recognised as significant contributors to hepatic steatosis, stellate cell activation, fibrosis, and hemodynamic modifications related to severe liver disease. Two cannabinoid receptors, CB1 and CB2, have opposite effects; while CB1 play a role in fibrogenesis, CB2 is anti-fibrotic. Likewise, opioids promote fibrogenesis via signal transduction in stellate cells, which an opioid antagonist, naltrexone, inhibits. Adrenergic neurotransmitters also participate in these activation pathways.

5.45.4 Contractility

In the early stages of fibrosis, the stellate cell contractility leading to worsening portal resistance during liver fibrosis is probably modifiable before the formation of thickened septae, intrahepatic shunts, and lobular distortion resulting in fixed increments in portal pressure. Like smooth muscle cells, the activated stellate cells display numerous contractile filaments (actin and myosin), producing both calcium-dependent and -independent contractile forces that provide cellular contractility. Physiologically, stellate cells also cause angiogenesis, regenerate, and respond to injury of liver cells. Post-partial hepatectomy, stellate cells form vascular links with hepatocytes and generate new sinusoidal branches.

5.45.5 Matrix Degradation and Resolution of Fibrosis

The findings showing that fibrosis and even cirrhosis are modifiable have renewed research interest in comprehending the modulation of the matrix degradation and fibrosis reversal, assuming that therapies manipulating these pathways can obverse the disease. Enzymes regulating matrix degradation consist of calcium-dependent matrix-metalloproteinases or *matrixins* that lyse collagenous and non-collagenous ECM substrates. Stellate cells are the primary source of proteases MMP-2, MMP-9, stromelysin (MMP-3), and interstitial collagenase MMP-13. The cross-linking of collagen by lysyl oxidase and tissue transglutaminase and the maturation of hepatic scar via disintegrin and metalloproteinase with thrombospondin—are critical contributors to hepatic fibrosis irreversibility. As advanced fibrosis is repaired, the micronodules that characterise active cirrhosis coalesce into macronodules.

Hepatic macrophages are resurfacing as pivotal modulators of matrix remodelling. During the development of hepatic fibrosis, macrophages increase fibrogenesis, while during resolution, they accelerate matrix degradation, largely via elevation MMP-13 synthesis. Proteases are inactivated on interacting with TIMPs, so this could be a major target of evolving therapy since these proteins trigger liver injury, and sustained production could inhibit the activity of interstitial collagenases, resulting in decreased degradation of the amassing matrix. Besides, TIMP-1 protects stellate cells from apoptosis, and thus its continued expression in liver injury may increase the population of activated stellate cells by avoiding their elimination. Stellate cells also display the uroplasminogen activator receptor, its inhibitor, and several constituents of the plasmin system. Taken together, these data hint that stellate cells comprise almost all components required for activating or inhibiting metalloproteinases. Eliminating activated stellate cells by apoptosis is still an attractive potential target for antifibrotic therapy since this would apply an endogenous pathway of fibrosis reversal.

5.46 Pathogenesis of Hepatic Encephalopathy

The pathophysiology of HE involves abnormalities of several factors such as ammonia (NH_3), inflammatory cytokines, manganese deposition in the basal ganglia, and benzodiazepine-like compounds [i.e. gamma-aminobutyric acid (GABA)]. Thus it is multifaceted. Newer evidence also suggests the involvement of microbiota and aromatic amino acids. However, ammonia remains the primary culprit. It is a toxin obtained from the bacterial metabolism of dietary proteins. It usually undergoes biotransformation by the liver followed by clearance via kidneys and, to some extent, by the muscles. In patients with cirrhosis, the urea cycle fails, and portosystemic shuts allow ammonia to enter systemic circulation without removal by the liver. In the brain, ammonia crosses the blood–brain barrier along the concentration gradient and is taken up by the astrocytes, where glutamine synthase converts it to glutamine. The rising glutamine concentration in astrocytes generates an osmotic gradient, leading to edematous astrocytes and the generation of reactive oxygen species, causing cerebral dysfunction observed in HE (Table 5.14).

Table 5.14 Pathogenesis of hepatic encephalopathy

Hypothesis	Evidence in favor of	Evidence against
Ammonia	• Evident neurotoxicity • Synthesized in copious amounts by gut bacteria • Liver failure enhances build-up • Use of L-ornithine-L-aspartate accelerates ammonia detoxification and HE symptoms • Association between *H. pylori* infection and HE • Triggers upregulation of BZRs and GABAergic enhancement	• No correlation between plasma levels and severity of HE • Eliminated via hemodialysis with no change in HE symptoms • No data supporting intoxication causes personality or sleep pattern abnormalities
Aromatic amino acids	• Levels elevated in HE patients • Levels proportionate to HE severity • May cause the synthesis of false neurotransmitters and change catecholaminergic transmission	• Studies associating levels of false neurotransmitters and HE is contentious
Mercaptans	• Evident neurotoxicity • Synthesized by gut microbiota	• No proven association with HE severity
Short and medium-chain fatty acids	• Proven potential to induce reversible coma • Levels elevated in chronic liver disease	• Evidence of elevated levels in HE is inconsistent
Phenols	• Evident neurotoxicity and hepatotoxicity • Levels elevated in HE	• Levels found in HE inadequate to cause neurotoxicity
Permeability alterations in the blood–brain barrier	• Enhanced by ammonia, mercaptans, and phenols • Levels of norepinephrine lowered in HE	• Inadequate evidence of chronic liver disease • Plausible facilitating factor for other toxins in HE
Change in brain monoamines	• Elevated levels of serotonin found in HE models • Levels of norepinephrine lowered in HE models	• Lack of proven association between neurotransmitter alterations and neurological symptoms • Similar serotonin tumour in the presence of cirrhosis with or without HE
Enhanced GABergic transmission (involvement of endogenous benzodiazepines)	• Elevated levels of circulating BZD ligands in animal models and patients with HE • Elevated brain levels of benzodiazepines and receptor density and binding • Positive association between HE symptoms and BZD ligand levels • Flumazenil demonstrates efficacy	• Elevated levels of BZD were found in a small fraction of patients • Levels associate better with the severity of liver disease compared with HE • Efficacy with flumazenil in a small number of patients • The likelihood of a non-specific effect of flumazenil cannot be excluded

The dysbiosis and increased gut permeability seen in cirrhosis also lead to enhanced production of pro-inflammatory cytokines, which contribute to an increase in blood–brain barrier permeability and, thereby, cerebral oedema. Ammonia can also bind to GABA receptors on astrocytes, leading to its activation and production of neurosteroids, which further aggravates HE. The neurotoxicity of ammonia results in elevated resting membrane potential and deactivation of neuronal chloride extrusion pumps, resulting in the suppression of both axonal conduction and excitatory postsynaptic potentials, inhibiting the postsynaptic potential generation and depolarising neurons. Raised mitochondrial ammonia increases levels of oxidative stress exacerbating cerebral edema.

Gut microbiota significantly contributes to the above chain of events, and dysbiosis may be linked with abnormal bile acid metabolism. Commensal bacteria obtain energy from primary bile acids and converted to secondary bile acids (i.e. cholic acid/deoxycholic acid). It destabilises bacterial membranes with a concomitant increase in intestinal permeability, leading to antimicrobial action. Dysbiosis may also correlate with the progression of cirrhosis to a decompensated state. It may be related to decreased hepatic bile acid synthesis with a consequent drop in intestinal levels, causing an overgrowth of pathogenic bacteria. A decrease in the ratio of secondary bile acids to primary bile acids, noted in cirrhotics, may result in altered gut microbiota composition and a reduction in beneficial flora. Cirrhotic patients also display an elevated Bacteroides/Firmicutes ratio and changes in other pathogenic bacteria with elevated levels of Enterobacteriaceae with a drop in Lachnospiraceae.

The key role of gut microbiota in HE was also shown in the experimental model of cirrhosis, in which, under sterile conditions, inflammation or hyperammonemia did not occur. Proton pump inhibitors (PPIs) frequently prescribed to these patients also increase the risk of dysbiosis, also noted in patients with cirrhosis. The further lower pH level in stomach acidity encourages intestinal bacterial overgrowth and increased risk of gut bacteria translocation. These observations argue that PPI poses a significant risk for the progression of HE (Fig. 5.15).

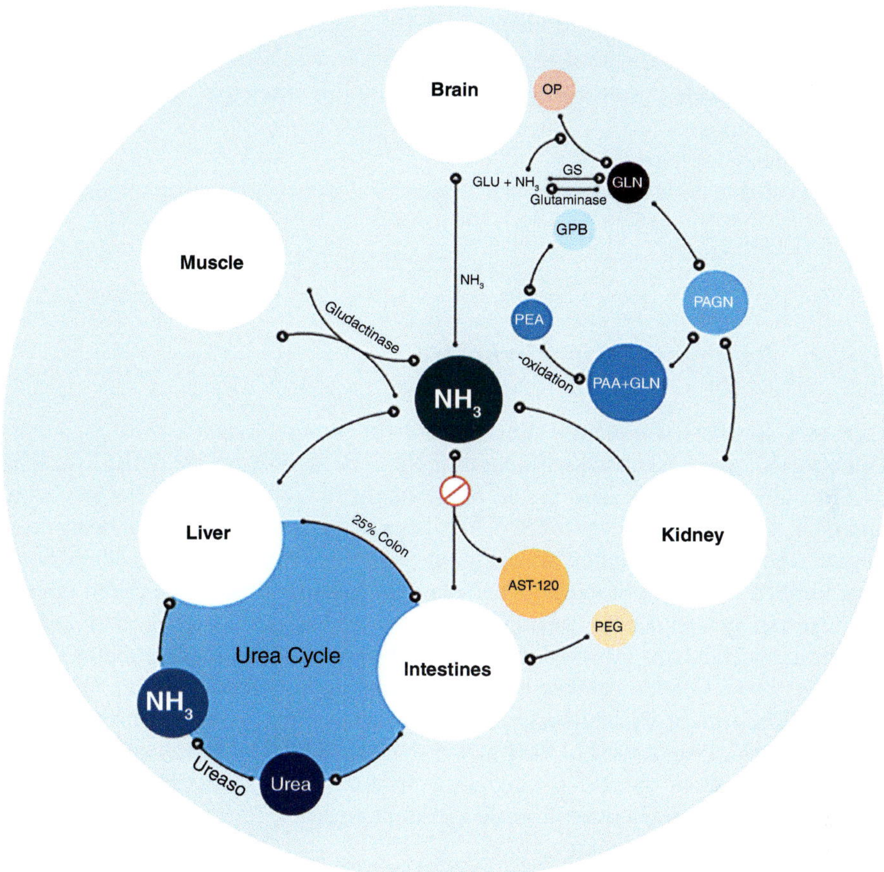

Fig. 5.15 Multiorgan ammonia pathways ammonia-lowering strategies used in cirrhosis. In Cirrhosis, the urea cycle fails, so serum ammonia (NH_3) levels are high. NH_3 crosses the blood–brain barrier, contributing to astrocyte swelling and hepatic encephalopathy. The alternative pathway consists of NH_3 binding with glutamate (GLU) to form glutamine (GLN) using glutamine synthetase (GS). Both ornithine phenylacetate (OP) and glycerol phenylbutyrate (GPB) are ammonia-lowering medications; they combine GLN and phenylacetate (PAA) to form phenylacetylglutamine (PAGN), which is excreted in the urine. AST-120 is a carbon microsphere adsorbent that binds NH_3 in the gut, thus lowering circulating ammonia levels. Polyethene glycol (PEG) is cathartic, which causes rapid clearance of gut bacteria synthesising ammonia to be excreted into the faeces. Skeletal muscle also contributes to the regulation of NH_3, as depicted. Although not shown in the figure, the presence of GS and glutaminase in most of the organ, contributing to NH_3 homeostasis. *PBA* phenylbutyric acid

5.47 Pathophysiology of Portal Hypertension in Cirrhosis

Portal hypertension is when the portal pressure elevates from the normal 5–10 mm of Hg to greater than 10 mm of Hg. The apparent clinical complications usually develop when portal hypertension exceeds 12 mmHg. Portal hypertension may develop due to obstruction to flow at three levels:

1. Prehepatic: portal vein thrombosis, EHPVO
2. Intrahepatic
   (a) Presinusoida: Schistosomiasis, vitamin A toxicity, sarcoidosis, and amyloidosis
   (b) Sinusoidal: cirrhosis
   (c) Post sinusoidal: venoocclusive disease and alcoholic hyaline sclerosis
3. Posthepatic: Budd–Chiari syndrome

5.47.1 Pathogenesis in Cirrhosis

Cirrhosis is most frequently the reason for portal hypertension in adults. Pathophysiological mechanisms were thought to be an obstruction in the liver from 1950 to the 1980s, and, as a result, one thought of ‘bypass’ shunt surgery as an approach to treatment. From 1980 to 2005, one realised that rather than being ‘congested’, the portal circulation was actually ‘hyperdynamic’ in advanced cirrhosis. This implied increased blood flow and shortened transit time in the portal circulation caused by splanchnic vasodilation. Many subsequent studies analyzed the molecular mechanisms involved in this hyperdynamic splanchnic circulation. It was found that there is an imbalance between the circulating vasodilators and vasoconstrictors. The role of vasodilatory chemicals such as nitric oxide and prostaglandin was elucidated. Progression of PHT may also be attributed to the transfer of bacteria or their products to the blood circulation from the gut. These intruders activate HSCs and KCs with resultant fibrosis and inflammation, further adding to the disturbance in portal pressure.

There was also increased cardiac index and decreased peripheral resistance, leading to ‘effective hypovolemia’ (or reduced central blood volume) is critical in exacerbating sodium retention and triggering renal impairment in patients with ascites. Studies showed that there was decreased NO production and increased responsiveness to endothelin at the sinusoidal level, resulting in increased sinusoidal vascular resistance. This led to the introduction of propranolol and terlipressin to treat portal hypertension. At the presinusoidal level and splanchnic bed, there are increased NO and adenosine concentrations, resulting in vasodilation and increased splanchnic blood flow. Other factors that contribute to the PHT are HSCs, the source of collagen, ECM, and procollagen. Later we realised that it also acts as myofibroblast with a contractile response to endothelin stimulation leading to dynamic increases in portal pressure. Carvedilol and simvastatin could help in portal hypertension.

5.48 Ascites in Liver Disease: Pathogenic Mechanism

Ascites is a standout among the most successive confusions of cirrhosis. More than three-fourths of the hospitalized ascites cases are due to cirrhosis, with the remaining attributable to tuberculosis, malignancies, congestive heart failure, pancreatitis, or other rarer causes. Cirrhotic patients with ascites are characterised by typical splanchnic and systemic haemodynamic changes, causing systemic vasodilation and reduced effective circulating blood volume leading to increased fluid and sodium retention.

5.48.1 Pathogenesis

Ascites formation in cirrhotic patients is related to two main pathogenetic mechanisms, namely portal hypertension (PHT) and renal sodium retention. PHT leads to considerable alterations in the splanchnic circulation. Mechanical obstruction to portal flow is only one of the causative problems; there is also elevated portal venous inflow due to splanchnic arterial vasodilation. In addition, they have significant reductions in systemic vascular resistance (SVR) and mean arterial pressure (MAP). This secondarily leads to elevated cardiac output and hyperdynamic circulation. The reduced SVR is most prominent in arterial splanchnic circulation. Partially it is due to the opening of portosystemic collaterals, which further decreases SVR. In addition, there are increased levels of circulating vasodilators. These include substance P, glucagon, platelet-activating factor, vasoactive intestinal peptide, and prostaglandins. Increased synthesis of systemic prostacyclin is observed in patients with cirrhosis even before ascites develop. It may be due to the increased synthesis of renal prostacyclin to maintain renal plasma flow following. This prostacyclin activation may be related to the presence of endotoxemia since selective intestinal decontamination with nonabsorbable antibiotics significantly decreases the synthesis of systemic prostacyclin.

The possible role of nitric oxide (NO) as a cause of vasodilatation cannot be ignored. Evidence shows that (a) the activity of endothelial NO synthase is elevated in the arterial vasculature of rodent cirrhosis models with ascites. (b) Compared to those without, those with cirrhosis display higher nitrite and nitrate levels. And (c) inhibition of the synthesis of NO experimental models of cirrhosis significantly elevates the arterial pressure, and SVR reduces the cardiac index and reverses the impaired response to vasopressors. It has been suggested that NO production may be triggered by endotoxin or bacterial DNA from the GIT, which is less efficiently eliminated due to portal-systemic shunting and diminished reticuloendothelial cell function. It is known that the NO concentrations in blood collected from portal veins are higher than those of peripheral veins, and serum NO concentrations parallel serum bacterial DNA concentrations. Oral administration of the antibiotic colistin to patients with cirrhosis significantly decreases not only plasma endotoxin levels but also lowers the serum concentration of nitrite and nitrate, and norfloxacin has been shown to increase systemic vascular resistance and decrease plasma renin in a subgroup of patients with ascites.

5.48.2 Sequalae of Vasodilatation

Vasodilation in the portal and systemic vasculature activates the release of endogenous vasoconstrictors, sodium and water retention, and increasing renal vasoconstriction to maintain the blood pressure (a) hypovolemia leads to stimulation of carotid and renal baroreceptors and thereby activation of the neurohumoral mechanisms that increase retention of sodium to reinstate normal perfusion pressure (Fig. 5.16). These include the renin–angiotensin–aldosterone system, adrenergic nervous system, and antidiuretic hormone (vasopressin). The overall effect is adequate sodium and water retention. (b) Sodium retention increases the plasma volume. In advanced disease, sodium excretion frequently drops to less than 10 mEq/day. Sodium retention suggests that hepatic function has dropped to half its capacity. Fluid overload is a crucial marker in the medical history of these patients, and the degree of sodium retention is inversely related to survival. (c) Water retention is predominantly due to the increased antidiuretic hormone (ADH) release described above. The dysfunctional water excretion leads to the development of hyponatremia and hypo osmolality. The severity of hyponatremia is a prognostic indicator of cirrhosis and correlates with worsening survival. (d) Renal vasoconstriction occurs later in the disease timeline and is related to overcoming compensatory mechanisms such as prostaglandins and perhaps nitric oxide. This exacerbates renal hypoperfusion, a slow drop in the glomerular filtration rate, and, in some patients, hepatorenal syndrome. Urea and creatinine production is low in cirrhosis, and its levels may be falsely low, masking renal dysfunction.

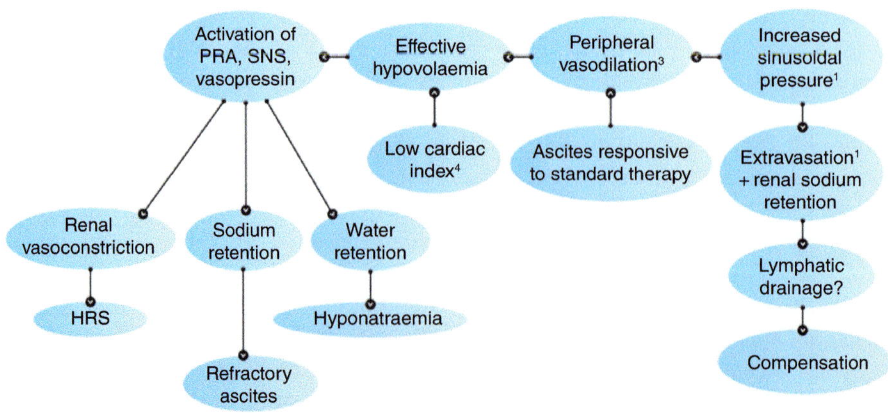

Fig. 5.16 Peripheral vasodilatation hypothesis

5.49 Pathogenesis of HRS in Cirrhosis

In patients with comorbid cirrhosis and ascites, acute kidney injury (AKI) is not synonymous with hepatorenal syndrome. Studies have indicated that only one-fourth of these patients have hepatorenal syndrome. A larger percentage is likely to have either prerenal azotaemia or acute tubular necrosis. A very small percentage of people may have obstructive uropathy.

5.49.1 Definition

AKI is defined as a rise in serum creatinine (SCr) of ≥50% from baseline or a rise in SCr by ≥0.3 mg/dl (26.5 μmol/l) in <48 h. Hepatorenal syndrome (HRS) type 1 is a peculiar type of AKI. If GRF remains reduced to 60 ml/min per 1.73 m^2 surface area or less, and a rise in creatinine persists by 50% or more for up to 3 months, it will be called acute kidney disease (AKD). Similarly, if these abnormalities persist beyond 3 months, the condition will be called chronic kidney disease (CKD).

The severity of AKI can be staged as shown in Table 5.15.

5.49.2 Pathogenesis of AKI in Cirrhosis

AKI HRS is a special form of AKI with multifactorial pathogenesis related to liver dysfunction and portal hypertension. Some of the factors involved are summarised below.

(a) Splanchnic vasodilatation and its pathogenesis have been described above in the section on Ascites. Mediators involved in splanchnic vasodilatation include prostacyclin, carbon monoxide, endogenous cannabinoids glucagon, adrenomedullin, epoxyeicosatrienoic acids, and nitric oxides. It causes a decrease in the circulating volume of blood and activation of compensatory sodium and water-retaining mechanisms. The decline of renal perfusion is due to altered renal blood flow autoregulation, ultimately leading to renal vasoconstriction. Therefore, the existing treatment relies on reversing responses using volume

Table 5.15 Severity of AKI

Stage	Rise in serum creatinine from baseline
1	1.5–2.5-fold (0.3 mg/dl)
2	2–3-fold
3	>3-fold (4.0 mg/dl with acute 0.3 mg/dl increase) + need renal replacement therapy

expansion (by 20% albumin infusion) and splanchnic vasoconstriction in HRS-AKI.

(b) Systemic inflammation is common in decompensated cirrhosis due to elevated levels of inflammatory cytokines in HRS-AKI. Compared to those with only advanced liver disease, patients with concomitant spontaneous bacterial peritonitis (SBP) and renal insufficiency involve high tumour necrosis factor (TNF)-α and interleukin (IL)-6 levels.

(c) Relative adrenal insufficiency is present in more than 25–30% of decompensated cirrhosis patients. It also may lead to cardiomyopathy in cirrhosis patients by downregulating β-adrenergic receptors and consequent lowering of adrenergic effects on vascular resistance.

(d) Cirrhotic cardiomyopathy is associated with declining cardiac compliance and contractility with electrophysiological signal abnormalities, particularly in response to stress. It is caused by sustained portal hypertension, associated with increased translocation of bacteria and portosystemic shunting. Bacterial translocation increases systemic inflammatory cytokines production resulting in endothelial dysfunction, while an elevated portosystemic shunt decreases systemic vascular resistance by elevating endogenous vasodilators nitric oxide, bradykinin, adenosine, and endocannabinoids. Reduced cardiac output and elevated right atrial pressure cause renal tissue hypoxia and venous congestion, worsening kidney injury and reducing glomerular filtration rate.

(e) The response to treatment with terlipressin plus albumin treatment in HRS type 1 patients with elevated serum bilirubin (≥10 mg/dl) indicates the effect of bile acid in chronic liver disease. Severely increased bilirubin levels trigger bile cast nephropathy, which is frequent in AKI pathology and detected by Hall histochemical staining. Bile cast nephropathy is associated with unalterable AKI in chronic liver disease. In less advanced renal injury, bile acid accumulation in cirrhosis causes proximal tubulopathy similar to Fanconi syndrome without bile cast nephropathy.

5.50 Non-cirrhotic Portal Hypertension

Nearly one-fourth of variceal bleeds in India used to be caused by Non-Cirrhotic portal Fibrosis (NCPF) about four decades back. Its contribution to variceal bleeding has been progressively falling as the disease is becoming uncommon. Its aetiology is not known yet. Its main features include the presence of fibrosis of small and medium branches of the portal vein causing portal hypertension (PHT), but hepatic anatomy and physiology are unremarkable.

5.50.1 Pathogenesis

Since the cause of this disease was unknown, several hypotheses were proposed (Fig. 5.17).

1. *Infective hypothesis*

   Non-cirrhotic portal fibrosis is widespread in people from mower economic strata. Neonatal abdominal infection or during infancy may be the reason. Occurrence of umbilical sepsis, bacterial infections, and diarrheal episodes in childhood increase the risk of portal pyelipblebitis or pyemia causing thrombosis and sclerosis of small and medium-sized portal vein radicles. With improvement in general hygiene, the disease is now becoming less common. A lesion similar to NCPF could be produced in experimental animals by causing portal pyemia.
2. *Trace element-chemical theory*

   Chronic consumption of arsenic has been suggested as a cause of NCPF based on histological studies on the liver of such patients. The liver shows fibrosis of the periportal area and fragmented septal cirrhosis, accompanied by or without the development of angiogenesis within the expanded portal zones. Seventeen percent of Belgian patients who consumed arsenic-containing Fowler's solution for psoriasis treatment developed NCPH with skin stigmata due to arsenicosis. However, such lesions were missing in Indian patients with NCPH. Liver studies in patients with NCPF also did not show an increase in hepatic arsenic levels. Other chemicals such as vinyl,

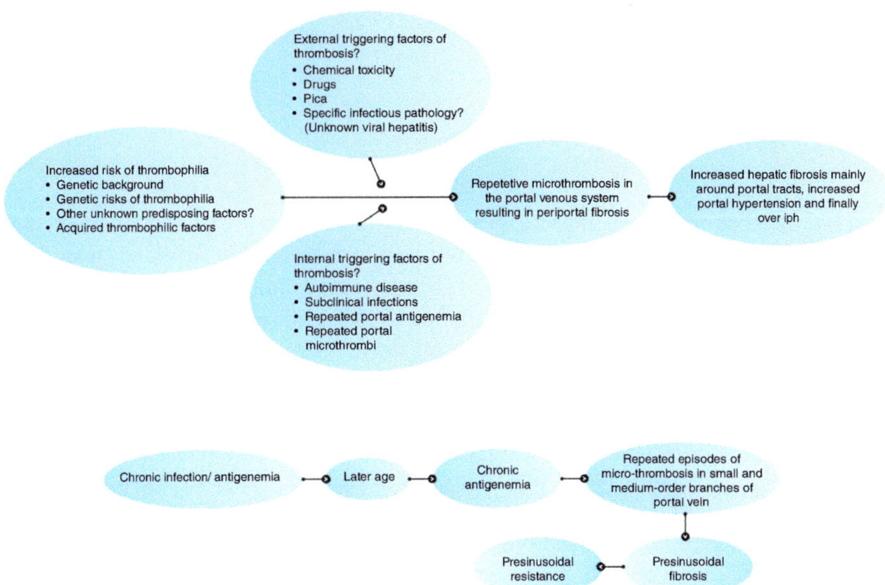

Fig. 5.17 Summary of proposed pathogenic mechanism for causation of NCPF

vitamin A toxicity, methotrexate, and 6-mercaptopurine are possible culprits. In the Iranian patient population, approximately 50% had a history of pica. Exposure to vinyl chloride, copper sulphate, prolonged therapy with methotrexate, excessive consumption of vitamin A, and certain immunosuppressants like azathioprine and 6-mercaptopurine transform the histology to that mirroring NCPF.

3. *Autoimmunity theory*

   Many patients with NCPF display (a) a decrease in the number of suppressor/cytotoxic T lymphocytes (T8) and a reduced T4/T8 lymphocyte ratio; (b) poor cell-mediated immunity; (c) a poor autologous mixed lymphocyte reaction (MLR). In the Japanese counterpart of this disease called Idiopathic pulmonary hypertension (IPH), several autoimmune disorders are also commonly associated, which include systemic lupus erythematosus (SLE), progressive systemic sclerosis (PSS), thyroiditis, and mixed connective tissue disease (MCTD). Almost 60% of women with IPH demonstrate anti-ds DNA antibody, and about 25% have positive for antinuclear antibody. Indian patients with NCPF have not shown such an association. There are, however, reports of familial aggregation and an excess HLA-DR3 in the Indian subpopulation.

4. *Thrombosis theory*

   Some studies have suggested an association between NCPH and prothrombotic disorders, which generate small and medium-sized portal radicles aggravating sclerosis and fibrosis. These include myeloproliferative disorders, protein C and S deficiency, and prothrombin gene mutation.

5. *Genetic theory*

   HLA-DR3 aggregation in family members with NCPF has been reported in India.

5.50.2 Pathology

Surgical studies and post-mortem evaluations have shown that the liver appears unremarkable or slightly larger or smaller with a smoother or wrinkled or nodular surface similar to that found in cirrhosis. Fibrous thickening of the hepatic capsule with excess vascularity and mild inflammation is observed. Histopathological examination again shows heterogeneous lesions. The common findings include obliterative portal venopathy, portal angiomatosis, dilated para portal shunt vessels, mega sinusoids, thin incomplete periportal/perisinusoidal fibrosis, and differential atrophy but unaffected lobular architecture. Another common finding is the dilatation of the main PV trunk with thick sclerosed walls. The obliterative portal venopathy interferes with the intrahepatic circulation, followed by parenchymal remodelling. Reduction in portal blood flow leads to atrophy of the hepatocytes and its hyperplasia in adequately perfused areas as compensation. This is particularly noted in patients with nodular hyperplasia, either partial or regenerative variants of INCPH.

5.51 EHPVO: Pathogenesis

Extrahepatic portal vein obstruction (EHPVO) has been the commonest cause of variceal bleeding in children during the first and the second decade of life. These patients have obstruction of the extra-hepatic portal vein, which may or may not involve large intra-hepatic portal veins, splenic, and superior mesenteric veins. The majority of reports are from the developing world, and this type of lesion is uncommon in Western countries. The disease has become less common in India over the last several decades.

5.51.1 Pathogenesis

The aetiology of EHPVO is unknown, and there are several theories for underlying causes. (a) Infections: in infants, umbilical sepsis or umbilical catheterisation may be a key factor for EHPVO, while in older pediatric patients, it could be attributed to extra-abdominal sepsis, appendicitis, and peritonitis. These can cause portal vein obstruction by causing portal pyeliphlebitis. (b) Dehydration: Among children, an episode of diarrhoea could lead to thrombosis. (c) Congenital agenesis or Atresia-meaning is another possibility. (d) Hypercoagulable states: protein C, protein S deficiency, Factor V Leiden deficiency, and polycythemia Vera may lead to thrombosis. And lastly, of course (e) idiopathic: cause not known.

The aetiology of EHPVO may be different in children and adults, as is outlined in Table 5.16.

5.51.2 Histopathology of EHPVO

Gross examination shows a cavernomatous transformation—at the hilum of the liver and the main portal vein is often not recognised. The liver is smooth, seldom finely granular on the surface. Histopathology shows that liver tissue structure is preserved. There may be circular condensation of reticulin fibres around portal veins.

Table 5.16 Aetiology of extrahepatic portal vein obstruction in children and adults

Etiology in children	Etiology in adults
• Infection: omphalitis, neonatal umbilical sepsis, umbilical catheterization, abdominal surgery, trauma • Congenital anomalies: PV stenosis, Atresia • Prothrombotic states • Idiopathic: 50%	• Prothrombotic disorders: more often occult • Local inflammatory lesions: pancreatitis, diverticulitis, appendicitis, duodenal ulcer • Injury to PV system: surgical shunting, splenectomy, colectomy, gastrectomy

5.52 Hepatitis B: Structure and Replication

5.52.1 Introduction

The hepatitis B virus is a DNA virus of the *Hepadnaviridae* family responsible for acute and chronic hepatitis B infection in humans. Similar viruses have been reported in certain animals, woodchucks, and ground squirrels. Baruch Blumberg was the first to discover Australian antigen (1976 Nobel Prize), later recognised as HBsAg. Nearly 33% of the global population has HBV infection, while 2–4% of individuals carry the virus (350–500 million). HBV causes nearly 800,000 deaths annually.

5.52.2 Structure of the Virion

Three types of viral particles displayed by electron microscopy include (a) smaller (20 nm) spherical structures and (b) filaments of 22 nm width but varying lengths. The spheres and filaments consist of the viral surface antigens or (HBsAg) and lipids from the host and are anucleate and non-infectious. (c) The infectious HBV virion (Dane particle) is a spherical, double-shelled 42 nm structure made of a lipid enclosure comprising of HBsAg that encloses an inner nucleocapsid which houses the hepatitis B core antigen (HBcAg) integrated with polymerase attached to the 5′ end of the minus strand and the DNA genome of the virus. HBV is classified into eight genotypes, A to H, each with unique geographic dominance.

The genome is a partially double-stranded circular DNA made of ~3.2 kilobase (kb) pairs covalently linked to the 5′ end of the long strand (Pol) and a short oligoribonucleotide at the 5′ end of the short strand. The '−' strand of fixed length encodes seven proteins from four overlapping Open Reading Frames (surface [S], core [C], polymerase [P], and the X gene [X]), shown as large arrows, and three upstream regions (preC, preS1, and preS2) (Fig. 5.18).

5.52.3 HBsAg

The envelope proteins are made in three forms, large (L), middle (M), and small (S) surface antigen. (a) S (24 kDa) is 226 amino acids (aa) in length and makes up a shared C-terminal region of the two longer envelope proteins. (b) The M protein (31 kDa) contains the S sequence with a 55aa N-terminal extension known as preS2. Expression of the M- and S-encoding mRNA is guided by the S promoter, with

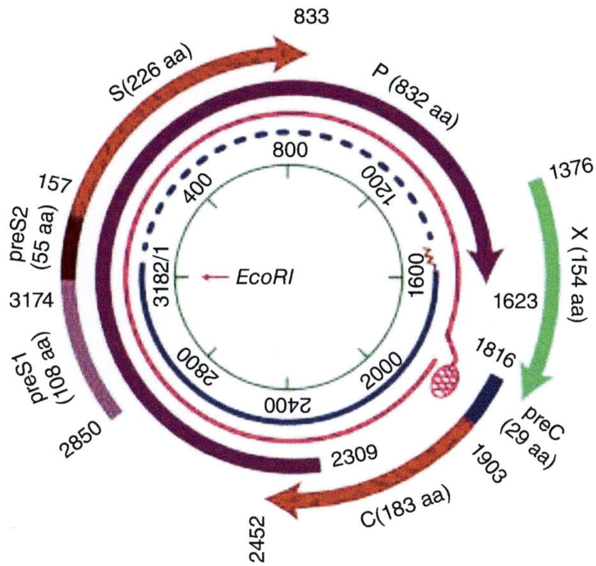

Fig. 5.18 Genome of HBV, showing the main four open reading frames

translation beginning from an upstream (M) or downstream (S) AUG. (c) The L protein (39 kDa), the largest of the envelope proteins, contains S, preS2, and an additional 108aa or 119aa (depending on the genotype) N-terminal extension known as preS1. L-HBsAg is encoded by a self-mRNA transcript regulated by the preS1 promoter.

The ER is where the envelope proteins are produced and reach their transmembrane configuration. The N-terminal signal sequence begins fitting S into the ER membrane, which is succeeded by another signal that moves the downstream peptide sequence into the ER lumen.

5.52.4 HBeAg

HBeAg is the final product of the post-translational processing of pre-core ORF. HBeAg ORF codes for an endoplasmic reticulum (ER) targeting sequence. It guides the peptide to the ER, transforming the protein into the final 15 kDa HBeAg secreted by the HBV-infected cells. The function of HBeAg is not completely understood. It facilitates HBV immune evasion by diverting the immune response elsewhere from infectious viral particles (devoid of eAg). It also inhibits the immune response to the HBV core protein and is a marker of HBV replication. HBeAg levels indicate viral titer. HBeAg seroconversion indicates transformation to the non-infectious carrier state.

5.52.5 HBcAg

The 21 kDa HBV core protein, or HBcAg, forms the fundamental structure for the virion. The core is translated from the pgRNA, and the initial 149aa of core forms the assembly domain, which is adequate for in vitro formation of capsids indistinct from capsids isolated from Dane particles. The remaining 34–36aa form the arginine-rich C-terminal domain (CTD); phosphorylation of various aa in the CTD modulates several stages of the HBV life cycle. The key function of core protein is gathering the nucleocapsid. It also interacts with HBV covalently closed circular DNA (cccDNA) to modulate the spacing of nucleosomes on cDNA; cccDNA is a nuclear-localized replication intermediate of hepadnaviruses that creates a mini-chromosome. CTD is also required for pgRNA packaging, initiating reverse transcription. It manifests on the surface of infected cells.

5.52.6 Polymerase/Reverse Transcriptase

Hepadnaviruses begin genome replication with several distinctive characteristics. HBV's 90 kDa, 838aa polymerase protein (reverse transcriptase/RT/Pol/P) constitutes three functional domains and a variable spacer region. The terminal protein (TP) domain is at the N-terminus, necessary for beginning the genome replication (P binding to the pgRNA, RNA packaging, and protein priming). A dynamic spacer separates the TP domain from the RT domain, and evidence indicates that almost every aa inside this spacer is likely to mutate without affecting the P function. Only three cysteine residues within the C-terminal end of the spacer region, together with a fourth in the N-terminal side of the RT domain, may be necessary for RT/pol function.

The RT domain accounts for genome replication by reverse transcribing the pgRNA to create the (−)-strand of the DNA genome, followed by its use for synthesising the (+)-strand of the DNA. Nucleoside analogues target this domain [as in human immunodeficiency virus (HIV) RT]. The last domain, P, is the RNase H domain, which accounts for the degradation of the pgRNA template during the synthesis of the (−)-strand of the DNA genome. Coordination of metal ion binding is performed by four conserved carboxylates for RNase H activity.

5.52.7 X Protein

HBx, a 154aa, 17 kDa protein, is the sole regulatory protein-encoded smallest HBV ORF and is important for HBV replication. HBx is combined with cccDNA and is necessary for transcription from cccDNA. HBx has oncogenic properties; however, whether it has true oncogenic potential or merely acts as a cofactor is unclear.

Hepadenavirus-linked HCC occurs only in mammalian hepadnaviruses via X protein expression. HBx is a protein with multiple functions involving the modulation of several hepatocyte signalling cascades and factors associated with cellular transformation.

HBx can regulate calcium, apoptosis, and proliferation signals and also stimulate several transcription factors, [activator proteins, nuclear factor of activated T cells (NFAT), and nuclear factor kappa-light-chain-enhancer of activated B cells (NF-κB)]. HBx also affects cellular signalling factors, such as Wnt/β-catenin, P53, and Akt, associated with HCC. It also modulates miRNA expression and may participate in immune regulation.

5.52.8 Replication

The outline of HBV replication in the hepatocyte is shown in Fig. 5.18. Sodium taurocholate cotransporting polypeptide (NTPC) has been identified as a functional receptor for human hepatitis B virion. HBV virion anchors to host cell membranes via the pre-S domain of the surface protein, followed by its internalisation. However, viral disassembly and intracellular transport of the viral genome into the nucleus are not elucidated and likely comprise nucleocapsid core protein modification. After the entry of the viral genome in the nucleus, the single-stranded gap region in the viral genome is modified by the viral pol protein, and the viral DNA is wound to the covalently closed circular (cccDNA) form. In this form, HBV DNA acts as the template for the transcription of numerous species of genomic and sub-genomic RNAs and is the stable component of the replication cycle that is comparatively resistant to antiviral action and immune elimination. The transcripts from the cccDNA are not spliced or polyadenylated and have a 5′ cap structure. The 3.5-kb genomic transcripts comprise two species with different 5′ ends: the pregenomic and the pre-core RNAs. The pregenomic RNA (pgRNA) acts as a basis for reverse transcription and the messenger RNA for core and polymerase; the pre-core RNA guides the translation of the pre-core gene product. The large HBsAg (L-HBsAg) protein is translated from the 2.4-kb subgenomic RNA, the middle (M-HBsAg) and small HBsAg (S-HBsAg) proteins from the various forms of 2.1-kb RNAs, and the HBxAg protein from the 0.7-kb RNA. The S-HBsAg is the primary *S* gene product, and the L and M proteins are minor.

The genome is encapsulated after the replication of HBV. The packaging signal is a *cis*-acting element called epsilon with a stem-loop structure. The terminal protein of the pol combined with the epsilon and the core protein creates the nucleocapsid. Followed by encapsidation, the pol intervenes in the reverse transcription of the pgRNA to minus-strand DNA and ensues positive-strand synthesis. Completing the circular form of the DNA involves complex means of strand transfer. The nucleocapsid and envelope proteins in the endoplasmic reticulum interact to form mature virions, followed by secretion into the extra-cellular environment.

5.53 Hepatitis C: Structure and Replication

5.53.1 Introduction

Hepatitis C virus (HCV) is a positive-strand RNA virus with an envelope belonging to the Hepacivirus genus of the Flaviviridae family, including yellow fever, West Nile, and dengue. The HCV particles are spherical and variable (40–80 nm in diameter). There are 7 HCV genotypes and 84 HCV subtypes. The intact HCV particle frequently exists with lipoproteins. A cross-sectional view of the HCV shows three main viral components: (a) the envelope glycoproteins (E1 and E2 heterodimers); (b) a lipid membrane and the nucleocapsid; and (c) the single-stranded RNA genome (Fig. 5.19).

Envelope

The HCV envelope glycoproteins are E1 and E2. This E1/E2 glycoprotein heterodimer is implanted in the host-derived lipid membrane, thus forming an HCV envelope. The envelope glycoproteins also create higher-order oligomeric structures like trimers and pentamers trimers. The hypervariable region in the E2 protein guards

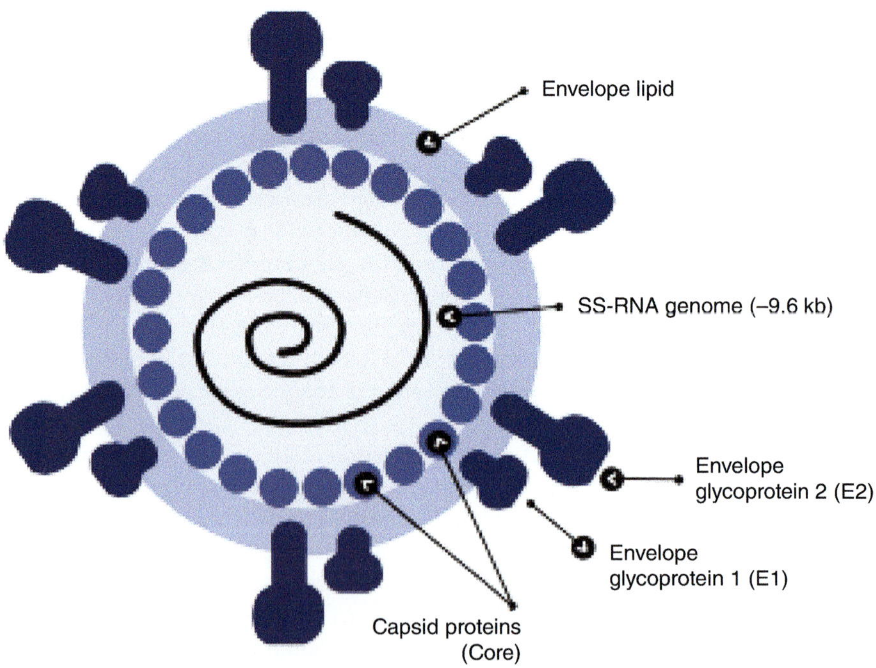

Fig. 5.19 Hepatitis C virus particle structure

the E1 protein from being detected by the immune system. The envelope proteins are necessary for binding to the host receptor, fusion with endosome-lipid membrane, and assembly. The E1 protein length has 192 amino acids, while the E2 protein is 363 amino acids long. The host-derived lipid membrane component of the envelope predominantly contains cholesterol, its esters, phosphatidylcholine, and sphingomyelin.

5.53.2 Capsid or Core

HCV core is a protein enclosure that encapsulates and shields the HCV RNA. It interacts with the viral membrane on the external surface, and the inner surface binds to several HCV RNA segments. The core proteins come together at the endoplasmic reticulum's cytoplasmic side; interactions with the HCV RNA and envelope E1 protein activate that. The core proteins use the HCV RNA to assemble as an arena upon which core proteins bind and encircle the HCV RNA. The HCV capsid is spherical and variable (30 nm diameter). The HCV capsid with the HCV RNA makes the HCV nucleocapsid.

5.53.3 HCV Genome

The HCV RNA (genome) used for both translation and transcription is a single-stranded, positive-sense RNA with nearly 9600 nucleotide bases along its length. The open reading frame (3006–3037 codons) is flanked by 5′ and 3′ untranslated regions (UTRs). The 5′ region, upstream from the open reading frame, is ~340 nucleotides long and consists of four highly structured domains: domain I have a short stem-loop and domains II–IV form the internal ribosomal entry site (IRES). The IRES is an organized structural regulatory component that facilitates the binding of the HCV RNA to the host ribosomal subunits, which is necessary to translate the open reading frame. The 3′ region is nearly 225 nucleotides long and has three constituents: variable, a long poly (U/UC) tract, and a 3′X region. The 3′X region plays an important role in switching between translation and transcription of the HCV RNA.

HCV RNA codes for structural proteins (Core, E1, and E2) and non-structural (NS) proteins [(p7 protein—for assembly), NS2 (a cysteine protease and as a cofactor in the assembly step), NS3 (serine protease and helicase activity), NS4A and B, and NS5A and B proteins].

NS4A is the most minute non-structural HCV protein that (a) attaches the NS3-4A complex to the external leaflet of the endoplasmic reticulum and mitochondrial external membrane, (b) acts as a cofactor for the NS3A serine protease, (c) enhances NS3A helicase activity, and (d) modulates NS5A hyper-phosphorylation and viral replication. The important interactions for viral assembly occur between NS4A and NS4B and NS3 and NS4A.

NS4B protein triggers modifications in the cytoplasmic membrane and intervenes in virus–host interactions, borders around the endoplasmic reticulum, together with non-structural proteins, to form a replication complex. NS4B is necessary to shape the membranous web, a microenvironment of the cytoplasm that aids viral replication.

NS5A is also necessary for the HCV replication complex. It interacts with non-structural proteins (NS4B, NS5B, and RNA) and host cell proteins (i.e. cyclophilin A and kinases) to direct viral replication and assembly. It is also necessary to form endoplasmic reticulum-derived double-membrane vesicles (DMVs), encouraging effective viral replication. It also aids in HCV pathogenesis, modulation of cell signalling pathways, virus propagation, and response to interferon via the interferon-alfa sensitivity-determining region. The direct-acting antiviral agent class NS5A inhibitors act on the NS5A protein.

NS5B is an RNA-dependent RNA polymerase (RdRp) that catalyses the polymerisation of ribonucleoside triphosphates (rNTP) during viral RNA replication. Two types of agents act on the NS5B: (a) nucleotide analogues induce chain termination and (b) non-nucleotide inhibitors interact with the allosteric sites and make it dysfunctional.

5.53.4 *Lipoviral Particles*

HCV consists of lipoproteins tightly associated with the HCV virion that can circulate in human blood as a hybrid lipoviral particle. These lipid components include triglycerides, HCV RNA, capsid protein, E1 envelope glycoprotein, E2 envelope glycoprotein, apolipoprotein B, and apolipoprotein E, which are ~100 nm in diameter. The lipoviral particle favours HCV entry into hepatocytes and shields HCV from being neutralised by the antibody. The density of these lipoviral particles and low-density lipoproteins (LDLs) and very-low-density lipoproteins (VLDLs) is similar.

5.53.5 *Replication*

Viral replication has been studied extensively to develop drugs to interfere with viral multiplication. Replication involves several steps.

(a) Binding: the HCV initially binds two low-affinity host receptors on the surface of a hepatocyte: LDL receptor and heparin sulphate proteoglycans (HSPGs), which activates the HCV outer E1/E2 heterodimer membrane protein to bind to the scavenger receptor B1 (SRB1) and the tetraspanin protein CD81. Subsequently, a wave in the lipid membrane is generated that propels the HCV particle into a tight junction between hepatocytes.

(b) Endocytosis: CD81 interactions with claudin-1 (CLDN1) lead to the folding of the virion and cell membrane into a pit-like region enclosed by clathrin. It evolves as an endosome of a virion enclosed by the host cell membrane, surrounded by a clathrin cage.
(c) Fusion and uncoating: The clathrin cage around the endosome soon dissolves. The low pH of the endosome results in the fusion of the viral and host membranes (endosomal fusion), releasing the HCV RNA into the cytosol.
(d) Translation: HCV RNA is translated into a polyprotein on the ribosomal subunits in the rough endoplasmic reticulum. The ribosome–RNA complex binds to the endoplasmic reticulum membrane to finish HCV polyprotein translation.
(e) Proteolytic breakdown of RNA polyprotein: Cellular proteases cleave the core, E1, E2, and p7 proteins, followed by the NS2 cysteine protease, which cleaves NS2 from NS3. Lastly, NS3, assisted by the membrane-bound NS4A, forms an NS3-4A protease complex that splits the balance proteins (NS3, NS4A, NS4B, NS5A, and NS5B). Overall ten mature HCV proteins are created.
(f) RNA replication: HCV proteins and host factors form a membranous web that collates to form a double-membrane vesicle. In the membranous web, the NS5B RNA-dependent RNA polymerase catalyses the progression of a negative-sense RNA intermediate (template) used to generate several copies of positive-sense HCV RNA. These newly synthesised HCV RNAs are either integrated into nucleocapsid particles or utilised for RNA translation and replication. The NS proteins facilitate the RNA replication process.
(g) Assembly: HCV particles gather in proximity to cytosolic lipid droplets (cells) and use the host enzyme to create the nucleocapsid built of core proteins adjacent to the HCV RNA. The proteins form a guarding shell around the HCV RNA. After the core is formed, the immature HCV particle integrates with a luminal lipid droplet (LuLD) filled with ApoE proteins to form a high-density HCV precursor. The ER also produces precursors of very low-density proteins (pre-VLDLs), which move to the Golgi apparatus along with the high-density HCV precursor. Here these undergo maturation before being packed in the vesicles and released.
(h) Maturation: HCV lipoviral particle exits the trans-Golgi network (TGN) in the form of vesicles by merging with the cell surface of the lipoviral particle.
(i) Release: HCV lipoviral particles in the vesicles fuse with the cell membrane near the surface. The generation of the HCV lipoviral particle occurs along with the cellular VLDL pathway; therefore, both are released simultaneously into the extracellular region.

5.54 HCV Replicon

The development of the HCV replicon cell culture is one of the most noteworthy examples of modern research capabilities. Repliconsare subgenomic RNA molecules that are capable of autonomously replicating in hepatoma cells. The first-generation replicons were obtained from a cloned HCV genome by eliminating the structural region and placing a selectable marker. A second internal ribosome entry site (IRES) was used to permit the translation of the HCV non-structural region. HCV replicons are tools to investigate viral RNA replication, find host cell factors necessary for HCV replication, decipher how HCV escapes innate immunity, and assess antiviral drugs. It is also useful in recognising mutations that grant resistance to antiviral drugs.

5.54.1 Development of Replicon

The first-generation HCV replicons Con1 were derived from a consensus genome cloned from the chronically infected liver with a genotype 1b virus. It was genetically modified by replacements of the core to NS2 or core to p7 region by a selectable marker (neo) and an internal ribosome entry site from a picornavirus (Fig. 5.20). Subsequently, subgenomic, selectable replicons were derived from two expression cassettes: the gene encoding for G418 resistance (neo) and the region encoding the viral replication enzymes NS2 to NS5B or NS3 to NS5B. Upon transfection of the human hepatoma cell line Huh-7, a small number of G418-resistant colonies could be formed, leading to stable replication of HCV RNAs. Compared to infection-based tissue culture systems, the replicons have the potential to amplify five times.

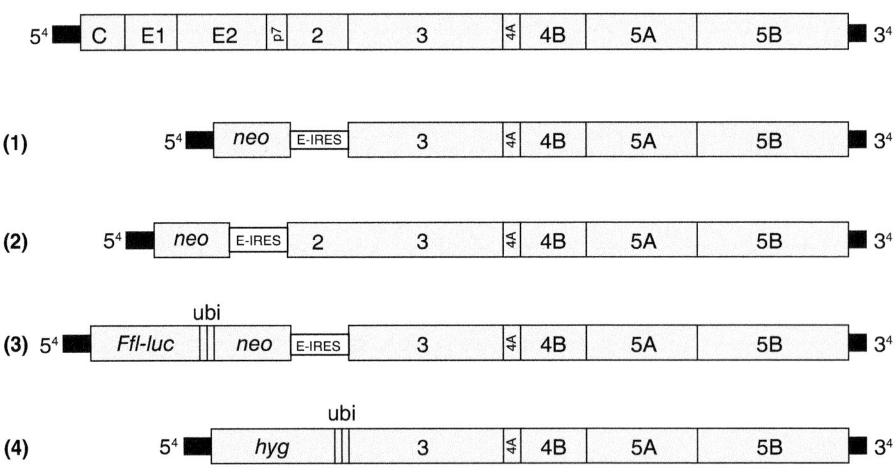

Fig. 5.20 Schematic presentation of the HCV genomic organisation and replicons derived therefrom. The genome is depicted in the top with 5′ and 3′ non-translated regions (thick black lines). Replicons are drawn below and identified on the left. (1) A subgenomic selectable, bicistronic replicon carrying the NS3 to 5B coding region; (2) a subgenomic selectable, bicistronic replicon carrying the NS2 to 5B coding region; (3) a subgenomic bicistronic replicon carrying in addition to the selectable marker neo the luciferase gene from the firefly (Ffl-luc); (4) a monocistronic selectable replicon carrying the hygromycin resistance gene

Further, the replicon RNAs can be retained with high replication potential and protein expression capacity for a long. Its potential relies on selecting highly permissive cells and cell culture adaptive mutations.

With evolving cell culture adaptive mutations and highly permissive cell clones, novel replicons such as monocistronic that carry a reporter gene for easy read-out of RNA levels or transient replication systems that do not need tedious G418 selection were created. Moreover, such HCV replicons can be used to develop human liver cell lines HepG2, IMY, and HuH6, the human embryonic kidney cell line 293, or the human cervical carcinoma cell line HeLa.

5.54.2 Applications

The most advantageous use of HCV replicons is their role development of antiviral drugs and their combinations for synergism or antagonism to avert the development of resistance. These can also be used for high-throughput screening assays to detect lead compounds in large collections of natural substances and combinatorial chemical libraries. These can also be used for assessing the antiviral activity of compounds produced by using isolated enzymes (NS3 proteinase or the NS5B RdRp).

Development of new drugs and in vitro testing of new drugs on HCV replicon led to record speed with which the new drugs were marketed to cure hepatitis C. HCV replicons also help in elucidating mechanisms of IFN used to treat chronic HCV.

Neo, neomycin phosphotransferase; hyg, hygromycin phosphotransferase; E-IRES, IRES from encephalomyocarditis virus; ubi, ubiquitin used as a proteolytic cleavage site to release neomycin phosphotransferase from the luciferase reporter.

5.55 Hepatitis E

5.55.1 Introduction

The hepatitis E virus is a naked virus that bears structural similarity to calciviruses and is, therefore, grouped under the Calciviridae family. It has an icosahedral symmetry measuring 27–30 nm diameter with a single-stranded RNA genome with positive polarity and 7.2 kb length.

5.55.2 Structure

Genomic RNA is polyadenylated and contains three ORFs. (a) ORF1 is the largest, containing 1693 codons, and it encodes the nonstructural proteins. These are responsible for the processing and replication of the virus, including a methyltransferase, an RNA helicase, a cysteine protease, and an RNA polymerase. (b) ORF2 (660 codons) encodes the capsid protein (CP), and (c) ORF3 encodes a small multifunctional protein that acts as a viroporin and facilitates the release of infectious virions from infected cells. The ORF2 and ORF3 proteins are translated from a single, bicistronic mRNA. ORF3 partially overlaps with the other two ORFs. At the HEV-CP N terminus is a signal peptide followed by an arginine-rich domain that potentially plays a role in viral RNA encapsidation during assembly. HEV-CP is a key antigen that stimulates the host immune response, and six antigenic domains have been identified. One neutralisation site has been mapped to the polypeptide region between amino acids 452 and 617.

5.55.3 *Genotypes*

The four genotypes of HEV have been recognised that fall into two noteworthy groups:

- Genotypes 1 and 2 are human infections recognised for causing acute hepatitis via waterborne and faecal–oral transmission.
- Genotypes 3 and 4 are primarily swine infections in domestic and wild pigs, deer, and bears; these genotypes can also infect humans as unintentional hosts (zoonoses), leading to severe acute hepatitis and often chronic hepatitis in immunosuppressed individuals.

HEV genotypes show geographic variation and transmission mode (e.g. waterborne versus zoonotic). Genotypes 1 and 2 are predominantly found in Asia, India, and North Africa; genotype 2 in Mexico and West Africa; genotype 3 in Western countries and in Asia and North America; genotype 4 in Asian and European countries. The prevalence rates of hepatitis E antibodies are higher in developing (10–70%) than in developed countries (1–21%). Asia and Africa demonstrate the highest prevalence due to the consumption of contaminated drinking water.

5.56 Test for Screening and Diagnosis of Virological Infections

HBV infection screening involves identifying HBsAg while screening for HCV infection relies on identifying anti-HCV antibodies by enzyme immunoassay. However, confirming the infection requires the detection of HBV DNA or HCV RNA by nucleic acid tests (NATs). A real-time PCR-based or transcription-mediated amplification-based assay is required for these tests. Several points of care (POC) or rapid diagnostic tests have also evolved, which can be used in the wards or physician offices. Most of the POC tests are so far based on immune essays.

5.57 ELISA

The enzyme-linked immunosorbent assay (ELISA) is regularly used to quantify antibodies, antigens, or proteins, e.g. HIV disease, pregnancy tests, and cytokines or solvent receptors estimation in cell supernatant or serum. ELISA assessments are mostly completed in 96 well plates, enabling different examples to be estimated in a solitary investigation. These should be uncommon absorbent plates (e.g. NUNC Immuno plates) to guarantee that the counteracting agent or antigen adheres to the surface. Every ELISA measures a particular antigen, and primers for an assortment of antigens are generally available.

The sandwich ELISA has two arrangements of antibodies utilised to recognise antigens (like secreted cytokines). Various steps of the test are given below.

1. Antibody coating: Using overnight incubation particular capture antibody is immobilised on high protein-binding plates. Plates are obstructed with unrelated proteins, e.g. albumin. One can also get pre-coated plates commercially.
2. Protein capture: Samples and standard dilutions are transferred to the wells, and the antigen is trapped by binding to the antibodies.
3. Detection antibody: Specific biotinylated detection antibodies in the wells enable the identification of the captured protein.
4. Streptavidin–enzyme conjugate: Streptavidin conjugated with alkaline phosphatase or horseradish peroxidase is put in the wells and binds to the biotinylated antibody.
5. Addition of substrate: Colorimetric substrate added will form a coloured solution when catalysed by the enzyme.
6. Analysis: Absorbance is determined in an ELISA reader, and the quantity of protein in the samples is estimated. The amount of colour obtained is proportional to the amount of antigen present.

5.58 PCR Versus bDNA Assays

Polymerase chain reaction (PCR) is usually done to recognise a tiny section of the target DNA, which is a prototype for generating the primers that trigger the reaction. The DNA is progressively amplified in number using polymerases which need a supply of nucleotides. A small fragment of DNA serves as a primer. For detection of RNA, it has to be reverse transcribed to DNA for processing by this technique; hence, it is called reverse transcriptase PCR or RT-PCR. Quantitative PCR in real-time or quantitative real-time PCR is used to the specific DNA or RNA level in a biological sample. Nested PCR is a modified PCR to enhance the sensitivity and specificity of the assay by using two primer sets targeted against the same target and two consecutive PCR reactions.

Thus, the conventional methods used to detect the expression levels of particular mRNA present in biological samples are Northern blots, ribonuclease protection analysis, microarrays, and quantitative polymerase chain reaction (qPCR). These methods are cumbersome, time-consuming, and labour-intensive. Two commonly performed tests are PCR and bDNA. The branched DNA (bDNA) assay offers reliable quantification of nucleic acid molecules. It differs in the underlying principle compared to PCR since it directly estimates nucleic acid molecules at physiological levels by amplifying the reporter signal rather than replicating target sequences for detection, and, therefore, devoid of the errors common to extraction and amplification of target sequences. A comparison of these two techniques is given in Table 5.17.

Table 5.17 Differences between bDNA assay and PCR

bDNA assay	PCR
It stands for branched DNA assay	It stands for polymerase chain reaction
The bDNA assay is a sandwich nucleic acid hybridisation method that uses bDNA molecules to amplify signals from captured target RNA	PCR relies on in vitro amplification of the target sequence
The bDNA technique permits the estimation of RNA gene expression directly from biological samples, precluding RNA contamination, and degradation risk	PCR, preparation of purified RNA can be challenging, running RT-PCR, and obtaining consistent results can be exigent and necessitates technical skills
Reverse transcription is not necessary	PCR requires reverse transcription and PCR amplification
Averts biases associated with cDNA synthesis and enzyme inhibition or fluorescent dye incorporation	PCR shows bias associated with cDNA synthesis and enzyme inhibition or fluorescent dye incorporation
Averts sample losses	PCR leads to sample losses
Assay workflow is uncomplicated compared to PCR	The workflow of PCR is complicated when compared to the bDNA assay
More cost-effective than PCR	RT-PCR technique is not economical since it provides only one result for one target gene per sample in one reaction, unlike bDNA, which can provide up to 80 different targets per sample per reaction
Fewer steps in RNA purification, reverse transcription, and other laborious real-time PCR steps make it less time and energy-consuming	The time to achieve real-time PCR data from the sample and primer is 1 month for one target gene

5.59 Genotyping Methods

Hepatitis B infection (HBV) increases the risk of liver cirrhosis and HCC. Eight diverse HBV genotypes are related to unmistakable geological circulations. Distinctive genotyping techniques are such as serotyping, sequencing, multiplex PCR, oligonucleotide microarray chips, INNO-LiPA, turn around dab smudge, restriction fragment polymorphism (RFLP) restriction fragment polymorphism (RFMP), intruder test, and continuous PCR. A few specialists have examined the impact of HBV genotypes on clinical results in incessant HBV patients.

More than ten diverse HBV genotyping techniques demonstrate different affectability, specificity, duration of process completion, and cost. The examination of these genotyping techniques is in Table 5.18.

Table 5.18 Genotyping techniques

Genotyping technique	Advantages	Limitations
Sequencing	• Gold standard, since it does not rely on a small nucleotide or amino acid sequences as other methods do, it is more consistent • Can detect new genotypes and recombination between genotypes • Simpler and more economical than sequencing	• Less efficient in detecting mixed genotypes • Tedious, not cost-effective and necessitates expertise
INNO-LiPA	• Simpler and more economical than sequencing • The overall success rate for identifying all eight HBV genotypes was 98% • Highly specific • Can recognise mixed genotypes	• Insensitive for HBV genomes with SNP or deletions in the sequences to which probes bind • Comparatively less cost-effective than PCR-based and serological techniques
RFLP	• Simple and cost-effective method • Can be used in large populations studies • Can determine subgenotypes	• Inconsistent results (<6%) • SNP at enzyme restriction sites affect the method sensitivity
Multiplex PCR	• Rapid, simple, and cost-effective method • Can be used in large population studies • Can determine subgenotypes • Accuracy reported to be 93% • Sensitive in detecting mixed genotypes (even if the minor genotype is as low as 10%	• Comparatively less cost-effective than PCR-based and serological techniqueses

(continued)

Table 5.18 (continued)

Genotyping technique	Advantages	Limitations
Serotyping	• No PCR is required • Simple and cost-effective • Can be used in large population studies	• May give indeterminate results (<6%) • SNP at enzyme restriction sites may influence method sensitivity • Indeterminate samples range from 2.9% to 4.5% • SNP at primer site may influence method sensitivity
Oligonucleotide microarray chips	• Sensitive • Determines genotype even at titers as low as 2000–5000 copies/ml • Superior to sequencing in determining mixed genotypes	• Less cost-effective than real-time PCR and sequencing • Sensitivity influenced by SNPs or deletions in the sequences to which the probes bind • Sensitivity influenced by SNPs or deletions in the sequences to which the probes bind
Reverse dot blot	• Can detect mixed genotypes • High sensitivity • Cost-effective, quick, and precise • Detection limit for genotypes B and D was found to be 10^2–10^3, and for A, C, and E and was 10^3–10^4 copies/ml	• Sensitivity influenced by SNPs or deletions in the sequences to which probes bind
RFMP	• High sensitivity • Can also be used to detect YMDD mutations with higher sensitivity and lower detection limit (100 copies/ml) compared to sequencing and can also detect mutations even in samples containing wild-type	• SNP at enzyme restriction sites may influence method sensitivity • Requires the availability of MALDI-TOF mass spectrometry (bulky, expensive, and requires trained operators)
Invader assay	• Extremely high sensitivity	• Sensitivity may be affected by SNPs or deletions in the sequences to which probes bind
Real-time PCR	• Low cross-contamination • Time-saving • High throughput • High sensitivity • Superior to sequencing and oligonucleotide microarray chips in detecting mixed populations	• SNP at primer site may affect method sensitivity • The lower ability to distinguish between genotypes with proximity between melting temperature

5.60 Malarial Hepatopathy

Malaria affects millions of individuals every year across the globe. The liver is the site for propagating the lifecycle of this parasite and is at risk of being affected. Hemolytic jaundice is quite common. Besides, sometimes an array of liver dysfunctions ranging from mild abnormalities of liver transaminases to fulminant liver failure has been reported. Malarial hepatopathy is the jaundice associated with malaria, the clinical significance of which is yet unclear.

5.60.1 Pathogenesis

Malarial hepatopathy can be diagnosed if the patient has (a) demonstrable *P. falciparum* infection; (b) a more than threefold rise in transaminases, persisting more than 24 h with or without conjugated hyperbilirubinemia; (c) viral hepatitis has been ruled out; and (d) patient improves with antimalarial drugs or autopsy shows of widespread falciparum infection and suggests severe infection. The organ damage occurs due to the attachment of RBC containing the parasites to the blood vessels and endothelium of the sinusoids leading to obstructions in capillaries or smaller portal vessels called hypoxemia or stagnant anoxemia. Liver function abnormalities reported are similar to those seen in severe systemic infections and endotoxemia. Jaundice in malaria can be due to severe abnormalities, namely malarial hepatopathy, intravascular hemolysis of parasitised RBC, septicemia, microangiopathic hemolysis associated with DIC/G6PD-related hemolysis, drug toxicity related to anti-malarial drugs, and also due to an unrelated cause.

Coma in a patient with cerebral malaria is largely due to cerebral malaria rather than hepatic encephalopathy. INR is rarely raised unless there is DIC present.

5.60.2 Clinical Manifestations

Paroxysms of fever, shaking, chills, and sweats occur every 48 or 72 h, depending on the species of plasmodium involved. Hepatosplenomegaly is common. Features of severe malaria include cerebral malaria, often presenting as coma, severe anemia, hypoglycaemia, metabolic acidosis, acute renal failure (serum creatinine >3 mg/dl), ARDS, shock or 'algid malaria', DIC, and hemoglobinuria may also be present.

5.60.3 Hepatic Manifestations

Malarial hepatopathy may manifest as a fulminant clinical illness (type A) with coma, severe jaundice, purpura, and renal dysfunction (severe malaria), or often indistinguishable from fulminant hepatic failure (i.e. coma and severe jaundice).

While type B is a milder illness with only fever, headache, and vomiting. Diagnosis is not difficult to make in hyperendemic areas, although sporadic cases may pose a challenge.

Roughly one-third of patients with *Plasmodium falciparum* or vivax may have hepatomegaly and splenomegaly. Jaundice has been reported in approximately 2–5% of patients with falciparum malaria. INR is rarely ever deranged in the absence of DIC. A rise in transaminases has been reported three to five times the upper limit of normal. Liver biopsy reveals hyperplasia of KCs with malarial pigmentation. Centrilobular necrosis reported from the autopsy series has been attributed to terminal shock.

5.60.4 Diagnosis

Severe malarial hepatopathy may be confused with ALF in non-endemic areas, but several features help to differentiate this presentation from ALF (see Table 5.19). Malaria can be diagnosed by microscopy of thick smears for parasite detection; and examination of thin smears for identifying species.

One can also do quantitative buffy coat tests and other rapid diagnostic tests (RDTs) such as histidine-rich protein, lactate dehydrogenase, and antigen-based immune-chromatography.

5.60.5 Treatment

The treatment for malaria depends upon the species and the risk of resistance among parasites.

For falciparum malaria, the drug of choice may be

- Artesunate 2.4 mg/kg i.v. Bolus at admission, repeated at 12 and 24 h, followed by once a day for 7 days + doxycycline 100 mg p.o. Every 12 h.
- or

Table 5.19 Differential diagnosis between malarial hepatopathy and ALF

	Malaria	ALF
Fever	From the onset, persistent	In prodrome
Liver size	Hepatomegaly	Small/normal
AST, ALT	< or ~5 times	↑↑↑ (>10×)
Bilirubin	Maybe unconjugated	Conjugated
PT	Normal unless DIC	↑↑↑
Anemia	Often	Absent
Hemolysis	Often	Absent
Specific tests	PBS, malaria antigen	Viral markers

- Quinine 20 mg/kg loading dose, followed by 10 mg/kg i.v. Infusion 8 hourly + doxycycline 100 mg p.o. Every 12 h. Doxycycline should be replaced by Clindamycin in pregnant females and pediatric patients. Exchange transfusion is a treatment option for parasitemia >10%.

5.61 Enteric Hepatopathy

Salmonella enterica serotype typhi causes typhoid, characterised by fever and abdominal symptoms. Approximately 16 million cases per year may be seen worldwide, of which 600 may end fatally. With the improvement of hygienic standards, the disease is becoming less common. The disease is endemic in the developing world and is mainly seen among travellers in Western countries.

5.61.1 Clinical Manifestations

Symptoms usually begin about 2 weeks after ingesting viable organisms through contaminated water or food. Fever is usually the first symptom and may be associated with abdominal and sometimes respiratory symptoms. Fever may have an insidious or abrupt onset, and diarrhoea and pain abdomen often follow the fever. Other findings may include headache, relative bradycardia, leukopenia, hepatomegaly, and splenomegaly. A minority of patients may have an evanescent truncal rash 2–4 mm in size and erythematous. Some patients may have constipation instead of diarrhoea, while the latter is more common among children and those with HIV infection.

Many untreated cases may have defervescence of fever by 4 weeks, although 5–10% may develop complications during the third week of the illness. Complications include intestinal perforation, bleeding, jaundice, DIC, and encephalopathy.

5.61.2 Hepatic Manifestations

The liver is involved by hematogenous spread from the intestines, and the disease primarily involves the reticuloendothelial system. Hepatomegaly, altered LFTs, or jaundice may be seen in nearly half the cases. Jaundice indicates a severe disease seen less often (reported in 0.4–26% of cases). AST is generally more elevated than ALT, and a rise may be seen three to five times of upper limit of normal. Alkaline phosphatase is usually mildly elevated.

5.61.3 *Diagnosis*

The diagnosis is made by positive blood cultures, which are positive in 60–80% of cases during the first week of illness. Bone marrow cultures may have greater yield and stay positive for longer. Serological tests often become positive after about 2 weeks. Typhidot has a sensitivity of 95–97% and a specificity of >89% (Widal test is non-specific). One study has shown that ALT:LDH levels are less than four in typhoid fever, while it may be four or more in viral hepatitis.

5.61.4 *Treatment*

The drug of choice is Ceftriaxone i.v. 50–75 mg/kg/day for 10–14 days.

5.62 Dengue Hepatopathy

Dengue is a severe, generally self-limited febrile zoonotic sickness often called 'breakbone fever'. It is caused by a flavivirus with four antigenically distinct genotypes. Infection with one type does not provide reliable immunity against other genotypes. Every year nearly 50 million individuals are affected around the world. The mosquito *Aedes aegypti* disseminates the infection and is common in the tropical and subtropical environments of Asia, Africa, the United States, Australia, and the Caribbean.

5.62.1 *Clinical Manifestations*

The incubation period ranges from a few days to up to 2 weeks. Usually, it manifests as a sudden onset of fever, extreme myalgias, arthralgias, migraine, retro-orbital pain, rash, and gastrointestinal symptoms. Serious complications usually develop in adults with second or subsequent infections due to antitoxic antibodies. Bleeding manifestations (dengue hemorrhagic fever) due to thrombocytopenia and shock (dengue shock syndrome) due to increased capillary permeability usually are seen towards the end of the 7-day fever or soon thereafter.

5.62.2 *Hepatic Manifestations*

Elevations of serum aminotransaminases are nearly universal (80–90%) and can be associated with features like viral hepatitis with hepatomegaly, jaundice, and right upper quadrant pain. The biochemical abnormalities usually peak on day 9 after the onset of fever and then gradually revert to normal levels within 15–21 days. Acute liver failure may occur, but fatality is generally attributed to dengue hemorrhagic fever or dengue shock syndrome. Liver involvement is usually due to the direct involvement of KCs and hepatocytes. Pathologic assessment of the liver shows finding similar to Yellow fever and may involve the following: necrosis of centrilobular structures, fatty change, hyperplasia of the KCs, and mononuclear infiltration of the portal tract and Councilman bodies.

5.62.3 *Treatment*

Currently, since no particular therapeutic regimen is available, management consists of symptomatic support. No vaccine has been developed yet.

5.63 Hydatid Illness of Liver (Echinococcosis)

5.63.1 *Introduction*

Human echinococcosis occurs due to *Echinococcus granulosus* (family: Taeniidae; class: cestode) and is a zoonotic infection spread by canines in livestock (sheep, cattle, and horse) shelter areas. These are classified based on genetic composition and infectious attributes. Of the six types, four are common: *E. granulosus* (hydatid cyst), *E. multilocularis* (alveolar echinococcosis), *E. vogeli*, and *E. oligarthrus* (polycystic echinococcosis).

There are two hosts in the life-cycle of the parasite, primary or definitive host and intermediate host. Adult forms occupy the intestines of primary hosts (feline and canines), causing only intestinal parasitosis. These 'Definitive hosts' disseminate millions of parasite eggs on defecation. Herbivorous animals and humans become 'intermediate hosts' for the parasite when they consume contaminated grass or food items. The embryo (or oncosphere) released in the GIT anchors itself with hooks to the intestinal wall and subsequently reaches the systemic circulation, followed by the liver. Thus, the liver is the most frequently affected site, responsible for more than half of the cases, while the lungs account for one-third of the cases, and the least affected include the spleen, kidneys, heart, bones, and brain. The embryo adopts the cyst form after the scolex is removed on settling in an organ, which consists of an exo- and endocyst named cuticula and germinal membrane, respectively. The cyst contains sterile, clear fluid enclosed in a fibrous capsule called a 'pericyst'. *E. granulosus* is prevalent in South America, Eastern Europe, and Asia, with an incidence rate of 50 per 100,000 person-years. Andhra Pradesh and Tamil Nadu have higher incidences in India than in other states.

5.63.2 Clinical Features

The patient may remain asymptomatic for several years. Some patients may present with epigastric and/or right hypochondriac pain, nausea, and emesis. Only a single organ is affected in the majority of patients (85–90%), with more than three-fourths of the patients harbouring only a single cyst. The liver is the commonest site of hydatid cysts, and one may get symptoms caused by local pressure of the cyst such as portal hypertension, biliary obstruction and jaundice and fistulae, ascites to abscess formation, and, finally, cholangitis with biliary ruptures. Cyst rupture can cause immunologic symptoms (pruritus, urticaria, and anaphylactic shock). The ruptured cysts release viable cystic constituents and larvae into the peritoneum, causing secondary hydatidosis. Bacterial super-infection was noted in >7% of patients with hydatid cysts. The cysts may rupture spontaneously or due to blunt trauma. Rupture of the cyst may be (a) Contained rupture in which only the endocyst is broken, and the cyst constituents stay within the pericyst. The size of the cyst remains the same on imaging. (b) A communicating rupture involves rupture of the endocyst, and cyst contents migrate via biliary radicals. Unlike contained rupture, such cysts become smaller on imaging with undulating membranes. (c) While in a direct rupture, both cysts rupture, leading to leakage of contents into the peritoneum and the spread of disease.

5.63.3 Classification

Sonographically the cysts can be classified by the Gharbi classification into five types (Table 5.20).

Table 5.20 Gharbi classification

Type	Description
I	Pure fluid collection
II	Fluid collection with a detached membrane
III	Fluid collection with multiple septa and/or daughter cysts
IV	Hyperechoic with high internal echoes
V	Cyst with reflecting calcified thick wall

Table 5.21 WHO classification of hydatid cyst

WHO class	Description
CL	• Unilocular anechoic cystic lesion without any internal echoes and septations
CE1 (type 1)	• It may be a uniformly anechoic cyst with fine echoes settled in it, representing hydatid sand
CE2 (type 2)	• A cyst with multiple septations gives it a multivesicular, rosette, or honeycomb appearance with a unilocular mother cyst. This stage is the active stage of the cyst
CE3 (type 3)	• Unilocular cyst with daughter cysts with detached laminated membranes (CE3a) appearing as a water lily sign. This is the transitional stage of the cyst. A later stage (CE3b) would indicate a predominantly solid cyst with daughter vesicles
CE4 (type 4)	• Mixed hypo- and hyperechoic contents with absent daughter cysts; these contents give an appearance of a ball of wool (ball of wool sign), indicating the degenerative nature of the cyst
CE5 (type 5)	• Arch-like thick, partially or completely calcified wall. This stage of cyst is inactive and infertile

5.63.4 WHO Classification of Hydatid Cyst

This classification differs from the Gharbi classification and is detailed in Table 5.21.

5.63.5 Management

The intervention required depends on the nature and location of the cyst and the patient's fitness for surgery. Small, densely calcified cysts only need monitoring. Cyst with viable daughter cysts that continue in the calcified shell, removal by excision may be considered in patients deemed fit for surgery. Smaller cysts (<4 cm) lying further within the hepatic parenchyma in the absence of complications may be treated conservatively, while superficial ones may be excised.

Medical therapy is with albendazole given in 10 mg/kg doses (usually 400 mg) twice daily and is indicated in the following cases: (a) surgically unfit patients with multiple cysts and peritoneal involvement; (b) to control cyst pressure, secondary seeding, and risk of repeat episodes in pre-surgical and pre-puncture cases.

Contraindications to medical therapy include (a) large cysts at risk of rupture; (b) inactive or heavily calcified cysts; (c) early pregnancy; (d) chronic hepatic conditions and bone marrow suppressive disorders since these increase the risk of adverse events.

Medical treatment is undertaken. (a) Assured cure—for univesicular cyst (types 1 and 2), 3–6 month treatment has a high success rate (82%) but is also at risk of relapse (>25%) within 2 years. Follow-up is recommended for life. (b) Drop in viability and cyst size can be attained in multi-vesicular cysts (type 3). Still, the germinal layer is quite resistant to treatment and achieving a cure is difficult, (c) pre-operative attenuation of viability of univesicular cysts prior to planning elective surgery or percutaneous drainage, (d) peri-operative or peri-interventional prophylaxis—ideally should be initiated at least 3 days prior to the surgical percutaneous treatment and continued for 3–8 weeks post-treatment in uncomplicated cases and 3–6 months in complicated cases.

Surgical management could be either partial or total cystectomy and offers a definitive cure in most cases. Surgery aims to completely remove the cyst without contaminating its contents and adverse events.

5.63.6 PAIR

PAIR is an acronym for a puncture, aspiration, injection, or reaspiration. This percutaneous approach degrades the cyst's germinal layer and empties the cystic contents. It involves ultrasound-assisted percutaneous needle puncture, with the aspiration of the cyst, then injection of an appropriate protoscolicide (20% sodium chloride or 95% ethanol) and cyst reaspiration is performed 15–20 min later. There is always a risk of anaphylaxis on exposure to cyst fluid. The cyst fluid aspirated is investigated for protoscoleces and bilirubin, and to attenuate the risk of secondary cysts, one also administers chemotherapy with albendazole. CE2 and CE3b stages are difficult because they have several sections necessitating individual puncture and are likely at risk of relapse PAIR. Larger, well-designed comparative studies are necessary to obtain more evidence to recommend one procedure over the other.

5.64 Interferons

Interferons are small proteins of the cytokine group. They modulate the immune system and suppress the tumour growth. They signal to the immune system during viral infections and elevate the antigens presenting to the T cells. They enable the healthy host to resist the new viral infection. They also stimulate the activity of macrophage B cells, alter the T cells, and enhance apoptosis. During viral infections, they may act as inducers, regulators, and effectors of innate and acquired immunity.

5.64.1 Types

Interferons are classified mainly into three types:

- IFN α: Primarily participates in the innate immune response against viral infections. There are 13 subtypes. Synthetic long-acting IFN α are Pegylated IFN α-2a and Pegylated IFN α-2b.
- IFN β: They are also important components of the innate immune response. The subtypes are IFN β1, IFN β2, and IFN β3. Its use in multiple sclerosis has shown a reduction in relapse rates.
- INF γ regulates immune and inflammatory responses and has an anti-tumour effect.

5.64.2 Induction of IFN

- Interferons are produced on exposure to pathogens. These bind to certain unique sites on the pathogens, such as viral glycoproteins, bacterial endotoxins, bacterial flagella, and CpG motifs, by pattern recognition receptors, such as membrane-bound TLRs or cytoplasmic receptors and are released. TLRs induce the IFNs and bind with dsRNA receptors activating IRF3 and NF-KB, which trigger the production of several inflammatory proteins. The activators of IFN synthesis include IL1-3, TNF, and CSF.

5.64.3 Clinical Application

1. IFN α
   (a) Hairy cell leukaemia and chronic myeloid leukemia
   (b) AIDS-related Kaposi's sarcoma and malignant melanoma
   (c) Condylomata acuminate
   (d) Chronic hepatitis B and C
2. IFN β: multiple sclerosis
3. IFN γ: chronic granulomatous disease

5.64.4 Toxicity/Side Effects

Common side effects: flu-like symptoms such as headache, muscle aches, joint aches, fever/chills, feeling sick, vomiting, loss of appetite, feeling tired, diarrhea, depression, mood swings, poor concentration, and vagueness.

Less common side effects: elevated liver function test and cytopenia.

5.64.5 Precautions

- Caution should be exercised for patients with pre-existing seizure disorders.
- Alcohol should be avoided.
- Should not undergo any immunisation or vaccination without the consult of the physician. Combining IFN with Ribavirin may develop certain blood problems like anemia and tooth and gum pain.
- Caution should be taken for administering to older adults due to the sensitivity of the drug's effect, effect on the heart, dizziness, or mood changes.
- Pregnant ladies should not be administered.

5.65 Ribavirin

Ribavirin is used along with interferon for the treatment of chronic hepatitis C virus infection.

5.65.1 Mechanism of Action

Ribavirin is a synthetic nucleoside analogue; however, its mechanism, combined with IFNs against HCV, is not completely understood.

5.65.2 Indications and Usage

Ribavarin, in combination with peginterferon alfa-2a, is indicated for the treatment of adults with chronic hepatitis C virus infection who have compensated liver disease and have not been previously treated with interferon-alpha and patients with cirrhosis (Child–Pugh class A).

5.65.3 Contraindications

- Patients with known hypersensitivity to ribavirin or any component of the drug.
- Women who are pregnant.
- Men whose female partners are pregnant.
- Patients with hemoglobinopathies (e.g. thalassemia major or sickle-cell anemia).
- Pancreatitis.

Ribavarin, in combination with peginterferon alfa-2a/2b, is contraindicated in patients with:

- Autoimmune hepatitis.
- Hepatic decompensation (Child–Pugh classes B and C) before or during treatment.

5.65.4 Side Effects

- Pregnancy: Ribavirin may cause congenital disabilities and/or death of the exposed fetus.
- Anemia: The primary toxicity of ribavirin is hemolytic anemia.
- Fatal and nonfatal myocardial infarctions have been reported in patients with anemia caused by ribavirin.
- Pulmonary: Pulmonary symptoms, including dyspnea, pulmonary infiltrates, pneumonitis and occasional cases of fatal pneumonia, have been reported during therapy with ribavirin and interferon. Sarcoidosis or the exacerbation of sarcoidosis has been reported.

5.66 Eltrombopag

- Eltrombopag is a nonpeptide, oral platelet growth factor
- Acts as a thrombopoietin-receptor agonist
- *Mechanism of action*
- Interacts with the transmembrane domain of the thrombopoietin receptor
- ↓
- Induces proliferation and differentiation of megakaryocytes
- ↓
- Increases in platelet production
- The most common side effects of eltrombopag are headache (14–36%), dry mouth, and abdominal pain.

5.66.1 Clinical Utility

- Thrombocytopenia may occur in patients with cirrhosis due to HCV; hence, the use of antiviral treatment with interferon may be difficult. In these patients, the use of eltrombopag enabled the completion of interferon treatment in 50–65%.

5.67 Sorafenib

5.67.1 *Mechanism of Action*

- Sorafenib suppresses tumour growth by inhibiting the serine/threonine and receptor tyrosine kinases found in both the tumour cell (c-Raf, b-Raf, V600E b-Raf, KIT, and Flt3) and in the tumour vasculature (c-Raf, VEGFR2, VEGFR3, and PDGFR-beta).
- The pivotal trials Asia-Pacific and Sorafenib HCC demonstrated improved survival (10 months) of patients with HCC at a late stage in child A.
- Lenvatinib is non-inferior to sorafenib as a first-line. Regorafenib, cabozantinib, and ramucirumab are the second-line agents.
- Due to the high rate and fast development (<6 months) of resistance, only 30% of patients can benefit from sorafenib.
- Adverse events with sorafenib mainly included gastrointestinal, physical, or skin diseases (e.g. hand and foot skin reactions, weight loss, and diarrhea). In serious cases, sorafenib can cause high blood pressure and abdominal pain, leading to treatment withdrawal.

5.68 Oral Antivirals in the Treatment of Chronic Viral Hepatitis B

The currently recommended first-line antiviral treatment for chronic hepatitis B is tabulated in Table 22.

Table 22 Oral antivirals in the treatment of chronic hepatitis B

	Entecavir	Tenofovir disoprovil fumarate (TDF)	Tenofovir alafenamide (TAF)
Mechanism	Nucleoside analogue—inhibits DNA polymerase	Nucleotide analogue	Nucleotide analogue
After 5 years of treatment			
HBV DNA viral suppression	99%	99%	Similar to TDF
HBeAg loss	53%	49%	
HBe Ag seroconversion		40%	
HBsAg loss		10%	1%
HBsAg seroconversion		8%	

Chronic treatment with oral nucleo(t)ide analogues have been shown to reverse necroinflammation and fibrosis histologically. It has been shown to improve clinical outcomes by decreasing the incidence of HCC and new onset decompensation

5.69 L-Ornithine L-Aspartate (LOLA)

- L-Ornithine L-aspartate (LOLA) is a compound salt of ornithine and aspartate.
- Increases in ammonia levels in the blood are pathogenetic in HE. Management of HE especially tends to reduce the amount of ammonia in the blood.
  - Catalyse the conversion of ammonia to urea or glutamine. LOLA reduces ammonia levels in patients with cirrhosis and hepatic encephalopathy.

5.69.1 *Mechanism of Action*

- Urea is synthesised in the periportal hepatocytes. Ornithine is an activator of ornithine transcarbamoylase and carbamoyl phosphate synthetase. Besides, ornithine acts as a urea genesis substrate (Fig. 5.9). Therefore, LOLA stimulates the periportal urea cycle.
- After conversion to α-ketoglutarate, aspartate and ornithine also serve as carbon sources for perivenous glutamine synthesis. LOLA upregulates glutamine synthesis in the skeletal muscle by providing a substrate for glutamine synthetase (GS).
- Ammonia is utilised during urea formation and glutamine synthesis, and consequently, LOLA reduces blood ammonia levels. Thus, LOLA stimulates the urea cycle and glutamine synthesis, resulting in ammonia detoxification.

5.69.2 *Dosage Schedule*

Oral: 5–10 g thrice a day
Intravenous: 20 g over 4 h infusion in 5% dextrose solution

5.69.3 *Studies*

Cirrhosis with Hepatic Encephalopathy

- There are only a few well-designed RCTs that have assessed LOLA therapy for HE. One compared LOLA with lactulose, while two more studies compared it with a placebo.
- In a meta-analysis, the RR of overt HE for LOLA versus placebo was 1.89 ([95% CI 1.32–2.71]; $p = 0.0005$), whereas RR for LOLA versus lactulose was 1.20 ([95% CI 0.54–2.67]; $p = 0.66$). The results indicated an association between LOLA therapy and improvement of HE and no significant difference in response to the treatment between LOLA and lactulose.

- Poo and colleagues conducted the first open-label clinical trial reporting the safety and efficacy of LOLA treatment versus lactulose treatment for managing hepatic encephalopathy. Patients treated with both study drugs have exhibited an improvement in quality of life according to the Visual Analogue Scale of the EuroQol Survey, and this improvement was significantly greater in patients given LOLA.
- LOLA is effective in reducing ammonia levels and improving psychometric performance in patients with cirrhosis.

Acute Liver Failure (ALF)

In the experimental model of acute liver injury, LOLA reduced the ammonia levels. LOLA was, therefore, a potential candidate for the treatment of ALF patients. But, a study by Acharya et al. has shown that it is ineffective in ALF patients.

5.69.4 Indications

- Hyperammonemia due to urea-cycle defects
- Hepatic encephalopathy, particularly due to cirrhosis

5.69.5 Contraindications

- Pregnancy
- Renal failure (serum creatinine of more than 3 mg/dl)

5.69.6 Limitations

- LOLA detoxifies ammonia to glutamine which circulates inside the body. Glutamine, in turn, gets broken down to ammonia later. Thus, there is no net ammonia removal from the body.
- Another compound, LOPA, is being evaluated in hyperammonemia states. This drug forms phenyl acetyl glutamine excreted in urine, resulting in net ammonia removal from the body.

5.69.7 *Future Studies*

- To be done in minimal hepatic encephalopathy.
- To be tested in HE of ACLF patients.

5.70 Lactulose and Other Ammonia-Lowering Agents

1. **Lactulose**
   (a) A non-absorbable disaccharide of galactose and fructose.
   (b) Stays undigested by the time it reaches the colon, where colonic bacteria degrade it to acetic acid and lactic acid.
   (c) **Mechanism of action**
   - These carboxylic acids (acetic and lactic acid) lower the intraluminal pH, due to which the growth of the intestinal urease bacteria (ammoniagenic bacteria) is inhibited.
   - Lactulose also has a cathartic effect, reducing the absorption of ammonia, thereby controlling the amount absorbed systemically and eliminating elevated faecal nitrogen waste.
   - It also attenuates the uptake of glutamine by the intestinal wall and subsequent metabolism into ammonia and consequent reduction in circulating levels.

   (d) **Side effects**
   - Abdominal cramping, bloating, nausea, vomiting, flatulence, and abdominal distension.
   - Lactulose treatment affects intestinal absorption, and this may worsen existing nutritional deficiencies in patients with end-stage liver disease, leading to a poorer prognosis post-liver transplantation.

2. **Oral antibiotics**
   (a) Orally administered antimicrobial agents aim at inhibiting urease-containing bacteria in the colon to reduce the production and prevent the absorption of ammonia through the GIT.
   (b) Antibiotics such as neomycin, metronidazole, and vancomycin lower blood ammonia in patients with end-stage liver disease. Owing to the associated serious adverse effects, these antibiotics find limited use.
   (c) **Neomycin**, an aminoglycoside antimicrobial, inhibits protein synthesis by binding to the 30S ribosomal unit. It is associated with nephrotoxicity and ototoxicity due to systemic absorption.

(d) **Metronidazole**, a nitroimidazole antimicrobial, is cytotoxic. It is excreted via the liver through hepatic oxidation, increasing the risk of toxicity in patients with chronic liver disease. Long-term use also increases the risk of neurotoxicity (peripheral neuropathy).
(e) **Vancomycin** is a glycopeptide antimicrobial that inhibits cell-wall synthesis, lowers blood ammonia, and attenuates HE in cirrhosis patients. However, it supports bacterial overgrowth and increases enteric bacterial resistance risk.

3. **Probiotics**

(a) Probiotic therapy enhances the properties of the intestinal microflora since it comprises monocultures or mixed cultures of live microorganisms.
(b) Clinical evidence demonstrates the benefits of probiotics in the treatment of HE by modifying intestinal bacteria by the colonisation of non-urease bacteria and decreasing blood ammonia concentrations. These therapeutic agents are well-tolerated, and, hence, adherence rates are high.

4. **Sodium benzoate**

(a) Glycine is metabolised by a four-protein glycine cleavage system, which produces ammonia as an end-product. Sodium benzoate combines with glycine to form hippuric acid, which is renally eliminated. Its administration prevents glycine metabolism and, consequently, the production of ammonia.
(b) It has been frequently used to alleviate hyperammonemia in paediatric patients with inborn urea-cycle disorders.

5. **Sodium phenylacetate/phenylbutyrate**

(a) Sodium phenylbutyrate rapidly oxidises into phenylacetate and has been shown to attenuate hyperglutaminemia in children with urea-cycle enzyme deficiencies by reducing hyperammonemia. It combines with glutamine in the hepatic and renal sites to form phenylacetylglutamine, which is eliminated in urine. Thus, glutamine-induced production of ammonia is prevented.
(b) The use of sodium phenylacetate/phenylbutyrate has never been investigated in the treatment of HE.

5.71 Rifaximin

- Rifaximin is a rifampin structural analogue. The addition of a pyrido-imidazole ring to the rifampin molecule makes it a water-insoluble, poorly absorbable antibiotic.

5.71.1 Mechanism of Action

- Exerts antibiotic effect by blocking RNA synthesis via interaction with DNA-dependent RNA polymerase.
- Has in vitro antimicrobial activity against Gram-positive, Gram-negative, aerobic and anaerobic bacteria (EAEC and Shigella).
- Alters epithelial binding and plasmid translocation.
- Provides cytoprotection by activating pregnane X-receptor.

5.71.2 Indication

The current indications and evidence of the beneficial effects of rifaximin in various GI disorders are tabulated in Table 5.23.

Table 5.23 Indications for rifaximin treatment and evidence

Disorder	Evidence of beneficial effect	Recommended regimen
IBS	Randomised controlled multicenter studies show improvement in global IBS symptoms	400–550 mg three times daily for 2 weeks. May require intermittent
Treatment of TD	Randomised controlled trials show the reduced duration of the illness	200 mg three times daily for 3 days
Prevention of TD	Randomised controlled trials in patients travelling to south and southeast Asia and Mexico	200 mg twice daily or 600 mg daily while in a high-risk area
SIBO	Rifaximin normalises the hydrogen breath test and improves symptoms	400 mg three times daily for 2 weeks may require retreatment
Hepatic encephalopathy	Randomised controlled trials prove efficacy and safety	550 mg 2 or 3 times daily chronically
Diverticular disease	Randomised controlled trials showed that rifaximin improves symptoms and prevents recurrence	400 mg twice daily for 7 days every month
IBD and pouchitis SBP prophylaxis	Observational data and small pilot studies: retrospective study	400–800 mg twice daily for 12 weeks may require retreatment or intermittent treatment 400 mg three times daily chronically
Recurrent *Clostridium difficile* infection	Small case series and retrospective studies	400 mg twice daily for 2 weeks

5.71.3 The Emergence of Drug-Resistant Bacteria

Real-life data on enteropathogens isolated from traveller's diarrhoea in different continents over a decade have shown no change in minimum inhibitory concentration (MIC) levels of the isolates to rifaximin as compared to increasing MIC levels of isolates to other antibiotics like fluoroquinolones, cephalosporins, and macrolides.

5.72 Somatostatin Analogues

Somatostatin is a 14 amino acid peptide that suppresses pancreatic exocrine and endocrine secretion. Owing to its brief half-life, it requires prolonged infusions, which limits its clinical utility.

5.72.1 Actions

- Somatostatin has numerous endocrine, paracrine, and exocrine effects. It also inhibits the release of growth hormones.
- Somatostatin inhibits the secretion of insulin, glucagon, and pancreatic polypeptide. It also inhibits pancreatic enzyme and bicarbonate responses to cholecystokinin and secretin.
- Several stimulatory gastrointestinal hormones, gastrointestinal motility, and blood flow are also inhibited.

5.72.2 Somatostatin Analogues

- **Octreotide (Sandostatin)** is a cyclic octapeptide dispensed as a clear sterile solution of octreotide acetate salt in a buffered lactic acid solution administered subcutaneously (SC) or intravenously (IV) injection.
  - It is a long-acting octapeptide with actions similar to the natural hormone somatostatin.
  - The predominant effects of octreotide include the inhibition of growth hormone (GH), glucagon, and insulin. Decrease in response of luteinizing hormone to gonadotropin-releasing hormone, decrease splanchnic blood flow, and suppression of release of serotonin, gastrin, vasoactive intestinal peptide, secretin, motilin, and pancreatic polypeptide.
  - **Sandostatin LAR** is a long-acting dosage form administered intramuscularly (every 4 weeks) that gradually releases octreotide from the injection site, tackling its limitation of multiple administrations.

- **Pasireotide (Signifor)** is an injectable cyclohexapeptide somatostatin analogue.
  - Pasireotide binds to somatostatin receptors (SSTRs) to inhibit ACTH secretion and decrease cortisol levels.
  - Corticotroph tumour cells from Cushing's disease patients display excessive expression of SSTR5 compared to SSTR1-4.
  - **Signifor LAR** is a long-acting release form of pasireotide.
- **Lanreotide (Somatuline Depot)** is a prolonged-release formulation of Lanreotide, a synthetic octapeptide for deep subcutaneous injection.

5.72.3 Indications

1. Endocrinological indications
   (a) Acromegaly
   (b) Cushing's syndrome
   (c) Carcinoid
   (d) VIPoma
   (e) Symptoms associated with functional GEP-NETs
   (f) Treatment of unresectable, locally advanced or metastatic, moderate, well-differentiated GEP-NETs
2. Digestive indications
   (a) Variceal haemorrhage in cirrhosis
   (b) Prevention of complications after pancreatic surgery
   (c) Dumping syndrome
   (d) GI and lymphatic fistulas
   (e) Hepatorenal polycystosis
   (f) Refractory chronic diarrhea

5.73 HEV Vaccines

- Hepatitis E is a global concern, as one-third of the population is at risk of infection. It is transmitted feco-orally. Two patterns of transmission are noted: from contaminated water in resource-limited settings (genotypes 1 and 2) and from infected animal products (pork) in developed countries (genotype 3/4).
- The majority of infections are mild acute hepatitis. There is a small risk of progression to acute liver failure in immunocompetent patients (~1%). Chronic hepatitis has been noted in transplant recipients, patients on chemotherapy, and co-existing HIV infection—usually caused by genotype 3.

- **At-risk population**: Pregnant females and patients diagnosed with chronic liver disease. The case fatality rate is high in pregnant females, in whom worsening to acute liver failure is common (~20%). Similarly, patients with preexisting chronic liver disease carry a greater risk of decompensation and develop acute or chronic liver failure with high fatality (~30%).
- **Vaccine development**: HEV is not easily cultivable in cell cultures. So, the development of live attenuated or killed vaccines is difficult. Efforts have been made to develop recombinant subunit vaccines. ORF 2 (open reading frame 2) encoded viral capsid proteins are highly conserved, immunogenic, have cross-reacting antibodies across all genotypes, and are a useful candidate for vaccine development.
- **HEV recombinant vaccines**: Due to the difficulty in culturing HEV, recombinant vaccines are being manufactured. Two systems—*E. coli* and insect cells of *Baculovirus* vector—express HEV viral proteins.

  - HEV 239 *E. coli* (Hecolin Innovax) vaccines have been authorised for clinical use in China for adults aged 15–65 years at risk of severe HEV infection—pregnant females, travellers to endemic zones, military recruits, and food handlers. It involves three intramuscular doses (30 μg) and provides 100% efficacy against genotypes 1 and 4. Its efficacy against genotypes 2 and 3 needs further evaluation.
  - *Baculovirus*-derived HEV vaccines, manufactured by GlaxoSmithKline (GSK), have been studied in military volunteers in phase-1 and -2 trials in the United States and Nepal. After three intramuscular doses (20 μg) at 0, 1, and 6 months, the vaccine efficacy was 95% after the third dose when monitored for 2 years after the vaccination schedule.
  - Enough safety data regarding these vaccines is unavailable; therefore, WHO has not recommended its inclusion in the national immunisation schedule.

5.74 Direct Acting Antivirals (DAA): Indian Perspective

Various oral antiviral medicines are authorised for the treatment of chronic hepatitis C acting on protease or RNA polymerase.

- NS3/4A protease inhibitors: Boceprevir, Gleacaprevir, Grazoprevir, Paritaprevir, Simeprevir, Telaprevir, and voxilaprevir.
- NS5A inhibitors: Daclatasvir, Ledipasvir, Velpatasvir, Elbasvir, and Pibrentasvir.
- NS5B inhibitors: Sofosbuvir and Dasabuvir.

In India, sofosbuvir has been made available at compassionate prices compared to the Western market. So, Sofosbuvir-based regimes form the backbone of treatment for chronic hepatitis C and are tabulated in Table 5.24. Most of these combination have efficacy >95%. The treatment duration recommended is 24 and 12 weeks in patients with and without cirrhosis, respectively.

Table 5.24 Treatment regimens for chronic hepatitis C

	Treatment option	Alternate option
Genotypes 1,4, 5, and 6	Sofosbuvir 400 mg + Velpatasvir 100 mg	Sofosbuvir 400 mg + Ledipasvir 100 mg
Genotypes 2 and 3	Sofosbuvir 400 mg + Velpatasvir 100 mg	Sofosbuvir 400 mg + Daclatasvir 60 mg

5.75 Antifibrotic Therapies in Liver Disease: Current Status

Liver fibrosis occurs in due course owing to the response of repair mechanisms to persistent tissue inflammation. Hepatic myofibroblasts are the key cells which mediate fibrogenesis. Hepatic myofibroblasts are not seen in normal liver. HSCs, placed in the space of Disse, are the main source of myofibroblasts. HSCs activated by cytokines (transforming growth factor-β and platelet-derived growth factor) lead to the formation of myofibroblasts which lay down extracellular matrix (collagen) and promote fibrogenesis.

The current approach to antifibrotic therapies focuses on the following:

1. Curing primary disease (hepatitis B and hepatitis C)
2. Controlling inflammation (CCR2, CCR5, and downregulating PPAR)
3. Inhibition of activation of HSCs (downregulate PPAR, FXR, and Cannabinoid receptors-mediated TGF β activation of stellate cells)
4. Clearance of activated HSCs (Cannabinoid receptor antagonists activate NK cells and downregulate NFKB)
5. Inhibition of the production of extracellular matrix (ECM) (LOXL2 antibody, pirfenidone, and tissue inhibitor of metalloproteinase antagonist)

Drugs in developmental stages and phases of a clinical trial are tabulated in Table 5.25.

Table 5.25 Promising antifibrotic agents in clinical trials

Drug	Class	Phase of trial	Disease
Obeitocholic acid	FXR agonist	Phase 3	NASH
Cenriviroc	CCR/CCR5 inhibitor	Phase 3	NASH
Selonsertib	ASK-1 inhibitor	Phase 3	NASH
Pirfenidone	Pyridinone derivative	Phase 2	HCV
Pegbelfermin/Aldafermin	FGF 21/FGF 19 analogue	Phase 2	NASH/PSC
Elafibranor	PPAR α/γ agonist	Phase 3/Phase 2	NASH/PBC

ASK1 apoptosis signal-regulating kinase 1, *CCR2/5* C–C chemokine receptor type 2, *FGF* fibroblast growth factor, *FXR* farnesoid X receptor, *HSP* heat shock protein, *NASH* non-alcoholic fatty liver disease, *OCA* obeticholic acid, *PBC* primary biliary cholangitis, *PPAR* peroxisome proliferator-activated receptor, *PSC* primary sclerosing cholangitis

5.76 TACE

- HCC ranks fifth among various malignancies in incidence and is the third leading cause of associated mortality globally. Of more than six lakh deaths annually, 75% occur in East Asian nations.
- The disease is usually in advanced stages at diagnosis; hence, even though surgical resection is an optimal treatment alternative total resection becomes challenging in these HCC patients, especially hypervascular HCCs that display dense staining on contrast images of the hepatic arteries.
- Transarterial chemoembolisation (TACE) is the standard therapy for HCC patients ineligible for surgical resection and is widely used. TACE concentrates on chemotherapeutic agents at the tumor site while obstructing the primary artery from supplying blood to the tumor. Thus, TACE is widely used to extend the survival of patients with HCC. Nevertheless, TACE can trigger local angiogenic factors that stimulate tumor regrowth and elevate the risk of metastasis.
- TACE, with a lipiodol-chemotherapeutic agent suspension and gelatin sponge (GS) particles, has been used as standard therapy in Japan and is also called conventional TACE (cTACE). It differs from TACE, which involves drug-eluting beads (DEB-TACE), commonly used in Europe and the United States.

5.77 Fibrolamellar Carcinoma

- Fibrolamellar HCC has been reported in <1% of patients with HCC, a rare subtype.

5.77.1 Pathological Features

- The fibrolamellar HCC manifests chromosomally stable tumour and genomic homogeneity. Yet no etiological element has been recognised.
- Over three-fourths of these patients reveal macroscopically central stellate scars and prominent fibrous tissue. Unique histologic characteristics are distinct from usual hepatic tumours clusters or sheets of large polygonal or spindle-shaped tumour cells with excess granular eosinophilic cytoplasm and conspicuous nuclei and macro nucleoli and paucicellular fibrous stroma organised in parallel lamellae encircling the tumour cells.
- Excessive expression of TGF-β may account for typical lamellar fibrosis. Round to oval cytoplasmic 'pale' anucleate bodies and intracytoplasmic hyaline droplets are present but not pathognomonic characteristics of the tumour.

5.77.2 Clinical Features

- Fibrolamellar HCC is present in adults at a mean age of 25 years, while HCC usually occurs at 65 years. Clinical manifestations include abdominal pain (21–74%), malaise and weight loss (25–30%), or a palpable abdominal mass or hepatomegaly (15–40%). Jaundice is infrequent and only reported in 2–5%. It seldom manifests as gynecomastia, recurrent deep vein thrombosis, Budd–Chiari syndrome, nonbacterial thrombotic endocarditis, fulminant liver failure, or encephalopathy.

5.77.3 Management

- Surgical resection with sufficient lymph node dissection is the primary management method.

5.77.4 Prognosis

- Compared to usual HCC, fibrolamellar HCC has a better clinical outcome, likely due to the younger age of the patients.

5.78 Major Hepatocyte Drug Transporters

The major hepatocyte drug transporters are depicted in Fig. 5.21.

Two major gene superfamilies have a key role in the passage of drugs across biological membranes (Table 5.26):

- Solute carrier (SLC) superfamily.
- ATP-binding cassette (ABC) superfamily.

5.78.1 SLC Superfamily

- This superfamily group includes *facilitated transporters or ion-coupled secondary transporters*, commonly associated with the transfer of compounds from the GIT to the systemic circulation, followed by the liver and kidney.
- The SLC superfamily includes 48 subfamilies, among which the ones that have been extensively studied are

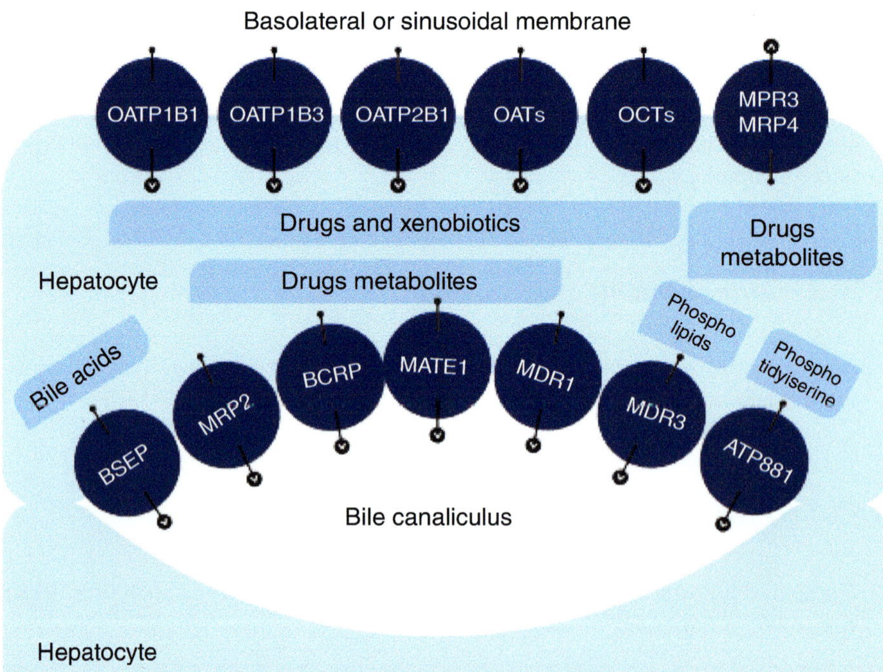

Fig. 5.21 Hepatocyte drug transporters. *OAT* organic anion transporter, *OCT* organic cation transporter, *MDR* multidrug resistance protein, *MRP* multidrug resistance-associated protein, *BCRP* breast cancer resistance protein, *BSEP* bile salt excretory protein, *MATE* multidrug and toxin extrusion

Table 5.26 Examples of substrates, inducers, and inhibitors of transporter involved in hepatocyte drug transport

Transporter (gene)	Substrates	Inhibitors	Inducers
OATP1B1 (SLCO1B1)	Estradiol-17β-D-glucuronide, estrone-3-sulphate, bosentan, enalapril, methotrexate, repaglinide, statins, valsartan	Estrone-3-sulphate, rifampicin, cyclosporine A, ritonavir	Rifampicin
OATP1B3 (SLCO1B3)	Estradiol-17β-D-glucuronide, cholecystokinin 8, digoxin, fexofenadine, statins, valsartan	Estrone-3-sulphate, cyclosporine A, rifampicin, ritonavir	Chenodeoxycholic acid
MDR1, -gp (ABCBI)	Digoxin, atorvastatin, fexofenadine, indinavir, loperamide, paclitaxel, vincristine	Cyclosporine A, quinidine, rifampicin, verapamil	Rifampicin, carbamazepine, phenytoin, ritonavir
NTCP (SLC10A1)	Taurocholate, estrone-3-sulphate, micafungin, rosuvastatin	Bromosulfophthalein, cyclosporine A, furosemide, ritonavir, rifampicin, propanolol	Dexamethasone
BSEP (ABCB11)	Taurocholate, micafungin, pravastatin, vinblastine	Ritonavir, cyclosporine A, rifampicin, troglitazone, bosentan, glibenclamide, vinblastine	Chenodeoxycholic acid

- Organic anion transporting polypeptide (OATP, SLCO) family.
- The organic cation transporters (OCT).
- Organic anion transporters (OAT).
- Multidrug and toxin extrusion (MATE, SLC47A).

5.78.2 ATP-Binding Cassette Family (ABC) Superfamily

- This superfamily comprises *primary active efflux transporters* using the energy released from ATP hydrolysis to transport substrates out of cells.
- The most important proteins involved in drug transport are
  - Multidrug resistance protein 1 (MDR1, ABCB1).
  - Breast cancer resistance protein (BCRP, ABCG2).
  - Multidrug resistance-associated protein (MRP, ABCC).

5.79 Statins in Liver Disease

- Statins are important drugs used to lower cholesterol levels. They form an integral part of treating metabolic syndrome-related non-alcoholic fatty liver disease. Patients with primary biliary cholangitis also develop hypercholesterolemia. Thus, patients with pre-existing liver diseases require drugs to treat hypercholesterolemia.
- Most statins are metabolised through cytochrome P 450. Many rely on transport mechanisms like organic anion-transporting polypeptide (OATP), BCRP, and p-glycoprotein. So, there is concern regarding the metabolism of statins in patients with liver disease. Pharmacokinetic studies with different statins have shown the greatest concern with ~6–16-fold rise in maximum concentration levels with atorvastatin. Pravastatin and rosuvastatin require minimal hepatic metabolism before biliary excretion. So, their drug levels are minimally affected in patients with decompensated liver disease.
- There was concern regarding the hepatotoxicity of statins, as liver enzyme elevation was noted in ~10% of patients who were started on statins in early clinical trials. Subsequent long-term prospective studies have shown this rise in transaminases to be transient. FDA advised periodic monitoring of liver function tests to monitor for drug-related liver injury.
- Long-term prospective studies and post-marketing surveillance have shown that the potential benefits of statins in mitigating cardiovascular diseases outweigh the slight risk of hepatotoxicity in patients with compensated cirrhosis and chronic hepatitis.
- Drug–drug interaction is a potential cause of concern as statins require transport proteins to facilitate their excretion in bile. Two main classes of drugs likely to interact in patients with hepatic disease are direct-acting antivirals and immunosuppressants.

5.80 Fibrates

Fibrates are a class of drugs used to treat dyslipidemia.

The authorised indications for the use of fibrates include:

- For use as an adjunct to dietary modifications (restricted in saturated fats and cholesterol) in adults with primary hypercholesterolemia or mixed dyslipidemia (Fredrickson type IIa and IIb). Fibrates help to reduce serum LDL, total cholesterol, triglycerides, and apolipoprotein B (Apo-B) and increase high-density lipoprotein cholesterol (HDL).
- To be used as an adjunct to dietary modifications in adults with severe hypertriglyceridemia (Fredrickson types IV and V).

5.80.1 Mechanism of Action

Fibrates act as lipid-lowering agents via the following mechanisms:

- Lowers the pool of dietary components to allow triglyceride synthesis in the liver.
- Overturns cholesterol transport and induces cellular fatty acid uptake.
- Fibrates enhance receptor-mediated elimination of LDL at an enhanced rate compared to normal, leading to an overall reduction in very low-density lipoproteins (VLDL).
- Small dense LDL, accounting for the high production of peroxide radicals, is decreased due to fibrates-mediated activation of lipoprotein lipase (LPL). Fibrates increase the activation of transcription factors for peroxisome proliferator-activated receptors (PPAR).
- Fibrates improve high-density lipoprotein (HDL) levels by upregulating the synthesis of Apo-AI and apo-AII in the liver.

5.80.2 Adverse Events

- Increased ALT/AST levels—to monitor periodically
- Reversibly increase serum creatinine levels—to monitor periodically
- Increased risk of myopathy (if co-administered with statins) and venous embolism
- Increased risk of gallstone formation
- Increased PT/INR (when co-administered with coumarin)

5.80.3 Contraindications

1. Severe renal dysfunction

5.81 Role of Albumin in Liver Disease

1. Albumin is the proportionally the most abundant serum protein (55–60%) and physiologically maintains intravascular volume and colloid osmotic pressure; binds and transports hormones, lipids, and drugs. It also has antioxidant and anti-inflammatory potential and endothelium-stabilising effects.

   The indications of giving albumin in liver cirrhotic patients are:

   (a) Spontaneous bacterial peritonitis (SBP)—These patients are susceptible to arterial and splanchnic vasodilatation and subsequent renal insufficiency.

Therefore, albumin is used to preserve intravascular volume and renal perfusion.

(b) Hepato renal syndrome—The primary findings include renal hypoperfusion due to systemic arterial vasoconstriction due to active renal vasoconstriction and reduced endogenous vasodilators such as prostaglandin, nitric oxide (NO), and natriuretic peptide.

(c) Plasma expansion in large volume paracentesis is used to manage refractory ascites in large volume paracentesis (LVP). Albumin is required to manage the complication of circulatory dysfunction post-LVP.

2. There have been varied results in various studies conducted with weekly administration of intravenous albumin to improve mortality and reduce complications of liver cirrhosis. It is still unclear if periodic administration of IV albumin is beneficial to cirrhotic patients.

5.82 Midodrine–Terlipressin

5.82.1 Midodrine

- Midodrine is an α-1 agonist acting via adrenergic receptors in arteriolar and venous vasculature, resulting in elevated vascular tone and blood pressure. It does not act on beta-adrenergic receptors in cardiac myocytes.
- Midodrine enhances renal excretion of sodium and reduces plasma renin and aldosterone.

Use

- Two small RCTs have shown the use of midodrine in decompensated cirrhosis patients with ascites is associated with complete/partial ascites control, reduction in abdominal girth, and a significant reduction in body weight.

Contraindications

- Severe heart disease, acute renal disease, urinary retention, pheochromocytoma or thyrotoxicosis, and hypertension.

Side Effects

The most common adverse reactions include

- Supine and sitting hypertension
- Paresthesia and scalp pruritus

- Goosebumps and chills
- Urinary urge, urinary retention, and urinary frequency

5.82.2 Terlipressin

- Terlipressin is a dodecapeptide containing three glycyl residues bound to the N-terminal of lysine vasopressin. It is metabolised to its active metabolite lysine vasopressin by enzymatic cleavage of glycyl residues.

Uses

- It is used for the treatment of bleeding esophageal and gastric varices.
- It is used for the treatment of hepatorenal syndrome along with albumin.

Contraindication

- Pregnancy
- Hypersensitivity

Precautions

- Terlipressin should only be used with caution and under strict monitoring of the patients in the following cases:
  - Uncontrolled hypertension.
  - Cerebral or peripheral vascular diseases.
  - Cardiac arrhythmias.
  - Coronary artery disease—recent angina or previous myocardial infarction.
- It may cause smooth muscle constriction, and, therefore, caution is necessary with strict monitoring in patients with severe asthma or chronic obstructive pulmonary disease.

5.83 Ursodeoxycholic Acid

Ursodeoxycholic acid is an oral, non-synthetic bile acid agent-7β epimer of chenodeoxycholic acid (less than 4% of the total bile acid pool). It was first introduced for the dissolution of radiolucent gallstones in the 1970s and subsequently authorised by the US FDA for the treatment of PBC.

5.83.1 Indications

1. Primary biliary cirrhosis (13–15 mg/kg/day)
2. Cholelithiasis (8–10 mg/kg/day)
3. Cystic fibrosis
4. Gallstone prophylaxis
5. Intrahepatic cholestasis of pregnancy (ICP)
6. Non-alcoholic steatosis-hepatitis (NASH)
7. Pruritus

5.83.2 Mechanism of Action

UDCA's therapeutic sites of action are in the liver, bile, and gut lumen. At chronic doses of 13–15 mg/kg/day, ursodiol comprises approximately 30–50% of the total biliary and plasma bile acids. The probable mechanism of action of UDCA includes

1. Suppress absorption of toxic, hydrophobic, and endogenous bile salts
2. Stabilisation of the hepatocyte membranes against toxic bile salts
3. Substituting the toxic endogenous bile acids with the nontoxic UDCA
4. Reducing expression of major histocompatibility complex (MHC) class I and II antigens

1. **Gallstone Dissolution**

   UDCA decreases the cholesterol proportion of bile by decreasing the hepatic secretion of cholesterol and fractional intestinal reabsorption. It also causes the distribution of cholesterol as liquid crystals in aqueous media. Thus, effectually, UDCA converts bile from a cholesterol-precipitating nature to a cholesterol-solubilising one, thus favouring gallstone dissolution.
2. **Primary Biliary Cirrhosis (PBC) and Other Cholestatic Liver Diseases**

   The precise pharmacological actions of UDCA that attenuate the worsening of the hepatic disease in patients with primary biliary cirrhosis are yet to be elucidated:
   (a) Accumulation of high levels of hydrophobic bile acids inside hepatocytes during cholestasis damages the membranes of intracellular organelles.
   (b) Enhanced hepatic bile flow, reduced bile viscosity, ductular proliferation, and portal inflammation are key effects of UDCA.
   (c) It lowers the intrahepatic concentration of hydrophobic bile acids within hepatocytes and increases the hydrophilicity of the bile acid pool. It also disturbs the enterohepatic circulation of bile acids by blocking the reuptake of endogenous bile acids in the terminal ileum.
   (d) Evidence also points to its immunomodulatory effect and suppression of certain cytotoxic cytokines that may attenuate lipid peroxidation in hepatocytes.

5.84 *S*-Adenosyl Methionine (SAMe)

- *S*-Adenosyl methionine is synthesised from methionine and ATP.
- It has multiple important metabolic roles. It is involved in transmethylation, trans-sulfuration, and aminopropylation reactions. It is the principal methyl donor in methyltransferase reactions.
- SAMe restores hepatic glutathione reserve and protects against liver injury. It is a key metabolite involved in hepatocyte growth, differentiation, and apoptosis.
- SAMe supplementation has been used in various rheumatological disorders and for the treatment of depression.
- Studies have evaluated the effect of SAMe on alcoholic and viral hepatitis. It is useful in improving biochemical parameters (total bilirubin and AST levels) when used in patients with chronic liver disease of varying etiologies. There is no strong evidence to show that SAMe improves hard clinical endpoints like mortality, decompensation, or fibrosis regression.
- SAMe is useful in the treatment of intrahepatic cholestasis of pregnancy.

5.85 Obeticholic Acid (OCA)

- OCA is a synthetic analogue of naturally occurring bile acid chino-deoxycholic acid. But this change leads to a 100 times higher binding potency with the nuclear receptor FXR (Farnesoid-X receptor).

5.85.1 Mechanism of Action

- Activation of FXR in the hepatocytes leads to the inhibition of the rate-limiting enzyme (CYP7A1) in the production of bile acids from cholesterol. In addition, it also stimulates the movement of bile acids out of hepatocytes and into the canaliculi. Eliminating potentially toxic bile acids from the hepatocytes leads to a reduction in liver cell necrosis.
- In the enterocyte, the primary effect of FXR activation is to activate fibroblast growth factor-19 (FGF-19). FGF-19 reaches the liver and further inhibits the CYP7A1. In addition, FXR activation in the enterocytes also inhibits the reuptake of bile acids from the apical bile acid transporter.

5.85.2 Indication

- **Role in PBC**: Based on phase 2 and 3 trials of OCA in primary biliary cholangitis patients, the FDA has approved the use of this drug in a dose of 5 and 10 mg among PBC patients who are intolerant or unresponsive to ursodeoxycholic acid

(UDCA). In the phase 3 POISE trial, 217 patients were randomised to receive a placebo or OCA 5 mg or OCA (starting 5 mg and increased to 10 mg). The primary endpoint of alkaline phosphatase reduction to $<1.67 \times$ ULN or 15% reduction, along with normal serum bilirubin, was achieved in 47% of treated patients versus 10% of placebo ($p = 0.0001$).

5.85.3 Role in NASH

- In preclinical studies, OCA dose dependently inhibits fibrosis in pig livers. This effect was associated with FXR activation in HSCs. OCA also inhibits gluconeogenesis, inhibits lipogenesis, and stimulates fatty acid beta-oxidation, thus improving hepatic steatosis.
- A pilot study of OCA on diabetic patients with NASH demonstrated a reduction in hepatic necroinflammation and weight and insulin resistance.
- Subsequently, there was a large phase 2 RCT (FLINT trial) in biopsy-proven NASH patients without cirrhosis randomized to either OCA (n = 141) or placebo (n = 142) for 72 weeks. Patients receiving OCA showed significantly greater histological improvement than placebo (45% vs. 21%; $p = 0.0002$).
- Recently the 18-month interim results (in >900 patients with biopsy-proven NASH with advanced fibrosis up to F3) from a large phase 3 trial (REGENERATE trial) have been published. The patients were randomised to receive either placebo or 10 mg OCA, or 25 mg OCA for 36 months. The primary endpoint of ≥1 stage reduction in fibrosis without worsening of NASH was achieved more frequently in patients receiving 25 mg of OCA per day (23.1% vs. 11.9%; $p = 0.0002$) and 10 mg/day (17.6% vs. 11.9%; $p = 0.04$), than placebo.

5.85.4 Side Effects

- Prominent adverse effects include significant itching and a rise in LDL cholesterol, both dose-dependent. While itching can be managed with emollients, antihistaminic use, and dose reduction, the rise in LDL can be treated by simultaneous use of statins.
- A USFDA black box warning was issued against high-dose OCA among patients with PBC with decompensated cirrhosis as it was found to increase mortality. So, in this group of patients, the recommended doses are 5–10 mg twice weekly.

5.86 Saroglitazar

- Saroglitazar is a newer, dual peroxisome proliferator-activated receptor (PPAR) alpha/gamma agonist. Activation of these receptors leads to many beneficial effects on carbohydrate and fat metabolism and liver inflammation.
- PPAR-alpha is involved in fatty acid transport and beta-oxidation. PPAR-gamma has a strong insulin-sensitising effect. Both regulate glucose and lipid metabolism in combination and show an anti-inflammatory effect.
- This molecule was earlier in use by endocrinologists to treat refractory hypertriglyceridemia in diabetic patients. However, with its beneficial effects demonstrated in patients with non-alcoholic fatty liver disease (NAFLD), it has been re-purposed for use in such patients.
- Pre-clinical studies in animals have suggested that saroglitazar significantly improves liver fat content, serum ALT levels, and markers of inflammation such as TNF-alpha and liver fibrosis. In a real-world analysis of the use of saroglitazar in 5824 patients with diabetic dyslipidemia, where it was prescribed from 12 to 58 weeks, it was shown to significantly reduce serum triglycerides, LDL cholesterol, and serum ALT levels. It was weight neutral and well-tolerated.
- The phase 2 study of Saroglitazar, called EVIDENCES IV, was an RCT carried out in patients with biopsy-proven NASH, and it compared the effect of placebo, 1, 2, and 4 mg doses of Saroglitazar on liver fat content and liver enzymes. It demonstrated that the 4 mg daily dose led to a significant decrease in liver fat content (30%) measured by MRI-PDFF and ALT levels. This improvement in liver fat reduction was noted in 40% of patients.
- A phase 3 study was conducted among 102 Indian patients with biopsy-proven NASH. The primary endpoint of ≥2 point reduction in NAS devoid of exacerbating fibrosis was achieved more frequently among the patients who received 4 mg/day of saroglitazar compared to placebo (52.3% vs. 23.5%; $p = 0.04$). It also demonstrated a reduction in ALT, triglyceride, and LDL cholesterol, and a significant rise in serum HDL levels. No mention was made of any change in fibrosis scores. The adverse events were mild to moderate.
- Saroglitazar is a dual PPAR alpha/gamma agonist that has shown significant activity in NASH patients regarding reducing liver fat, ALT, and necroinflammation. It also improves dyslipidaemia and insulin resistance. Long-term studies are evaluating its efficacy in reducing hepatic fibrosis and improving cardiovascular outcomes in patients with NAFLD.

5.87 Newer Molecules for the Treatment of HCC

HCC diagnosed in the early stages of BCLC classification benefit from curative therapies like resection and transplant. In intermediate and advanced BCLC stages, systemic and locoregional therapies are used. Sorafenib was the first systemic therapy approved for HCC in 2007. From 2016 to 2018, five new molecules have been approved for treatment for HCC: Lenvatinib, Regorafenib, Cabozantinib, Ramucirumab, and Nivolumab (Table 5.27).

Table 5.27 Newer molecules for the treatment of HCC

Drugs	Mechanism of action	Current status	Key evidence
		First line	
Sorafenib	Multikinase inhibitor (RAF, BRAF, KIT), anti-VEGF and PDGFR β	First-line therapy in advanced-stage HCC	SHARP, Asian Pacific trial
Lenvatinib	Multikinase inhibitor (VEGFR, PDGFRα, KIT, FGFR)	First-line therapy in advanced-stage HCC	REFLECT trial
		Second line	
Regaorafenib	Similar to sorafenib with broader activity against VEGFR, KIT, and angiotensin 1 receptor	Second-line therapy who progressed on first-line treatment	RESORCE trial
Cabozantinib	Small molecule tyrosine kinase inhibitor—inhibits VEGFR	Second-line therapy who progressed on first-line treatment	CELESTIAL trial
Ramucirumab	Anti-VEGFR-2 monoclonal antibody	Second-line therapy who progressed on first-line treatment	REACH trial (HCC patients with baseline AFP >400 ng/ml)
Nivolumab	Monoclonal antibody against negative immunoregulatory cell surface receptor programmed death-1 (PD-1)	Accelerated approval by the FDA for patients with advanced HCC who progressed on sorafenib	Based on the Phase II trial CHECKMATE-phase III trial awaited

5.88 Caspases, Apoptosis, and Liver

- Apoptosis is the physiological process that involves a cascade of molecular steps in a cell resulting in cell death. It is a necessary process to eliminate worn-out or anomalous cells from the system. The fading cell is subject to condensation of the nucleus and cytoplasm, fluid-filled projections of the plasma membrane, and ultimately cleaving as apoptotic bodies enclosed in a membrane that contains organelles and the nucleus. Subsequently, phagocytes, HSC, or innate immune cells degrade these apoptotic bodies and, in the liver, KCs perform this function.
- Unlike apoptosis, necrosis is a pathological cell death involving non-reversible cytoplasmic edema and deformed organelles.
- Caspases are a family of proteases (cysteine proteases that are aspartate specific) that play a key role in proteolytically dismantling cellular structures of dying or dead cells. Activation of caspases is the primary pathway for initiating hepatic cell apoptosis. Caspases 1–11 have been reported in humans. The process begins with activating Caspases-2, -8, -9, and -10, with subsequent activation of a cascade of downstream effector caspases such as -3, -6, and -7. Caspase activation and lysis of specific cellular proteins are typical features of apoptosis. Thus, apoptosis is a caspase-mediated form of cell death.
- Caspase-3 and caspase-7 are activated with the worsening of the disease. It lyses a major filamentous protein known as cytokeratin-18 (CK-18). Cytokeratin-18 is an intermediate filament indicating ~5% of total hepatic protein, exocrine pancreas, intestine, and other epithelial tissues. In vitro, studies have shown that due to caspase digestion during the intermediate stage of apoptosis, cytokeratin 18 fragments are released into the extracellular space and can act as apoptosis biomarkers.

5.88.1 Caspases in Liver Disease

- Caspase-cleaved CK18 fragments are significantly higher in NAFLD patients than in the non-NAFLD group. Therefore, Caspase inhibition can be a target for the prevention of inflammasome activation and thus avoid inflammation-associated liver diseases.
- Emricasan is an oral pan-caspase inhibitor found to be effective in acute hepatitis and chronic models of non-alcoholic steatohepatitis (NASH). It has lowered apoptosis, inflammation, fibrosis, cirrhosis, and death. In experimental NASH models, emricasan decreases caspase activity, apoptosis, and fibrogenesis and improves survival.
- Early clinical studies in patients with HCV hepatitis and cirrhosis have demonstrated effective reduction of ALT levels and anti-apoptotic and anti-inflammatory actions.

- In patients with NASH and elevated aminotransferases, emricasan decreases caspase-3/7 activity, cCK18, flCK18, and serum ALT levels after a month's treatment. Thus, pan-caspase inhibition reduced the pro-inflammatory cytokine IL-1β synthesis, apoptosis, and reduced liver inflammation and fibrosis.
- However, a large 72-week clinical study with emricasan (5 mg or 50 mg) failed to demonstrate a reduction in fibrosis and NASH without worsening the other.
- Suicide-substrate types of caspase inhibitors are potent and show significant efficacy in averting apoptotic cell death. However, the other caspase-independent form of cell death, necroptosis, may be triggered by the inhibition of caspases. Thus, switching to necrosis, and triggering a more prominent inflammatory response, caspase inhibitors, may be less effective than previously anticipated when used long term and may even be detrimental.

5.89 Acute Kidney Injury and Hepatorenal Syndrome

A serious complication of cirrhosis, hepatorenal syndrome is associated with increased mortality and morbidity. The definition and criteria have been refined in the past decade in line with the evolving acute kidney injury network (AKIN) definition. There are two types of hepatorenal syndrome: Type 1 and Type 2 HRS (Table 5.28).

Acute kidney injury in cirrhosis patients may be related to hypovolemia secondary to diuretics. The approach to acute kidney injury in cirrhosis patients is outlined in Fig. 5.22. Renal perfusion is directly correlated with mean arterial blood

Table 5.28 Previous and current definitions of hepatorenal syndrome (HRS)

Old definitions	New definitions
HRS type 1 • Doubling of serum creatinine to a concentration ≥2.5 mg/dl within 2 weeks • No response to diuretic withdrawal and 2-day fluid challenge with 1 g/kg/day of albumin 20–25% • Cirrhosis with ascites • Absence of shock • No current or recent use of nephrotoxic drugs (NSAIDs, contrast dye) • No signs of structural kidney injury • Absence of proteinuria (>500 mg/day) • Absence of hematuria (>50 RBCs per high power field) • Normal findings on renal ultrasonography	HRS-AKI • Increase in serum creatinine of ≥0.3 mg/dl within 48 h OR • Increase in serum creatinine ≥1.5 times from baseline (creatinine value within previous 3 months), when available, may be used as a baseline, and value closest to presentation should be used • **No response to diuretic withdrawal and 2-day fluid challenge with 1 g/kg/day of albumin 20–25%** • **Cirrhosis with ascites** • **Absence of shock** • **No current or recent use of nephrotoxic drugs (NSAIDs, contrast dye)** • **No signs of structural kidney injury** – **Absence of proteinuria (>500 mg/day)** – **Absence of hematuria (>50 RBCs per high power field)** – Normal findings on renal ultrasound

Table 5.28 (continued)

Old definitions	New definitions
HRS type 2 • Gradual increase in serum creatinine, not aligning with criteria described above	HRS-NAKI HRS-AKD • **Estimated glomerular filtration rate <60 ml/min/1.73 m^2 for <3 months in the absence of other potential causes of kidney disease** • **Percentage increase in serum creatinine <50% using the last available value of outpatient serum creatinine within 3 months as a baseline value** HRS-CKD • **Estimated glomerular filtration rate <60 ml/min/1.73 m^2 for ≥3 months in the absence of other potential causes of kidney disease**

Adapted from: BMJ 2020;370:m2687. https://doi.org/10.1136/bmj.m2687
AKD acute kidney disease, *AKI* acute kidney injury, *CKD* chronic kidney disease, *NAKI* non-acute kidney injury, *NSAID* non-steroidal anti-inflammatory drug, *RBC* red blood cell

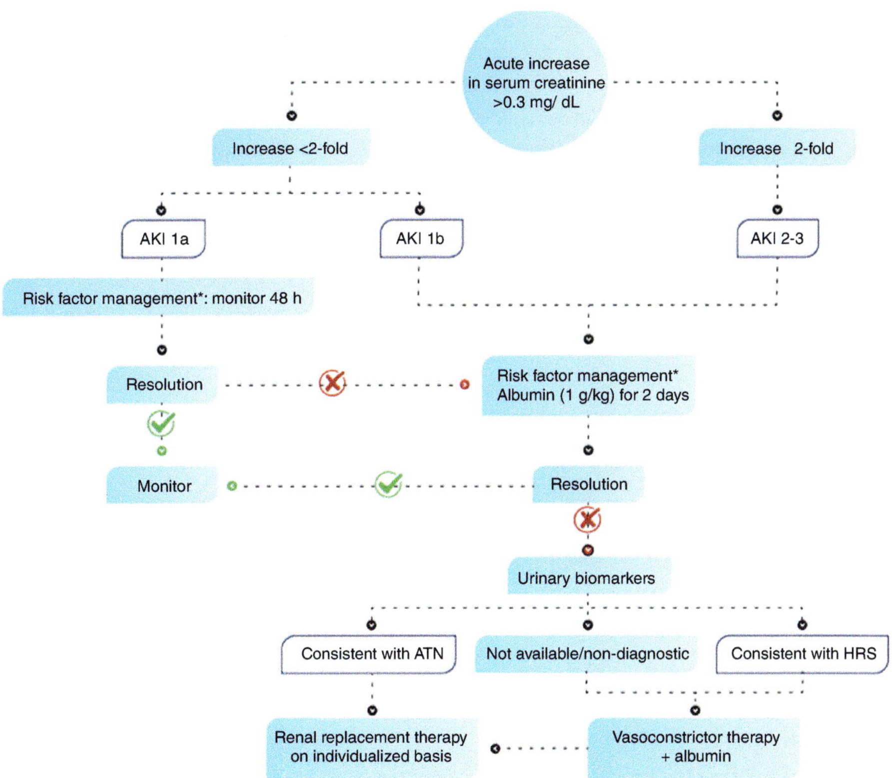

Fig. 5.22 Algorithm for management of acute kidney injury in cirrhosis. *ATN* acute tubular necrosis, *HRS* hepatorenal syndrome. *Risk factor management: hold nephrotoxic drug blockers, withdraw diuretics, treat infections, plasma volume expansion as needed

pressure. Vasoconstrictors reverse splanchnic vasodilation, increase circulatory blood volume in the systemic circulation, and enhance mean arterial blood pressure, thereby improving renal perfusion. Terlipressin, norepinephrine, and midodrine with octreotide have been used along with albumin in various RCTs to treat HRS and have shown to reverse HRS in ~20–60%.

5.90 Clinical Relevance of HVPG

- Hepatic venous pressure gradient (HPVG) is the difference between wedged and free hepatic venous pressures indicating a gradient in the portal vein and intra-abdominal portion of the inferior vena cava. The normal HPVG value is 1–5 mmHg pressure higher than this represents portal hypertension.
- The natural history of chronic liver disease involves the progression to cirrhosis, first compensated, then to decompensated cirrhosis and ultimately to death. HVPG >10 mmHg was the strongest predictor of the development of varices.
- Most complications resulting in decompensation are associated with portal hypertension. Thus, portal pressure is a marker of decompensation. HVPG, MELD, and albumin per se predict clinical decompensation in patients with compensated cirrhosis. Patients with an HVPG <10 mmHg have a low probability of developing decompensation as per a study of median 4-year duration. Clinical evidence suggests that for each 1 mmHg increase in HVPG, there is a ~11% greater risk of clinical decompensation.
- Non-responders with HVPG per se are associated with a greater incidence of portal hypertension, complications, and have fatal outcomes. Therefore, a decrease in HVPG with medical therapy is recognised as a negative predictor of variceal hemorrhage and spontaneous bacterial peritonitis.

5.91 Prognostic Models for Acute Liver Failure

Acute liver failure is characterised by the acute onset of encephalopathy and coagulopathy. Some patients respond to medical management, while others worsen and die. It is an emergent indication for a liver transplant. There are various prognostic models which help select appropriate patients for liver transplants. None of them is used universally. The components of various scoring systems and their utility and limitations are tabulated in Table 5.29.

Table 5.29 Comparison of different criteria used for the selection of patients with acute liver failure for liver transplantation

Prognostic mode/marker	Parameters included	NAALF		AALF		Remarks/drawbacks
		Sn	Sp	Sn	Sp	
KCC	Age, INR, serum bilirubin, icterus-encephalopathy interval, drug toxicity	68%	82%	58.2%	94.6%	Poor sensitivity, performs better with advanced HE. Combining lactate with the KCC improves sensitivity but reduces specificity
Meld score (>30)	Serum bilirubin, serum creatinine, and INR	76	73	80	53	Inconsistent results
Clichy criteria	Advanced HE with Factor V levels <20% in patients <30 years and <30 in patients ≥30 years	69 PPV 64%, and NPV 55%	50	75 PPV 50% and NPV 79%	56	Inferior to KCC and MELD in validation studies. Poor Sp and PPV
APACHE II	Multiple parameters APACHE II > 15	–	–	82	98	Non-specific Not validated Cumbersome
SOFA	SOFA scores of >6 by 72 h post-acetaminophen overdose	–	–	90 PPV 42% and NPV 96%	69	Non-specific challenging to calculate neurological components in intubated patients
		Sn	*Sp*			
Arterial ammonia	Baseline arterial ammonia >124 mol/l	78.6	76.3			Ammonia levels affected renal impairment, sepsis, bleeding, haemolysis, and drugs Not validated at LT centres. Persistent hyperammonemia is a better predictor, but the strategy is delayed
Blood lactate	Post-resuscitation arterial lactate cut-off 3.0 mmol/l in AALF	76	97			Inconsistent results

(continued)

Table 5.29 (continued)

Prognostic mode/marker	Parameters included	NAALF		AALF		Remarks/drawbacks
		Sn	Sp	Sn	Sp	
Serum phosphate	Level of 1.2 mmol/l at 48–96 h after acetaminophen overdose	89 PPV 100%, and NPV 98%	100			Not reproducible
Serum Gc globulin	A cut-off level of 80 mg/l in the NAALF	49 PPV 85%, and NPV 43%	90			Lacks validation studies
Cytokeratin 18-based modification of the MELD	CK18 M65, INR, MELD. A baseline cut-off of 53.5 modified MELD	81 PPV 65%, and NPV 91%	82			Better than MELD and KCC, but lack validation studies
Monocyte HLA-DR expression	Monocyte HLA-DR expression 15% or less in AALF	96	100			Lacks validation studies No correlation between a reduction in monocyte HLA-DR expression and NAALF outcome
BiLE score	Bilirubin, lactate, and aetiology	79	84			Lacks validation study
ALFED mode ≥4	Over 3 day values of arterial ammonia, serum bilirubin, INR, and advanced HE	PPV 85% and NPV 87%	Needs further validation. The strategy may be delayed since advanced HE is a late feature			
ALFSG index	Coma grade, INR, serum bilirubin, and phosphorus levels and log10 M30	85.6	64.7			Less cost-effective and needs validation but better than MELD and KCC

AALF acetaminophen-associated acute liver failure, *ALFED* acute liver failure early dynamic, *ALFSG* acute liver failure study group, *APACHE* acute physiology and chronic health evaluation, *AUROC* area under receiver operating characteristic, *DA* diagnostic accuracy, *DOR* diagnostic odds ratio, *INR* international normalised ratio, *KCC* King's college criteria, *MELD* model of end stage liver disease, *NAALF* non-acetaminophen-associated acute liver failure, *NPV* negative predictive value, *PPV* positive predictive value, *Sn* sensitivity, *Sp* specificity, *SOFA* sequential organ assessment score, *HE* hepatic encephalopathy

5.92 Pregnancy and Liver Disease

5.92.1 Physiological Changes in Pregnancy and Effect on Liver

- During gestation, the maternal heart rate and cardiac output are elevated by 40% and the plasma volume by 30%, with a drop in peripheral vascular resistance. These alterations lead to hyperdynamic circulation development and resemble decompensated chronic liver disease patients.
- Palmar erythema and multiple spider naevi are observed in up to 70% of pregnant females.
- Reduced gallbladder motility increases susceptibility to gallstones.
- Maternal alkaline phosphatase (ALP) is elevated in the third trimester when ALP is produced in the placenta. The alpha-fetoprotein (AFP) level is elevated in pregnancy as the fetal liver synthesizes it.
- Due to hemodilution, the blood urea, haemoglobin levels, and the prothrombin time either remain constant or somewhat lower.
- However, elevations in transaminases, bilirubin, or the prothrombin time are not part of physiological gestational changes and should be evaluated.
- Pregnancy is also found to be a pro-coagulant state with elevation in levels of clotting factors (I, II, V, VII, X, and XII) and fibrinogen.

5.92.2 Liver Disease in Pregnancy

Liver dysfunction in pregnancy may be associated with

1. Pregnancy-associated liver diseases
2. Worsening of pre-existing liver disease, or
3. Disease states not associated with pregnancy.

5.92.3 Pregnancy-Associated Liver Diseases

The liver diseases specific to the pregnant state can be classified into the following:

- Hyperemesis gravidarum is early pregnancy.
- Acute fatty liver of pregnancy (AFLP), pre-eclampsia with hepatic involvement including haemolysis, elevated liver enzymes and low platelets (HELLP) syndrome, liver rupture/infarction, and intrahepatic cholestasis of pregnancy (ICP) occurring in late pregnancy.

5.92.4 Hyperemesis Gravidarum

- It involves unmanageable emesis, leading to dehydration, ketosis, and weight loss of >5%.
- Complications occur in a maximum of 2% of pregnant women, and symptoms generally begin before 9 weeks of gestation.
- The human chorionic gonadotropin (HCG) hormone is the highest in the first trimester and correlates with the severity of HG. HG is more frequent in molar and twin pregnancies where HCG levels are very high. HCG can physiologically trigger the thyroid stimulating hormone (TSH) receptor causing inhibition of TSH and elevation in T4. Thus, HCG, TSH, and HG severity are interrelated.

5.92.5 Intrahepatic Cholestasis of Pregnancy

- Intrahepatic cholestasis of pregnancy (ICP) is the most frequent liver disease specific to pregnancy. It is a reversible form of cholestasis that manifests as pruritus in pregnancy and elevated fasting or post-prandial serum bile acids, which resolves spontaneously within 6 weeks post-partum. However, the patients should be evaluated if symptoms persist for >3 months.
- ICP primarily occurs in the last trimester but is also reported around 7 weeks of gestation. It occurs in multiple gestations and assisted pregnancies.
- ICP involves genetic, endocrine, and environmental factors. Probably, altered sex hormone levels uncover certain conditions in genetically predisposed females. Approximately 15% of cases have genetic variation in one of the hepatocanalicular transport proteins: ABCB11 (bile salt export pump) or ABCB4 (phosphatidylcholine flippase).
- It generally manifests as pruritus, particularly worse on the palms and soles; however, it could also be generalised.
- The first-line treatment for ICP is ursodeoxycholic acid (UDCA), which resolves symptoms in three-fourths of pregnant females. Besides, UDCA also improves the maternal and fetal effects of ICP. It improves the biliary transport of bile acids, is anti-apoptotic, and probably eliminates the pruritus-causing metabolites (progesterone sulphates).
- ICP also occurs at a high frequency in subsequent gestations. There is an increased risk of future hepatic diseases in these women, the commonest being gallstones due to the ABCB4 gene common to both hepatobiliary malignancies and immune-mediated and cardiovascular diseases. Combination OCPs increase the risk of hepatic dysfunction. However, progesterone-only OCPS are less likely.

5.92.6 *Pre-eclampsia, Eclampsia and Haemolysis, Elevated Liver Enzymes, Low Platelets (HELLP) Syndrome*

- Hypertension (blood pressure 140/90) in pregnancy after 20 weeks of gestation or 4 weeks postpartum, along with proteinuria (>300 mg/day) and renal insufficiency during pregnancy, are grouped as a multisystem disorder—pre-eclampsia. It also may involve hepatic impairment, neurological, haematological complications, uteroplacental dysfunction, or fetal growth restriction. In a patient with pre-existing essential hypertension, superimposed pre-eclampsia is diagnosed if the above features are present.
- Pre-eclampsia occurs in ~3–5% of all pregnancies. If seizures occur in females with coexisting pre-eclampsia, it is then classified as eclampsia. HELLP syndrome is a severe form of pre-eclampsia.
- History of pre-eclampsia or hypertension in pregnancy, chronic kidney disease, essential hypertension, diabetes, and autoimmune disorders increases the risk of pre-eclampsia.
- Premature delivery of the fetus aids in the management of eclampsia/HELLP syndrome.

5.92.7 *Acute Fatty Liver of Pregnancy*

- AFLP is a rare emergency state that can occur in the last trimester. It can have serious maternal and fetal negative outcomes in the absence of timely identification and treatment.
- It occurs in approximately 1 in 20,000 pregnancies. However, the incidence rate could be higher since it can be subclinical and unrecognised. Nulliparity, male infants, and twin pregnancies increase the susceptibility of AFLP.
- The clinical spectrum is similar to that of mitochondrial cytopathies, and an abnormality in mitochondrial b-oxidation could be an underlying reason for AFLP in a subset of cases. The long-chain 3-hydroxyacyl coenzyme A dehydrogenase (LCHAD) is a necessary component of mitochondrial trifunctional protein. Nearly 20% of neonates born to mothers with AFLP may have defective β-oxidation and lack LCHAD due to mutation of any component of trifunctional protein.
- Early delivery is the treatment of choice. The majority of AFLP resolves with the delivery of the fetus. Rarely, liver failure persists after delivery and may require an urgent liver transplant.

5.92.8 Pregnancy in the Setting of Pre-existing Liver Disease

- In women with cirrhosis who become pregnant hepatic decompensation occurs in 10%, and the MELD score can be used as a predictor pre-pregnancy.
- Existing treatment with immunosuppressive agents (azathioprine, tacrolimus, cyclosporine, and steroids) should not be withdrawn in women post-transplantation or with autoimmune hepatitis.

5.93 Non-invasive Markers of Liver Fibrosis

The non-invasive modalities for diagnosing liver fibrosis are classified into serum markers and imaging modalities.

1. **Serum Markers**

   (a) The biomarkers are either direct or indirect markers.

   - Direct or class biomarkers correlate with the alterations in the ECM structure, elevated markers of ECM turnover, fibrogenesis, and fibrolysis.
   - Indirect (Class II) biomarkers indicate liver damage and impairment of hepatic function associated with fibrosis and cirrhosis.

   (b) **AST/ALT ratio**: ALT and AST are released from damaged cells and, when considered in combination, provide more information than either alone.
   (c) **APRI**: stands for AST-Platelet Ratio Index. It is calculated as APRI = [AST level (/ULN)/Platelet counts (10^9/l)] × 100. It is one of the simplest marker panels to diagnose significant fibrosis and cirrhosis with reliable accuracy in HCV.
   (d) **FIB-4**: consists of four variables: AST, ALT, age, and platelet count. It is calculated with the following formula: The FIB-4 index = [age (years) × AST (IU/l)]/[platelet count (10^9/l) × ALT (IU/l)]1/2. FIB-4 provides results equivalent to FibroTest in diagnosing advanced fibrosis and cirrhosis in HCV patients.
   (e) **Fibrotest** is an extremely validated test among the various panels. It is a combined battery of five serum biochemical markers (alpha2-macroglobulin, apolipoprotein A1, haptoglobin, γ-glutamyl transpeptidase, and bilirubin) modified in liver fibrosis. Its score is proportional to the severity of hepatic damage.
   (f) **Collagen**: Collagen is synthesised as a precursor with propeptide extensions at both the N- and C-terminal ends. Higher rates of collagen production suggest that increased HSC activation rates reflect increased fibrogenesis.
   (g) **PIIINP**: The N-terminal pro-peptide of collagen type III (PIIINP) has been evaluated as a marker of liver fibrosis in several studies. It may detect cirrhosis with a sensitivity of ~94% and a specificity of ~81%. Levels of PIIINP

increase in hepatitis and are proportional to those of aminotransferases. Therefore, these, in all likelihood, indicate inflammation rather than fibrosis.

Other biomarkers:

(h) PINP, YKL-40, TIMP, CK18, malondialdehyde (MDA), and superoxide dismutase (SOD)

2. **Imaging Techniques**:

   (a) **Elastography**: These techniques differ in the physical approaches used and can be grouped into three major types: (1) transient elastography (TE) (Fibroscan), which uses a mechanical external push; (2) acoustic radiation force impulse (ARFI) techniques, which use an acoustic internal push; and (3) strain elastography (SE) technique, which uses frame-to-frame differences (tissue deformation) with stress, caused by pressing the body surface or by internally occurring physiologic motion. The ARFI techniques can be divided into point shear wave elastography (p-SWE) and 2-D shear wave elastography (2-D SWE) techniques.

   (b) The shear wave-based techniques (TE and ARFI techniques) determine the speed of shear waves in tissues generated by an external mechanical push in TE or by the push pulse of a focused ultrasound beam in the ARFI techniques. The shear wave speed calculated correlates with liver stiffness. It can be converted into kilopascals because the tissue is purely elastic, incompressible, its elastic response is linear, and the tissue density is always 1000 kg/m^3.

   (c) **MRI** is commonly used to evaluate cirrhosis and its complications. Nevertheless, detecting mild-moderate severity is difficult, and many novel MR imaging techniques have been developed.

   (d) **MR elastography**: Similar to sonographic transient elastography (TE), the principle of MR elastography is that the velocity and wavelength of the wave propagating in the tissue are directly proportional to the stiffness of the tissue. Specific software and hardware are necessary to perform MR elastography.

   (e) **Reticuloendothelial-specific contrast agents**: Some studies with reticuloendothelial system-specific contrast agents have been conducted. Superparamagnetic iron oxide (SPIO) leads to a signal drop in the hepatocyte containing liver parenchyma, and as a result, it elevates the conspicuousness of fibrotic tissue that is less affected by this contrast agent.

   (f) **Susceptibility-weighted MRI**: Susceptibility-weighted imaging is a gradient echo sequence with greater sensitivity to the presence of iron, haemoglobin, and calcifications. Measurement of liver-to-muscle signal intensity ratio correlates with liver fibrosis with high inter-observer agreement.

   (g) **MR spectroscopy**: Assessments of liver fibrosis using MR spectroscopy achieved inconsistent results in different studies. Phophodiester (PDE) can be estimated by MR spectroscopy with sensitivity and specificity of 81% and 69%, respectively, for distinguishing advanced from mild fibrosis.

(h) **Computed tomography**: Morphological liver alterations, signs of cirrhosis, and signs of portal hypertension can be identified by computed tomography (CT, splenomegaly, collateral venous circulation, and enlarged portal vein), but CT is less sensitive for less advanced cirrhosis.

(i) **Fibro CT** is an experimental processing method of conventional CT scan images analysed by additional software. Optical analysis of CT images of the liver applying this technique identified the stage and location of liver fibrosis in patients with chronic hepatitis.

Other modalities: Ultrasonography, contrast-enhanced US, elastography, T1 mapping of the liver, supersonic shear wave elastography or 2-D SWE, ARFI, and real-time elastography—Hi-RTE or SE.

5.94 Autoimmune Liver Disease and Overlap Syndromes

Autoimmune hepatitis (AIH) is a progressive inflammatory disorder of the liver in genetically predisposed individuals with a breakdown of immune tolerance mechanism due to unknown aetiology driven by T-cell-mediated injury to hepatocytes.

Diagnosing autoimmune hepatitis relies on supportive biochemical, serological, and histopathological evaluation with the exclusion of other etiologies of liver disease. There is no specific diagnostic marker for autoimmune hepatitis.

5.94.1 Biochemical Markers

ALT/AST are key biological markers of autoimmune hepatitis. Elevated ALP/GGT and raised transaminases raise the possibility of overlap syndrome.

5.94.2 Serological Markers

Raised serum immunoglobulin G (IgG) level is key to diagnosing autoimmune hepatitis. Antinuclear antibody (ANA > 1:80), anti-smooth muscle antibody (ASMA > 1:80), and liver kidney microsomal (ALKM > 1:40) antibodies are key serological markers for autoimmune hepatitis. Type 1 autoimmune hepatitis is characterised by ANA, ASMA antibodies, and soluble liver antigen (SLA), while ALKM antibodies characterise Type 2 autoimmune hepatitis. Typical serological markers may be absent in up to 30% of patients with acute onset AIH.

5.94.3 Histopathological Pattern

Lymphoplasmacytic interface hepatitis and hepatocyte rosettes are characteristic histopathological findings in active autoimmune hepatitis. KC hyaline granules (eosinophilic PAS-resistant granules) are typically seen in pediatric AIH. Centrilobular necrosis is typically seen in acute onset AIH and ALF.

Close mimics of a histopathological pattern of AIH is drug-induced autoimmune hepatitis-like liver injury. The absence of advanced fibrosis and the rare presence of bridging fibrosis may favour DI-AIH, although this may not help in acute AIH.

5.94.4 Diagnostic Scores

Diagnosing autoimmune hepatitis is established by combining clinical, biochemical, serological, and histopathological features. There are no single diagnostic criteria for AIH. The currently simplified autoimmune hepatitis score uses a combination of serological markers (ANA, ASMA, ALKM, SLA, and IgG levels) with supportive histological features on liver biopsy in the absence of an alternate diagnosis (negative workup for viral hepatitis). The final score gives the likelihood of autoimmune hepatitis as possible, probable, and likely.

5.94.5 Clinical Presentation

Autoimmune hepatitis presents in different forms ranging from asymptomatic transaminase elevation to acute liver failure. Presentation could be acute hepatitis, severe AIH, or liver failure AIH. Acute hepatitis could be a 'true' acute onset of disease or flare of pre-existing subclinical disease. Acute severe AIH includes patients with jaundice and INR >1.5 without encephalopathy or previously diagnosed liver disease. ALF-AIH includes patients with jaundice, INR > 1.5, with encephalopathy occurring within 26 weeks of the onset of illness without any preexisting liver disease. Up to 3% of patients with acute or chronic liver failure (according to APASL ACLF criteria) are due to autoimmune hepatitis.

5.94.6 DI-AIH, Like Liver Injury

The majority of patients with DIAIH-like injury are acute-onset, and up to 30% of the cases are accompanied by hypersensitivity reaction; the latency periods of minocycline and nitrofurantoin (the two most commonly implicated drugs) can exceed 12 months. HLA-DR3 or -DR4 and cirrhosis at presentation are unusual.

5.94.7 *Overlap Syndromes*

There may be an overlap between AIH and primary biliary cirrhosis (PBC) (AMA positivity and florid bile duct lesions on liver histology), PSC (p-ANCA positivity, strictures on MRCP/ERCP, and onion peel fibrosis). This distinction is important as patients with overlap syndromes may not respond well to conventional AIH treatment alone and may require treatment for overlapping entities.

5.94.8 *Treatment*

Steroids (oral prednisolone, budesonide, or intravenous methylprednisolone) form the first line of treatment for acute and acute severe AIH. It is recommended to check for response to steroids in 1–2 weeks and decide on the need for the second-line agents (mycophenolate) if no significant response is noted in acute severe AIH. The normalisation of liver biochemistry and immunoglobulin levels define a biochemical response. Maintenance treatment with azathioprine is needed in the majority of patients to prevent relapse.

5.94.9 *Liver Transplant*

AIH–ALF is treated with a liver transplant. The 5-year survival after liver transplant for autoimmune hepatitis is lower (78%) than that for other etiologies due to infective complications (fatal fungal infections in the postoperative period).

5.95 Chronic HEV Infection

- Chronic hepatitis E occurs mainly in immunosuppressed individuals such as transplant recipients, human immunodeficiency virus patients with low CD4 counts, and patients with hematological malignancies receiving chemotherapy. A few cases have also been reported in immunocompetent individuals.
- These patients may present with unexplained elevation in transaminases or less frequently with neurological or renal manifestations.
- The diagnosis is confirmed by a persistent elevation of hepatitis E RNA in serum or stool.
- Reduction of immunosuppression to achieve spontaneous viral clearance should be attempted. However, it is effective only in about a third of patients. Therefore, most patients require treatment with antiviral agents like ribavirin.

5.95.1 Virology

- Hepatitis E virus (HEV) is a non-enveloped, single-stranded RNA virus with a 27–34-nm diameter. It belongs to the family Hepeviridae. Five genotypes of HEV have been described, of which genotypes 1–4 infect humans, while genotype 5 has only been known to infect birds.
- More recently, a new taxonomic scheme for the classification of HEV has been proposed based on analysis of existing genomic sequences. The HEV belongs to the orthohepevirus A, from the genus Orthohepevirus which includes isolates from human, mammalian, and avian HEV.
- There are no known animal reservoirs for HEV genotypes 1 and 2, but genotype 3 can also affect animals such as deer, pigs, rodents, mongeese, and shellfish.

5.95.2 Clinical Manifestations

- Fatigue, diarrhoea, arthralgia, abdominal pain, and jaundice are the most common symptoms.
- The ALT elevation in chronic hepatitis E is modest (usually less than 300 IU/l) compared to acute hepatitis E (more than 1000 IU/l).
- While chronic hepatitis E can lead to spontaneous resolution in some patients without any specific intervention, it can lead to rapid progression to cirrhosis and death in others.
- There is no correlation between serum HEV concentration and progression to liver fibrosis.
- Neurologic complications are the most frequent extra-hepatic manifestations. Bilateral pyramidal signs, ataxia, proximal myopathy, encephalitis, cognitive dysfunction, peripheral demyelinating polyneuropathy, and painful sensory peripheral neuropathy were the wide range of neurologic manifestations noted in these immunocompromised patients with chronic HEV infection.
- Renal complications have also been reported in patients with HEV infection. Membranoproliferative glomerulonephritis and relapses in IgA nephropathy were noted in solid organ transplant recipients with acute and chronic HEV infection. Cryoglobulinemia is also noted in patients with chronic HEV.
- Rheumatologic manifestations such as arthralgia, myalgia, skin rashes, and cryoglobulinemia have been noted.
- Chronic hepatitis E has also been reported in patients with hematological malignancies like non-Hodgkin's lymphoma and HIV.

5.95.3 Diagnosis

Diagnostic tests for hepatitis E can be classified as indirect and direct tests.

- Indirect tests to diagnose HEV involve the detection of anti-HEV antibodies IgM and IgG in the blood.
- Direct methods involve the detection of the virus itself or its constituents, such as the viral nucleic acids.

5.95.4 Management

- Reducing immunosuppressive drugs targeting T cells could help eradicate HEV in transplant recipients.
- The use of calcineurin inhibitors stimulates and mycophenolic acid inhibits the replication of HEV.
- A combination of ribavirin and mycophenolic acid had a greater ability to inhibit HEV replication than ribavirin or mycophenolic acid alone.
- Treatment of chronic hepatitis E in HIV-positive patients with pegylated interferon (IFN)-α alone and a combination of pegylated IFN-α and ribavirin has been reported.

5.96 Hepatopulmonary Syndrome (HPS) and Portopulmonary Hypertension

HPS is a disease process with a triad of:

- Liver disease/portal hypertension.
- Widespread intrapulmonary vasodilatation.
- Gas exchange abnormality presents with increased alveolar–arterial oxygen gradient ($\Delta P\,(A - a)O_2$) while breathing room air, ultimately resulting in hypoxemia.

The most common liver disease responsible for HPS is liver cirrhosis. Other liver diseases may contribute like non-cirrhotic portal hypertension, extrahepatic portal vein obstruction, chronic active hepatitis, and fulminant hepatic failure.

5.96.1 Pathophysiology

Vasodilatation

The imbalance of vasodilator and vasoconstrictor agents favours vasodilation. This could be due to the (a) Overproduction of the vasodilators from the injured hepatobiliary system & decrease in their clearance by the liver, (B) Either production of a vasoconstrictor inhibitor or normal sensitivity of the pulmonary vessels to vasoconstrictors in response to hypoxemia is blunted in HPS. Numerous vasodilators are suspected, but nitric oxide (NO) is the most appreciated. Other mediators include vaso-active intestinal peptide (VIP), calcitonin-related peptide, glucagon, substance P, and platelet-activating factor.

Hypoxemia

The main pathophysiologic event underlying hypoxemia is widespread pulmonary precapillary and capillary vasodilatation. The pulmonary capillary diameter is normally about 8–15 μm, which could rise to 500 μm in HPS. In addition, there are distinct arterio-venous (AV) malformations and direct AV communications. Pleural spider angiomas may also form. These changes lead to the following:

(a) Ventilation–perfusion (V/Q) mismatch: This hypoxemia is correctable by breathing 100% oxygen.
(b) Right to left shunting of the blood: If numerous, they can give rise to severe hypoxemia, unresponsive to breathing 100% oxygen.
(c) Diffusion impairment: Excessive vasodilatation causes O_2 molecules not to reach the centre of dilated capillaries readily. On the other hand, increased cardiac output and decreased transition time of blood through the pulmonary vascular bed impairs diffusion; this is called diffusion perfusion defect or alveolar capillary oxygen disequilibrium.
(d) Response to breathing 100% O_2: In response to breathing 100% oxygen, if PaO_2 rose to levels ≥600 mmHg, shunting of blood is unlikely. If it fails to exceed 500 mmHg, the shunt cannot be ruled out, and if it does not rise to levels above 150–200 mmHg, the shunt is most probably the main mechanism of hypoxemia.

5.96.2 Clinical Manifestations

- Three Cs are important in diagnosing hepatopulmonary syndrome: cirrhosis, clubbing, and cyanosis.
- More than 80% of patients present with symptoms and signs of liver disease. Less than 20% of the presenting symptoms and signs are related to lung disease. These include dyspnea, cyanosis, clubbing, platypnea, and orthodeoxia.
- There is controversy on the correlation between the severity of liver disease and HPS. Some studies have shown a correlation between liver disease severity and HPS severity, but others have failed to show so.
- Mortality is high among HPS patients and is reported to be around 40% within 2–3 years after presentation. Curious enough, the causes of mortality are most commonly non-respiratory (e.g. GI bleeding, sepsis, and renal failure).

5.96.3 Diagnostic Procedures

- Arterial blood gas analysis
- Chest X-ray and chest CT
- Pulmonary function tests
- Two-dimensional contrast-enhanced echocardiography (CEEC)

5.96.4 Treatment

1. Medical therapy:
   (a) Almitrin bimesylate
   (b) Indomethacin
   (c) Methylene blue
2. Interventions other than liver transplantation:
   (a) Embolotherapy
   (b) Portal decompression with TIPS
3. Orthotopic liver transplantation (OLT)

5.96.5 Portopulmonary Hypertension

Definition

Haemodynamically, portopulmonary hypertension is defined by the presence of the following features in patients with portal hypertension:

- Raised pulmonary arterial pressure (mean pressure determined by right-heart catheterisation of >25 mmHg at rest and >30 mmHg during exercise).
- Raised pulmonary vascular resistance (>240/dynes/cm^5) in the presence of a pulmonary arterial occlusion pressure; or a left-ventricular end-diastolic pressure of less than 15 mmHg.

According to the current WHO classification, portopulmonary hypertension is no longer classified as secondary pulmonary hypertension but as pulmonary arterial hypertension associated with liver disease or portal hypertension.

5.96.6 Pathogenesis

- The development of portopulmonary hypertension seems to be independent of the cause of portal hypertension. This syndrome has been described in patients with portal hypertension due to liver cirrhosis and non-hepatic causes, such as portal venous thrombosis in the absence of chronic hepatic disease.

- The mechanisms by which portal hypertension causes pulmonary hypertension remain incompletely understood.

5.96.7 Clinical Manifestation

- The most common presenting symptom in patients with pulmonary hypertension, irrespective of its cause, is progressive dyspnoea on exertion.
- Other symptoms, such as fatigue, palpitations, syncope, or chest pain, are less frequent.
- Physical findings indicating pulmonary hypertension are generally subtle and may be completely absent.
- The most common findings are an accentuated pulmonary component of the second heart sound and a systolic murmur, indicating tricuspid regurgitation. Extended jugular veins, oedema, and ascites can be signs of either decompensated cirrhosis or overt right-heart failure.

5.96.8 Diagnosis

- The diagnosis of portopulmonary hypertension requires raised pulmonary arterial pressure and resistance combined with excluding other causes of pulmonary hypertension.
- Electrocardiography, radiography, or pulmonary function testing can provide some clues as to the presence of pulmonary hypertension.
- In most cases, echocardiography provides the first and most accurate sign leading to the diagnosis.
- Right-heart catheterisation can conclusively show or rule out pulmonary hypertension if uncertainty remains. The presence of pulmonary hypertension in a patient with cirrhosis and portal hypertension per se is insufficient to diagnose portopulmonary hypertension.
- Other common causes of pulmonary hypertension, especially left-heart disease, valvular heart disease, interstitial or obstructive lung disease, sleep-related breathing disorders, and chronic thromboembolism, must be excluded first.

5.96.9 Management

- Anticoagulants
- Pharmacotherapy: intravenous epoprostenol and other prostanoids
- Liver transplantation

Table 5.30 Target population for screening for HCC (EASL 2018)

Target population	Level of evidence	Grade of recommendation
Child A and B cirrhosis	Low	Strong
Child C cirrhotics awaiting a liver transplant	Low	Strong
PAGE-B score >9 (platelet count, age, and gender) in non-cirrhotic HBV	Low	Weak
Non-cirrhotic F3 fibrosis, based on individual risk factor	Low	Weak

5.97 Surveillance for HCC in CLD

Surveillance strategies depend on disease incidence in the target population, the widespread availability of good tools that help identify the disease early, and effective treatment availability. HCC incidence varies widely across the world.

EASL has recommended certain high-risk groups for surveillance (Table 5.30).

5.97.1 Method of Surveillance

Six-month ultrasound is the recommended screening tool. The utility of biomarkers like Alpha-fetoprotein and PIVKA for surveillance is uncertain.

5.98 TIPSS

5.98.1 TIPS: Indications and Complications

Introduction

TIPS creation is the percutaneous formation of a tract between the hepatic vein and the intrahepatic segment of the portal vein to reduce the portal venous pressure. The blood is shunted away from the liver parenchymal sinusoids, thus reducing the portal pressure. TIPS, therefore, represents a first-line treatment for complications of portal hypertension, typically in patients with decompensated liver cirrhosis.

Indications

Accepted indications for TIPS include the following

- Uncontrolled variceal hemorrhage from esophageal, gastric, and intestinal varices that do not respond to endoscopic and medical management

- Refractory ascites
- Hepatic pleural effusion (hydrothorax)

Less common indications include
- Hepatorenal syndrome
- Chylous ascites
- Budd–Chiari syndrome

Indications under evaluation
- TIPSS prior to major abdominal surgeries
- Treatment of portal vein thrombosis

Contraindications

- Right heart failure
- Recent left ventricular failure
- Child C cirrhosis with hepatic encephalopathy

Complications

Peri-procedural
- Carotid artery puncture, right atrial puncture, and liver capsular laceration
- Non-target insertion of a stent
- Stent migration

Post-procedure
- Stent occlusion
- Hepatic encephalopathy
- Worsening of heart failure
- Mortality

5.99 Molecular Adsorbent Recirculating System and Bioartificial Devices for Liver Failure

The only long-term therapy in most cases is orthotopic liver transplantation unless the liver can regenerate. Many patients, especially those not listed for high-urgency transplantation, may not survive until a suitable donor organ is available since donor organs are rare. In other cases, contraindications do not permit liver transplantation. For these indications, extracorporeal liver-assisted devices have been developed to either bridge the patient to transplantation or temporarily support the failing organ until it can regenerate, therefore, leading to the development of bioartificial liver support devices.

These are categorized as:

- Artificial systems
- Bioartificial systems

5.99.1 *Molecular Adsorbent Recirculating System*

The molecular adsorbent recirculation system (MARS) is an artificial system that banks on the recycling of albumin solution via an anion exchanger and active charcoal. The patient's blood is led through the hollow fibre capillaries of a high-flux dialysis filter. Circulated in the extracorporeal circuit, albumin solution passes the membrane counter-directionally, allowing albumin-bound toxins in the blood to cross the membrane and bind to the albumin of the MARS circuit. The membrane is, however, impermeable to albumin. When passing the adsorber and filter cartridges, the filter clears the toxins and albumin is regenerated and able to accept new toxins when passing the membrane again. Additionally, the albumin circuit is dialysed in continuous veno-venous haemodialysis (CVVHD) or continuous veno-venous hemodiafiltration (CVVHDF), diminishing the load of water-soluble toxins (Fig. 65).

Bioartificial Systems

Bioartificial systems were developed to take over partially the synthetic and regulatory function of the liver besides detoxifying the patient's plasma. They use liver cells as biological components to accomplish this task. Different cell sources have been utilised in several systems. In general, there are the following options: using primary cells, either human or of xenogeneic derivation, cell lines (tumor cell lines or immortalized cell lines), or developing expandable progenitor cell populations.

1. The *CellModule* is part of the modular extracorporeal liver support (MELS) device and can be combined with continuous veno-venous haemodiafiltration and albumin dialysis. The *CellModule* bio-reactor comprises three interwoven hollow fibre bundles embedded in a polyurethane housing. Two bundles of hydrophilic polyethersulfone membranes with a pore size of 0.5 ± 0.1 μm serve for culture medium or plasma perfusion, while one bundle of hydrophobic multilaminate fibres is used for decentralised oxygenation. By mimicking the vessel structure of the liver, those hollow fibres form small repetitive units and ensure the supply of the cells with oxygen and nutrients. The system can be operated with up to 600 g of liver cells inoculated in the inter-capillary space via 24 open-ended silicone tubes. During therapy, the patient's plasma is separated from the blood cells via a plasma filter and recirculated through the hollow fibres at 200–250 ml/min.

2. The bioartificial liver of the Academisch Medisch Centrum Amsterdam (AMC-BAL) does not separate the cells from the patient's plasma by capillary membranes. A spirally wound mat of nonwoven polyester fibres inside a housing provides an attachment area for the liver cells. The matrix incorporates oxygenation capillaries to provide local oxygenation. 10×10^9 primary porcine liver cells are seeded in the matrix, where they can adhere. During therapy, the patient's plasma is directly perfused through the matrix, so this system features only one membrane barrier, one fewer than most other bioartificial livers, and enables direct cell–plasma contact.

   MELS and AMC-BAL were compared in an in vitro study with porcine cells. This was the first attempt to directly compare two bioartificial liver support systems under similar conditions, showing that the cell performance was similar in both bioreactor types, only showing minor differences in some parameters.
3. The extracorporeal liver-assisted device (ELAD) utilises C3A cells, a cell line derived from the human hepatoblastoma cell line HepG2. The cells are localised in the extra capillary space of a modified dialysis cartridge. The membrane cut-off of 70 kDa was chosen to prevent the crossing of immunoglobulins, blood cells, and C3A cells. Two additional cell filters ensure that no tumour cells can reach the patient's bloodstream.

5.100 Nodular Regenerative Hyperplasia

- Nodular regenerative hyperplasia (NRH), known as nodular transformation, is characterised histologically by diffuse involvement of the liver by hyperplastic nodules composed of cells resembling normal hepatocytes.
- The nodules range from smaller than a hepatic acinus to conglomerate nodules forming large masses. Portal areas may be trapped within the nodules. No significant fibrosis is found in or around the nodules. This is an important feature distinguishing NRH from cirrhosis and focal nodular hyperplasia of the liver.
- Large nodules, viewed out of context of the diffuse nodularity, can mimic hepatocellular adenoma histologically. Thus, limited sampling by needle biopsy may yield a false diagnosis of hepatocellular adenoma or normal liver. Although the prevalence of NRH may be as high as 0.6% based on autopsy series, it is probably underdiagnosed because the entity is not easily recognised and because needle biopsy of the liver can be misinterpreted.
- Clinically, portal hypertension is found in up to one-half of the cases; these patients are often presumed to have cirrhosis or idiopathic portal hypertension, and cross-sectional imaging studies of the liver are not always done. Only a few case reports have appeared in the radiologic literature. Various systemic diseases and drugs are often associated with NRH and provide a clue to its diagnosis, particularly when portal hypertension is present.

- The correct diagnosis prevents an unnecessary hepatic lobectomy should NRH be mistaken for hepatocellular adenoma. It prevents an unnecessary search for a primary neoplasm should the multiple nodules of NRH be mistaken for metastases. Correct diagnosis also provides an appropriate prognosis; the prognosis of NRH differs from focal nodular hyperplasia because of the associated portal hypertension, systemic diseases, and risk of hemorrhage or rupture of a large nodule in NRH, and differs from cirrhosis since portal hypertension in cirrhosis carries a poorer prognosis.
- The radiologic features of NRH reflect its composition of cells resembling normal hepatocytes and KCs (present within or between the nodules), the tendency for large nodules to bleed, and the presence of portal hypertension. Patients with NRH may be asymptomatic or may present with idiopathic portal hypertension with varices, splenomegaly, or ascites evident radiologically.
- If a liver spleen scan shows discrete masses, particularly if uptake of sulfur colloid is seen within a mass, then NRH should be considered, and the radiologist should recommend sonography and CT to look for other findings suggesting NRH. These include hypoechoic or isoechoic masses on sonography, particularly with anechoic centres (suggesting hemorrhage) and a hypodense pattern on CT with minimal or no enhancement.
- NRH may have radiologic features resembling focal nodular hyperplasia, hepatocellular adenoma, or hepatic metastases. When multiple, other tumours (HCC or sarcomas, for example) should be considered in the differential diagnosis. As with NRH, focal nodular hyperplasia, hepatocellular adenoma, and metastases may be multiple, hypodense on CT, have variable echogenicity, and contain areas of hemorrhage centrally that may manifest as central sonolucency.
- A stellate central scar, characteristic of focal nodular hyperplasia, has not been seen in NRH. Unlike hepatocellular adenoma, our cases of NRH did not enhance significantly, except for one patient in whom the NRH enhanced and had a central hemorrhage. These entities are not usually accompanied by portal hypertension.
- Uptake of Tc sulphur colloid in a hepatic mass has been suggested as virtually pathognomonic of focal nodular hyperplasia. Recent pathologically correlated data indicate that this may occur infrequently in hepatocellular adenomas (Lubbers, unpublished data) and rarely in hepatoblastoma. A spoke wheel pattern on angiography with filling from the periphery, seen in one of our cases, has also been described as a feature of focal nodular hyperplasia and hepatocellular adenoma
- When multiple liver nodules are present, NRH and metastases should be considered in the differential diagnosis; however, most multiple hepatic metastases show some enhancement and are usually not incidental findings. When the nodules are microscopic in size, sonography or CT may be normal or show portal hypertension findings. Thus, even if hepatic masses are not evident radiologically, the radiologist should include NRH in the differential diagnosis of noncirrhotic portal hypertension, for example, when splenomegaly, varices, and/or ascites are present.

- The pathogenesis of NRH is obscure; vascular, immune, and hormonal factors and drug effects have been hypothesised. One theory is that a vasculitis or vascular obstruction of small hepatic arteries or portal veins occurs at the level of the acinus. This causes ischemic atrophy of the acinus, resulting in mild elevation of liver enzymes and bilirubin. Adjacent acini with an intact blood supply then undergo compensatory hyperplasia causing micronodularity. Coalescence of small nodules leads to large macroscopic nodules and masses, evident radiologically.
- Portal hypertension may be related to the compression of terminal hepatic venules, portal veins, or sinusoids by vasculitis or small nodules. Large nodules near the porta hepatis may compress the main portal vein, thus aggravating portal hypertension. Sometimes the nodules are limited to the portal area, a variant of NRH termed 'partial nodular transformation'. As with other forms of noncirrhotic portal hypertension, the prognosis is usually better than that of patients with portal hypertension due to cirrhosis. Portal diversion is useful in relieving the symptoms of portal hypertension in NRH.
- NRH occurs in men and women equally and is reported in all ages and races. Many cases are found incidentally on physical examination, radiologic examination, autopsy, or during workup for associated disease. There is experimental evidence that NRH may be a precursor of HCC. NRH may be fatal because of portal hypertension complications such as bleeding varices or from rupture of a large nodule. If a patient with a large nodule shown radiologically develops hypotension or other signs of acute blood loss, a bleeding nodule can be confirmed sonographically and treated with surgical resection.
- Diseases associated with NRH include myeloproliferative syndromes, lymphoproliferative syndromes, and miscellaneous immune-related disorders. Various drugs have also been implicated, although most are probably not causative; these include steroids (particularly oral contraceptives), antineoplastics, and others. The aetiologic role of these diseases and drugs and the exact incidence of NRH in these diseases are uncertain. When NRH is diagnosed, one should look for associated diseases that are not clinically evident.
- In summary, NRH is an underdiagnosed disease with a variable radiologic presentation ranging from a grossly normal liver to single or multiple masses, often in a patient with portal hypertension. In patients with idiopathic portal hypertension, the radiologist should recommend scintigraphy and cross-sectional liver imaging studies to look for NRH. Suppose multiple hepatic masses characteristic of NRH are detected, particularly in a patient with idiopathic portal hypertension or a disease associated with NRH. In that case, the radiologist should recommend multiple needle biopsies or a laparoscopically guided needle biopsy or wedge biopsy of the liver rather than initiating a search for occult neoplasm. Resection is unnecessary unless a nodule bleeds or ruptures.

5.101 miRNA in Liver Disease

- MicroRNAs (miRNAs) are small RNAs that do not encode proteins but regulate gene expression. MiRNAs are transcribed by RNA polymerase II or RNA polymerase II. The resulting 500–3000 nucleotides long transcripts (pri-miRNAs) are cleaved in a second step by the 'microprocessor complex' into approximately 70 nucleotides long precursor miRNAs (pre-miRNA), which are actively exported from the nucleus into the cytoplasm. Finally, pre-miRNAs are processed by the RNase III endonuclease 'Dicer' into approximately 22 nucleotides long double-stranded miRNAs, which bind to the Argonaute protein and are integrated into the 'RNA-induced silencing complex'. Within this complex, miRNAs bind the 3′ or 5′ untranslated region of the target mRNAs, leading to transcriptional or translational repression of the target mRNA.
- Alterations in miRNA expression profiles were described in organ development, ageing, and cell death, as well as in the pathophysiology of complex diseases such as inflammation, fibrosis, and cancer.

5.101.1 Role in Liver Diseases

- Besides their role in regulating gene expression, miRNAs have been described in body fluids, where they might serve as biomarkers. Based on their extraordinary stability, less complex chemical structure, and lack of post-processing modifications, circulating miRNAs were suggested as 'optimal' serum-based biomarkers.
- Circulating miRNAs can be bound to serum proteins and lipoproteins or encircled in extracellular vesicles, including exosomes, microvesicles, or apoptotic bodies. As exosomes can be released by various hepatic cells (e.g. hepatocytes and KCs) and can be transferred to other recipient cells to regulate expression profiles in these cells, they were suggested to play an important role in hepatic cell–cell communication and the pathophysiology of different liver diseases.
- Findings that miRNAs encircled in these vesicles are well protected from degradation further highlight exosomal miRNAs' potential to serve as potent biomarkers. With respect to the concept of 'liquid biopsy', which has recently been suggested as a novel detection tool for malignant diseases, miRNA might thus function as a potential 'liquid biopsy' for malignant and benign liver disease.
- Circulating miRNAs might represent diagnostic and prognostic biomarkers in liver fibrosis or cirrhosis patients.
- The role of circulating miRNA in autoimmune liver disease has only been analysed in limited studies with comparatively small cohort sizes. Thus, further studies are needed to clearly state the potential role of serum miRNAs as a biomarker for AIH, PSC, and PBC.
- Circulating miRNAs might also help assess patient outcome and their likelihood to benefit from different treatment options (surgery, systemic treatment, and

locally ablative treatment) to find a personalised therapy for individual patients. For instance, serum levels of miR-221 correlated with tumour size and tumour stage, and patients with high levels of circulating miR-221 showed a significantly reduced overall survival compared to patients with lower miR-221 levels. Likewise, high serum levels of miR-122 independently predicted poor overall survival in a cohort of HCC patients.

- Circulating miRNAs are of increasing interest for diagnosing and prognosis liver cancer. Although reliable data on serum/plasma miRNAs in the field of CCA are limited, circulating miRNAs are likely to play a decisive role in early detection and the prediction of survival for both analysed types of liver cancer in the future.

5.102 Surgical Risk Assessment in Cirrhosis

Patients with the liver disease face a higher risk of surgical complications, including death from surgery and anaesthesia. Many studies can guide in assessing risk. The goal is to avoid surgery in high-risk patients without denying the benefits of surgery to those at reasonable risk.

5.102.1 Assessment

Nearly all candidates for surgery receive a panel of blood tests that include markers of liver disease, such as the serum concentrations of bilirubin, aminotransferases, alkaline phosphatase, and albumin. Although these individual variables (and others, such as a low platelet count and variceal bleeding) are still valid and important, composite measures have been developed specifically to predict whether patients with advanced liver disease may survive the surgery. Two scoring systems are in widespread use: Child score and MELD score.

Risk of Surgery in Cirrhosis

Abdominal surgery: Most of the risk of surgery in patients with cirrhosis comes from studies of abdominal surgery. The risk appears to be much higher with 'open' abdominal surgery than with laparoscopic surgery. Reported mortality rates range from 20% to 54% for various types of open abdominal surgery, and multiple factors are associated with mortality (Table 5.31).

- Most studies found that the mortality rate was higher in patients with one or more: elevated bilirubin, prolonged prothrombin time, ascites, decreased albumin, encephalopathy, portal hypertension, and emergent surgery. The first five of these factors are the components of the Child–Pugh score.

Table 5.31 Risk of mortality associated with surgery in patients with liver disease

Liver disease	Type of surgery	Mortality (%)	Prognostic factors
Cirrhosis	Non-laparoscopic biliary surgery	20	Ascites, prothrombin time, albumin
	Peptic ulcer surgery	54	Prothrombin time
	Umbilical herniorrhaphy	13	Systolic blood pressure, haemoglobin
	Colectomy	24	Urgent surgery
	Abdominal surgery for trauma	47	Hepatic encephalopathy, ascites, albumin, haemoglobin
	Emergency abdominal surgery	57	Child–Pugh class urgent surgery
	Laparoscopic cholecystectomy	0.9, 6	
	Emergency cardiac surgery	80	Child–Pugh class
	Elective cardiac surgery	3–46	Child–Pugh score
	Knee replacement	0	
	Transurethral resection of the prostate	6.7	
	Abdominal surgery	5–60	Haemoglobin, bilirubin, malignancy
Chronic hepatitis	Various types	0	
Hepatitis C	Laparoscopic cholecystectomy	0	
Acute hepatitis	Exploratory laparotomy	100	
Obstructive jaundice			

- Child–Pugh class was useful in stratifying the risk of death: patients in class A (5 or 6 points) had mortality rates of about 10%, those in class B (7–9 points) had mortality rates of around 30%, and those in class C (>9 points) had mortality rates above 70% in two studies. Child–Pugh score, therefore, emerged as a broadly accepted criterion for predicting the risk posed by abdominal surgery in patients with cirrhosis.
- The MELD score was also evaluated for predicting survival after nontransplant surgery in patients with cirrhosis in a recent large retrospective analysis. The investigators did not offer any cut-off MELD score for avoiding surgery, but in patients with MELD scores lower than 10, the survival rate was 99% at 7 days, 96% at 30 days, and 92% at 90 days—numbers deemed acceptable. Survival rates were significantly lower with MELD scores of 10 or more. Another recent retrospective analysis suggested that a MELD score of 14 or more and a haemoglobin concentration of less than 10 g/dl predicts a poor outcome of abdominal surgery in patients with cirrhosis.

5.103 Coffee and Liver

Coffee consumption has recently been associated with health benefits. Apart from caffeine, coffee contains diterpenes, cafestol, kahweol, and other ingredients, which have been associated with health benefits. The putative mechanism by which coffee benefits the liver is listed in Table 5.32.

Clinical studies (largely retrospective and case–control or cohort) have shown beneficial effects of coffee consumption on the level of liver enzymes, NAFLD, viral hepatitis, fibrosis, and development of HCC. Coffee consumption of >2 cups/day is associated with

- Lower ALT/AST levels
- Better SVR rates in chronic HCV patients treated with Peg IFN and ribavirin
- Lower fibrosis scores in patients with NAFLD and
- Decreased risk of HCC (one cup of coffee equals 10 g of whole green coffee or 5 g of instant coffee)

Table 5.32 Mechanism of the beneficial effect of coffee on liver

Effect on liver	Site of action	Chemical involved	Mechanisms
Antifibrotic	HSC Hepatocyte	Caffeine Caffeine	Inhibit focal adhesion kinase (FAK) and actin synthesis Increase HSC apoptosis and intracellular F-actin and cAMP expression Inhibit procollagen type 1C and alpha-SMA expression Decrease transforming growth factor beta (TDF-B) Stimulate ARE-regulated signalling
Cancer prevention	Hepatocyte	Cafestol and Kehweol	Inhibit phase I activating enzyme expression and activity Induce phase II detoxifying enzymes (i.e. glutathione *S*-transferase) Stimulate antioxidant responsive element (ARE) regulate signalling Induction of gamma-glutamylcysteine synthetase (GCS)
Antioxidant effect	Hepatocytes	Hydrophilic (caffeine and polyphenols, such as chlorogenic acids) Hydrophobic (cafestol, kahweol, and trigonelline), including MRP	Preventing inflammatory reaction down-regulation of immune and inflammatory markers, such as interferon-gamma (IFN-γ) chemokine coded by CX3CL1 or fractalkine, chemokine, ligand4 or CCL4 also called macrophage inhibitory protein (MIP-1b), fibroblast growth factor-2, (FGF-2), and tumour necrosis factor receptors (sTNFRII)

Wadhawan M, Anand AC. Coffee and liver disease. J Clin Exp Hepatol. 2016;6(1):40–6. https://doi.org/10.1016/j.jceh.2016.02.003

5.104 Post-exposure Prophylaxis for HBV

The approach to post-exposure prophylaxis depends on the exposure type and the source's viral status. CDC recommendations on the approach to post-exposure prophylaxis have been summarised in Table 5.33.

Table 5.33 Guidelines for post-exposure[a] prophylaxis[b] for hepatitis B

	Treatment	
Exposure	Unvaccinated[c]	Previously vaccinated[d]
HBAg[e]-positive source		
Percutaneous (e.g. bite or needlestick) or mucosal exposure to HBsAg-positive blood or body fluids	Administer hepatitis B vaccine series and hepatitis B immune globulin (HBIG)	Administer hepatitis B vaccine booster dose
Sex or needle-sharing contact of an HBAg-positive person	Administer hepatitis B vaccine series and HBIG	Administer hepatitis B vaccine booster dose
Victim of sexual assault/abuse by a perpetrator who is HBsAg positive	Administer hepatitis B vaccine series and HBIG	Administer hepatitis B vaccine booster dose
A source with unknown HBAg status		
Victim of sexual assault/abuse by a perpetrator with unknown HBsAg status	Administer hepatitis B vaccine series	No treatment
Percutaneous (e.g. bite or needlestick) or mucosal exposure to potentially infectious blood or body fluids from a source with unknown HBAg status	Administer hepatitis B vaccine series	No treatment
Sex or needle-sharing contact of a person with unknown HBAg status	Administer hepatitis B vaccine series	No treatment

[a] These guidelines apply to nonoccupational exposures. Guidelines for managing occupational exposures have been published separately and can be used for managing non-occupational exposures, if feasible

[b] When indicated, immunoprophylaxis should be initiated as soon as possible, preferably within 24 h. Studies are limited on the maximum interval after exposure during which postexposure prophylaxis is effective. But the interval is unlikely to exceed 7 days for percutaneous or 14 days for sexual exposures. The hepatitis B vaccine series should be completed

[c] A person who is in the process of being vaccinated but who has not completed the vaccine series should complete the series and receive treatment as indicated

[d] A person who has written documentation of a complete hepatitis B vaccine series and did not receive postvaccination testing

[e] Hepatitis B surface antigen

5.105 Prevention of Vertical Transmission for HBV

Children born to mothers with chronic viral hepatitis B and C risk acquiring these infections either in utero or during labour. The risk of transmission and mechanism of transmission is summarised in Fig. 5.23.

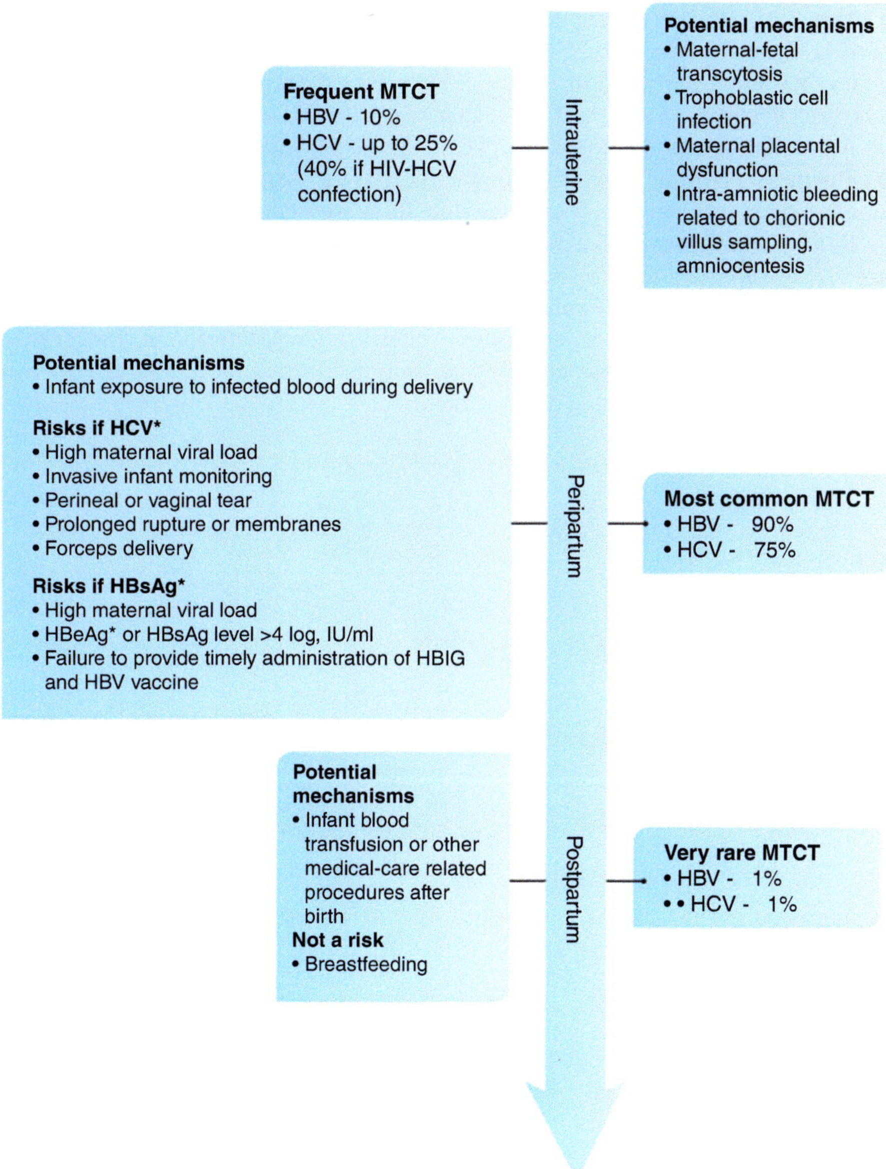

Fig. 5.23 Risk of mother-to-child transmission of chronic hepatitis and proposed mechanisms

Universal screening of hepatitis B during pregnancy is recommended. If maternal indications for the treatment of hepatitis B exist during pregnancy, tenofovir disoproxil fumarate is the preferred antiviral agent. If no maternal indication for initiation of treatment is noted, it is advised to do HBV viral load. If HBV DNA viral load <200,000 IU/ml, the risk of standard newborn immunoprophylaxis is recommended. The risk of failure of standard immunoprophylaxis is low when the maternal HBV viral load is <200,000 IU/ml. If maternal HBV viral load is >200,000 IU/ml or >8 log copies/ml, the risk of failure of standard immunoprophylaxis is higher. So, this subset of mothers may need antiviral therapy during pregnancy, starting from the 28th to the 30th week of pregnancy till delivery. The approach to chronic hepatitis B in pregnancy is summarised in Fig. 5.24.

All neonates born to HBsAg-positive mothers should receive the hepatitis B vaccine within 24 h of delivery. If the mother is HBeAg positive, the risk of transmission is higher, and hepatitis B immunoglobulin is advised.

In a systematic review of the risk of MTCT, the mean estimated rates in the absence of vaccination, with vaccination alone and with vaccination plus the HBIG birth dose, were 75%, 21%, and 6% in women positive for HBeAg, respectively, and 10%, 3%, and 1% in women negative for HBeAg, respectively. In women positive for HBeAg, MTCT was notably low (only 2%) when the HBIG and vaccine birth doses were administered early.

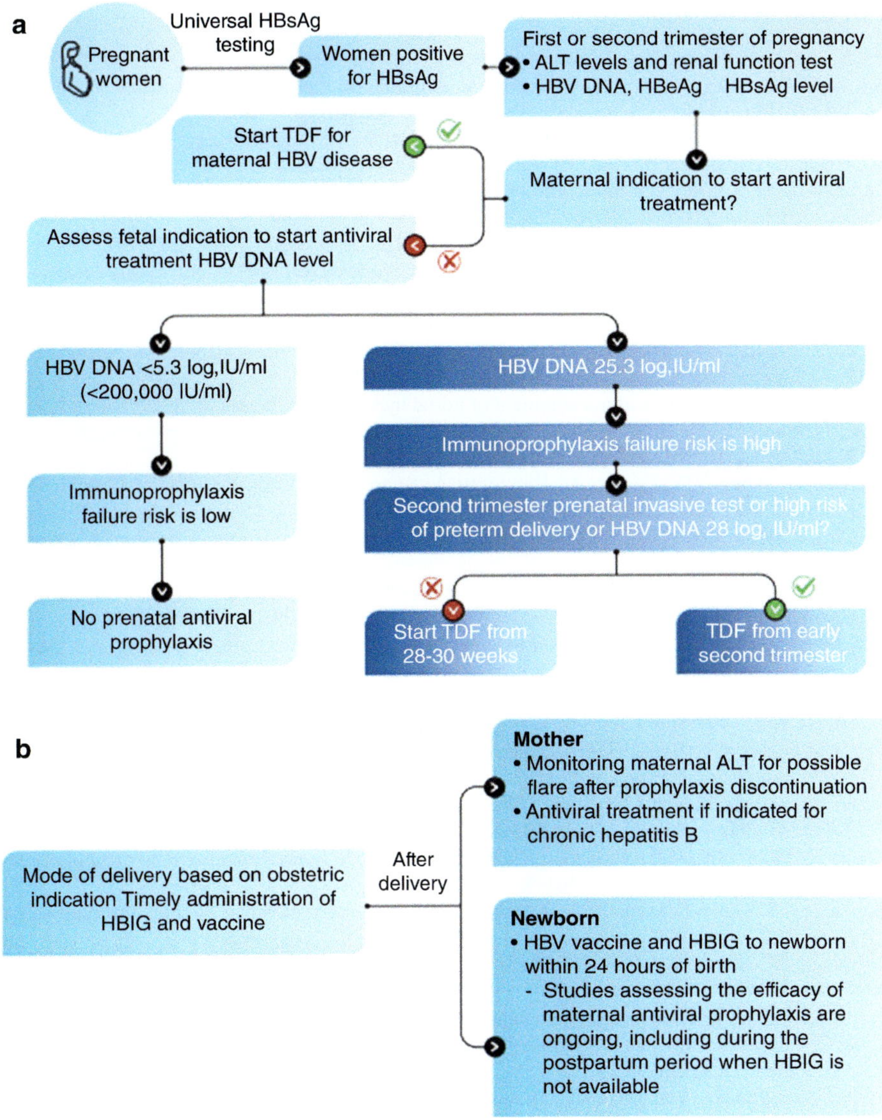

Fig. 5.24 Approach to management of HBV in pregnancy. (Adapted from: Terrault, N.A., Levy, M.T., Cheung, K.W. et al. Viral hepatitis and pregnancy. Nat Rev Gastroenterol Hepatol 18, 117–130 (2021). https://doi.org/10.1038/s41575-020-00361-w)

Further Reading

Adeva MM, Souto G, Blanco N, Donapetry C. Ammonium metabolism in humans. Metabolism. 2012;61:1495–511.
Bacon BR, et al. Hemochromatosis. In: Feldman M, et al., editors. Sleisinger and Fordtran's gastrointestinal and liver disease. 9th ed. Philadelphia, PA: Saunders; 2010.
Bartenschlager R. The hepatitis C virus replicon system: from basic research to clinical application. J Hepatol. 2005;43:210–6.
Beg M, Gupta A. Hepatopulmonary syndrome: a clinical view. Med Update. 2008;99:748–51.
Bharadwaj S, et al. Malnutrition: laboratory markers vs nutritional assessment. Gastroenterol Rep. 2016;4(4):272–80.
Bowlus CS, Pockros PJ, Kremer EA, et al. Long term obeticholic acid improves histological endpoints in patients with primary biliary cholangitis. Clin Gastroenterol Hepatol. 2020;18:1170–8.
CDC. CDC recommendations. MMWR. 2006;55(RR16):30–1.
Collins JC, Sarfeh IJ. Surgical management of portal hypertension. West J Med. 1995;162:527–35.
Dhakshinamoorthy G, et al. Imaging features of fibrolamellar hepatocellular carcinoma. Am J Roentgenol. 2014;202:544–52.
Ferraioli G, Wong VW, Castera L, Berzigotti A, Sporea I, Dietrich CF, Choi BI, Wilson SR, Kudo M, Barr RG. Liver ultrasound elastography: an update to the World Federation for Ultrasound in Medicine and Biology guidelines and recommendations. Ultrasound Med Biol. 2018;44(12):2419–40.
Guu TSY, et al. Structure of the hepatitis E virus-like particle suggests mechanisms for virus assembly and receptor binding. Proc Natl Acad Sci U S A. 2009;106(31):12992–7.
Hancharoenthana W, Leelahavanichkul A. Acute kidney injury spectrum in patients with chronic liver disease: where do we stand? World J Gastroenterol. 2019;25(28):3684–703.
Harris AG. Somatostatin and somatostatin analogues: pharmacokinetics and pharmacodynamic effects. Gut. 1994;Suppl 3:S1–4.
Hemochromatosis. National Heart, Lung, and Blood Institute. http://www.nhlbi.nih.gov/health/dci/Diseases/hemo/hemo_all.html.
Hemochromatosis. National Institute of Diabetes and Digestive and Kidney Diseases. http://digestive.niddk.nih.gov/ddiseases/pubs/hemochromatosis/index.htm.
Hoeper MM, Krowka MJ, Strassburg CP. Portopulmonary hypertension and hepatopulmonary syndrome. Lancet. 2004;363:1461–8.
Hohenester S, Oude-Elferink RPJ, Beuers U. Primary biliary cirrhosis. Semin Immunopathol. 2009;31(3):283–307.
Hoofnagle JH, Nelson KE, Purcell RH. Hepatitis E. N Engl J Med. 2012;367(13):1237–44.
Jain MR, Giri SR, Bhoi B, et al. Dual PPAR alpha/gamma agonis Saroglitazar improves liver histopathology and biochemistry in experimental NASH models. Liver Int. 2018;38:1084–94.
Jenne CN, Kubes P. Immune surveillance by the liver. Nat Immunol. 2013;14(10):996–1006.
Kaul U, Parmar D, Manjunath K, et al. New dual peroxisome proliferator activated receptor agonist Saroglitazar in diabetic dyslipidemia and non alcoholic fatty liver diseases: integrated analyses of real world evidence. Cardiovasc Diabetol. 2019;18:80.
Kim CW, Chang K-M. Hepatitis C virus: virology and life cycle. Clin Mol Hepatol. 2013;19:17–25.
Kordes C, Haussainger D. Hepatic stem cell niches. J Clin Investig. 2013;123(5):1874–80.
Kowdley KV, Vuppalanchi R, Levy C, et al. A randomised, placebo controlled, phase II study of obeticholic acid for primary sclerosis cholangitis. J Hepatol. 2020;73:94–101.
Kubes P, Jenne C. Immune responses in the liver. Annu Rev Immunol. 2018;36:9.1–9.31.
Kubes P, Mehal WZ. Sterile inflammation in the liver. Gastroenterology. 2012;143:1158–72.
Kumar A, Sharma P, Sarin SK. Hepatic venous pressure gradient measurement: time to learn! Indian J Gastroenterol. 2008;27(2):74–80.
Kumar R, Anand U, Priyadarshi RN. Liver transplantation in acute liver failure: dilemmas and challenges. World J Transplant. 2021;11(6):187–202.
Landes Bioscience. Artificial and bioartificial liver support. Organogenesis. 2007;3(1):20–4.

Leida Z, Peng H, Xi C, Ping B. Transarterial chemoembolization (TACE) plus sorafenib versus TACE for intermediate or advanced-stage hepatocellular carcinoma: a meta-analysis. 2014;9(6):e100305.

Liang TJ. Hepatitis B: the virus and disease. Hepatology. 2009;49(5 Suppl):S13–21.

Loosen SH, et al. Role of circulating miRNAs in liver diseases. World J Hepatol. 2017;9(12):586–94.

Lurie Y, et al. Non-invasive diagnosis of liver fibrosis. World J Gasroenterol. 2015;21(41):11567–83.

Mannan AASR. Chronic hepatitis—general. Pathology. http://www.pathologyoutlines.com/topic/liverchronichepgeneral.html. Accessed 17 Apr 2018.

Marconi L, Haibo B, et al. Functional and structural features of cholangiocytes in health and disease. Cell Mol Gastroenterol Hepatol. 2015;1:368–80.

Masakatsu T, Takamichi M. Surgical and locoregional therapy of HCC: TACE. Liver Cancer. 2015;4:165–75.

McManus DP, Gray DJ, Yang Y. Diagnosis, treatment, and management of echinococcosis. BMJ. 2012:344, e3866. https://doi.org/10.1136/bmj.e3866.

Murali AR, Kotwal V, Chawla S. Chronic hepatitis E: a brief review. World J Hepatol. 2015;7(19):2194–201.

Nagral A, et al. Wilson's disease: clinical practice guidelines of the Indian National Association for Study of the Liver, the Indian Society of Pediatric Gastroenterology, Hepatology and Nutrition, and the Movement Disorders Society of India. J Clin Exp Hepatol. 2019;9(1):74–98.

Nagy LE, Ding WX, Cresci G, Saikia P, Shah VH. Linking pathogenic mechanisms of alcoholic liver disease with clinical phenotypes. Gastroenterology. 2016;150:1756–68.

Neuschwander-Tetri BA, Loompa R, Sanyal AJ, et al. Farnesoid X nuclear receptor ligand obeticholic acid for non-cirrhotic non alcoholic steatohepatitis (FLINT): a multi centre, randomised, placebocontroleld trial. Lancet. 2015;385:956–65.

Nevens F, Andreone P, Mazzella G, et al. A placebo-controlle trial of obeticholic acid in primary biliary cholangitis. N Engl J Med. 2016;375:631–43.

Palta S, Saroa R, Palta A. Overview of the coagulation system. Indian J Anaesth. 2014;58(5):515–23. https://doi.org/10.4103/0019-5049.14464.

Priyanka R, et al. Interferons and interferon therapy. J Pharm Sci Res. 2014;6(12):400–3.

Rena NMRA, Wibawa IDN. Albumin infusion in liver cirrhotic patients. Acta Med Indones. 2010;42:162–7.

Ripoll C, Groszmann R, Garcia-Tsao G, et al. Hepatic venous pressure gradient predicts clinical decompensation in patients with compensated cirrhosis. Gastroenterology. 2007;133(2):481–8.

Rizvon MK, et al. Surgery in patients with liver disease. Med Clin North Am. 2003;87:211–27.

Roche Pharmaceuticals. Medication guide has been approved by the U.S. Food and Drug Administration. 2002. https://www.accessdata.fda.gov/drugsatfda_docs/label/2002/21511_Copegus_lbl.pdf.

Russo RA, Vernuccio F. One, no one and one hundred thousand: the wide spectrum of hepatic cystic lesions—tips and tricks for differential diagnosis. EPOS. https://doi.org/10.26044/ecr2019/C-0923.

Sarin SK, Sharma M, Koradia P, et al. A prospective, multi-centre, double-blind randomised trial of Saroglitazar 4 mg compared to placebo in patients with non-alcoholic steatohepatitis. Abstract #1427; APDW 2020.

Shayto RH, Abou Mrad R, Sharara AI. Use of rifaximin in gastrointestinal and liver diseases. World J Gastroenterol. 2016;22(29):6638–51. https://doi.org/10.3748/wjg.v22.i29.6638.

Solá S, Aranha MM, Steer CJ, Cecília MP. Rodrigues game and players: mitochondrial apoptosis and the therapeutic potential of ursodeoxycholic acid. Curr Issues Mol Biol. 2007;9(2):123–38.

Sureka B, Rastogi A, Bihari C, Bharathy KGS, Sood V, Alam S. Imaging in ductal plate malformations. Indian J Radiol Imaging. 2017;27(1):6–12. https://doi.org/10.4103/0971-3026.202966.

Tajiri K, Shimizu Y. Branched-chain amino acids in liver diseases. Transl Gastroenterol Hepatol. 2018;3:47–58.

Tan HK. Molecular adsorbent recirculation system (MARS). Ann Acad Med Singap. 2004;33(3):329–35.

Terrault NA, Levy MT, Cheung KW, et al. Viral hepatitis and pregnancy. Nat Rev Gastroenterol Hepatol. 2021;18:117–30. https://doi.org/10.1038/s41575-020-00361-w.

UDCA. Drug monograph. http://www.mdconsult.com/das/pharm/body/192066773-5/975800014/full/1617.

United Healthcare Commercial Medical Benefit Drug Policy. Somatostatin analogues. 2018. https://www.uhcprovider.com/content/dam/provider/docs/public/policies/comm-medical-drug/somatostatin-analogs.pdf.

Venkatanarshimha N, Thomas R, Atmstrong EM, Shirley JF, Fox BM, Jackson SA. Imaging features of duct plate malformations in adults. Clin Radiol. 2011;66:1086–93.

Yin C, Evason KJ, Asahina K, Stainier YR. Hepatic stellate cells in liver development, regeneration, and cancer. J Clin Investig. 2013;123(5):1902–10.

Younossi ZM, Ratziu V, Loomba R, et al. OBeticholic acid for the treatment of non-alcoholic steatohepatitis: interim analysis from a multi centre, randomised, placebo-controlled phase 3 trial. Lancet. 2019;394:2184–96.

Chapter 6
Pancreas, Gallbladder, and Bile Ducts

A. C. Anand

6.1 Pancreas: Embryological Development and Anomalies: Annular Pancreas and Pancreas Divisum

- The pancreas' initial appearance is around 4–5 weeks of gestation when the embryo is about 4 mm. The endodermal lining of the embryo forms two protrusions, each developing into dorsal and ventral sections of the pancreas.
- The dorsal pancreas grows faster than the ventral forming a nodular unit lengthening into the dorsal mesentery, where further growth continues. The ventral pancreas continues to be comparatively smaller. It rotates towards the dorsal pancreas and away from the small intestine since it is linked to the common bile duct.
- The ductal pathways of the ventral and dorsal pancreas merge so that the secretions of the latter can drain into the common passage between the ventral part and the bile duct.
- The dorsal and ventral parts contribute to the head of the pancreas, while the tail and body are formed entirely by the dorsal component.

6.1.1 *Pancreatic Ducts*

In the primitive stage, the dorsal and ventral parts each arise from the duodenal wall and the common bile duct, respectively.

The ventral duct later unites with the dorsal duct to form the main pancreatic duct. The liver and pancreas share a common exit path for the bile duct and pancreatic duct due to their shared origin.

A. C. Anand (✉)
Department of Gastroenterology and Hepatology, Kalinga Institute of Medical Sciences, Bhubaneswar, India

G. Choudhuri et al. (eds.), *Gastroenterology and Hepatology*,
https://doi.org/10.1007/978-981-99-9287-4_6

- In adults, the proximal part of the dorsal duct transforms to form the accessory duct and remains patent in 70%.

  Around the third gestational month, the pancreatic acini become conspicuous and are derived from the side ducts and termini of the primitive ducts. The pancreas is divided into lobes and lobules by the mesenchymal tissue enclosing it.
- The endocrine and exocrine cells of the pancreas undergo differentiation almost simultaneously. Acini and ductal cells appear slightly earlier than the islet cells.

6.1.2 Signalling and Growth Factors

A variety of signalling and transcription factors influence pancreatic development.

6.1.3 Hedgehog (Hh) Proteins

- Certain proteins regulate cell formation, multiplication, and differentiation by acting as signalling ligands. These are known as Hedgehog (Hh) proteins.
- There are three hedgehog genes found in mammals, namely, Indian (Ihh), sonic (Shh), and desert (Dhh)
  - The Shh protein expression blocks the pancreatic development and, therefore, is usually not present in the developing pancreas, unlike the Indian and desert hedgehog.
  - The Indian hedgehog protein activates pancreatic growth.
- There are two transmembrane receptors: patched (Ptc) and smoothened (Smo) for the Shh. Ptc negatively regulates Smo. When Shh interacts with Ptc, Smo is activated and transcriptional factors are translocated.

6.1.4 Experimental Evidence of the Role of Hedgehog Proteins in Pancreatic Development

- **Indian hedgehog**: Experimental studies have demonstrated that the suppression of Ihh causes a reduction in pancreatic size and impedes endocrine cell formation.
- Excessive Shh expression considerably attenuates exocrine and endocrine pancreatic tissue formation and has been associated with neoplastic intraepithelial lesions. Therefore, the excessive presence of Shh may be of diagnostic help.
- Inactivation of Ihh and Shh has been associated with anomalous structural deformations such as annular pancreatic development, which obstruct the passage of food in the intestine, and incomplete fusion of the dorsal and ventral pancreas.

  Thus, hedgehog protein abnormalities may cause an annular pancreas, the presence of extra-pancreatic tissues, and pancreas divisum.

6.1.5 Annular Pancreas

- The annular pancreas is a ring of abnormal pancreatic tissue around the descending duodenum ahead of the ampulla, causing an obstruction. It originates from the ventral pancreas.
- It occurs when the pancreatic tissue encroaches into the duodenal wall.
- Its incidence is 1 in 20,000.
- Bimodal presentation: (Infancy and fourth to fifth decades)
- Other associated congenital anomalies are trisomy 21 (Down syndrome), malrotation, esophageal and duodenal atresia, biliary atresia, pancreas divisum, tracheoesophageal fistula, and cardiorenal abnormalities.
- In some cases, symptoms of the annular pancreas may appear for the first time in an adult. Presenting findings in adults include
  - Duodenal stenosis
  - Peptic ulceration
  - Chronic pancreatitis
  - Biliary obstruction is a rare complication
- The most frequent manifestation is abdominal pain, which could also be an auxiliary finding.
- In symptomatic presentations, a surgical bypass would be preferred to resection since sometimes the annular tissue may be part of the large pancreatic duct.

6.1.6 Pancreas Divisum

- When the dorsal and ventral ducts do not merge appropriately, pancreas divisum occurs, although patients largely remain asymptomatic. This results in exocrine secretions draining through the comparatively small duct of Santorini and minor papilla. Only a fraction drains through the duct of Wirsung through the larger main ampulla of Vater.
- Pancreas divisum has been observed in
  - 5–10% of autopsy series
  - 2–7% of patients undergoing ERP
  - 25% in a series of patients diagnosed with idiopathic pancreatitis
- Whether the pancreas divisum represents, an anatomic variant of no pathologic significance or a congenital anomaly responsible for recurrent acute pancreatitis in some patients is controversial.
- It has been proposed that both pancreas divisum and a relatively stenotic accessory papilla must be present for clinically evident pancreatic disease to occur. Pancreas divisum is usually diagnosed by ERP, although EUS and MRCP can aid in its diagnosis.
- Most series suggest endoscopic sphincterotomy with the placement of stents through the accessory papilla has a clinical benefit for patients with recurrent idiopathic pancreatitis with accessory papilla stenosis and pancreas divisum.

6.2 Choledochal Cysts

Cystic dilatation of the extrahepatic and intrahepatic bile ducts attributed to congenital abnormalities are known as Choledochal cysts.

6.2.1 *Incidence*

- The incidence rate is 0.00007–0.001%, greater in Asian countries than in Western ones.
- This anomaly is not hereditary and is more predominant in women than men.
- These cysts are not familial.
- Females are more commonly affected than males.
- More than 60% of the cases are diagnosed before completion of 10 years of age.

6.2.2 *Todani's Classification (Fig. 6.1)*

1. Type I cysts: segmental or diffuse cystic [Ia], saccular [Ib] and fusiform [1c] dilatation of the common bile duct (type I cysts account for 80–90% of cases).

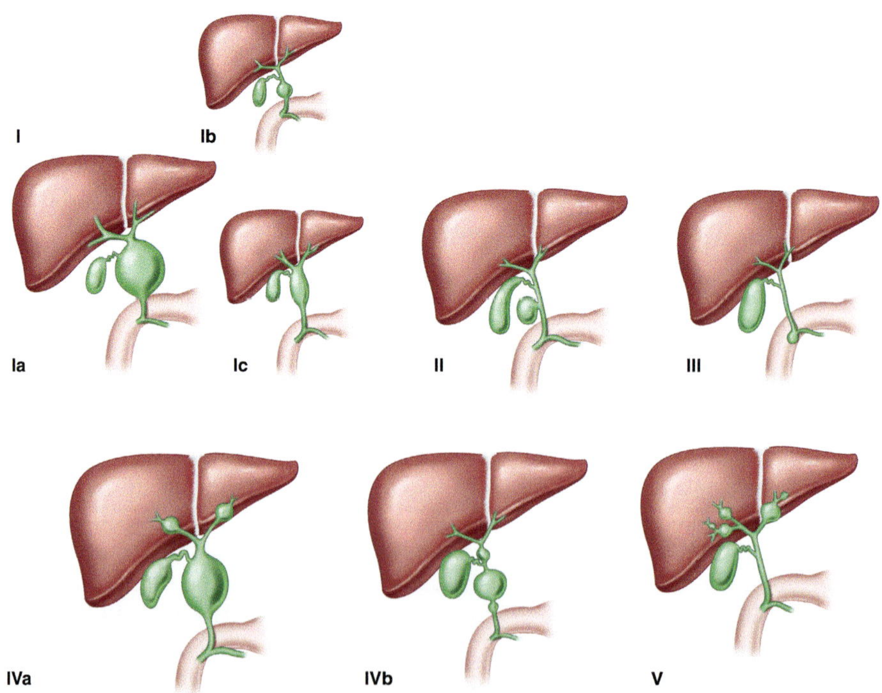

Fig. 6.1 Choledochal cyst—Todani's classification

2. Type II cysts: a true choledochal diverticulum.
3. Type III cysts: dilatation of the intraduodenal portion of CBD, or choledochocele.
4. Type IV cysts may be subdivided into the following:
   (a) Type IVa: multiple intrahepatic and extrahepatic cysts.
   (b) Type IVb: multiple extrahepatic cysts (the type IVb variant is infrequent or may coexist with type I).
5. Type V or Caroli's disease: A solitary or multiple dilatations of the intrahepatic ductal system.

6.2.3 Aetiology

Although the definite aetiology is not identified yet, the probable ones are listed below:

- Weakness of the bile duct wall due to congenital causes.
- The anomalous epithelial proliferation of the duct during embryologic morphogenesis.
- Obstruction of bile ducts due to congenital causes.
- Reovirus RNA has been detected in 78% of patients.
- Pancreaticobiliary maljunction (PBM) occurs frequently (40%), an anomalous junction of the pancreatic and common bile ducts near the intestinal wall. It causes the reflux of pancreatic secretions into the biliary tract.

6.2.4 Pathology

- Comprises of a fibrous wall without epithelial lining or a low columnar epithelium.
- Liver histology in the affected neonate shows characteristic features of large duct obstruction.
- Portal tract oedema, bile ductular proliferation, and fibrosis may be conspicuous.
- Carcinoma of the cyst wall may occur by adolescence.

6.2.5 Clinical Features

During the initial days after birth, more than three-fourths of the patients suffer from cholestatic jaundice, acholic stools, vomiting, irritability, and failure to thrive.

- Hepatomegaly: a palpable abdominal mass in nearly half of the patients
- The choledochal cyst may perforate spontaneously
- Worsening hepatic injury due to biliary obstruction

In older patients, epigastric pain is the commonest manifestation.

- Patients may experience febrile illness with episodes of jaundice and fever due to repetitive bouts of cholangitis.
- Abdominal pain, jaundice, and a palpable abdominal mass are observed in less than 20% of patients.

6.2.6 *Diagnosis*

- Best established with ultrasonography (USG) in children and adults and antenatal USG in foetuses.
- In older paediatric patients, cholangiography aids in elucidating the anatomic characteristics of the cyst, its site of biliary origin, an abnormal junction of the pancreaticobiliary ducts, and the proportion of both extrahepatic and intrahepatic disease, formation of intraductal strictures and calculi.
- Many paediatricians prefer an operative cholangiogram to assess the involvement of intrahepatic and extrahepatic disease.

6.2.7 *Treatment*

- Surgical resection of the cyst with the restoration of the extrahepatic biliary tree is preferred since it attenuates the bile stasis and lowers the susceptibility to cholangitis and cancerous development.
- Biliary drainage is generally performed by a choledocho-jejunostomy with a Roux-en-Y anastomosis.
- In cases which prevent complete resection due to complex anatomic features, simple decompression, and internal drainage should be opted for.
- Considering the high risk of recurrent cholangitis, lithiasis, anastomotic stricture, and pancreatitis subsequent to the surgery, a periodic, long-term follow-up is important.

6.3 Lesser Sac

- The lesser sac, or the omental bursa, is a hollow space/pocket enclosed by the lesser and greater omentum. The epiploic foramen or Foramen of Winslow links the omental bursa to the greater sac.
- The stomach, pancreas, and splenic artery share a part of the wall of the lesser sac; therefore, any damage to these results in seepage of their contents into the lesser sac (Fig. 6.2).

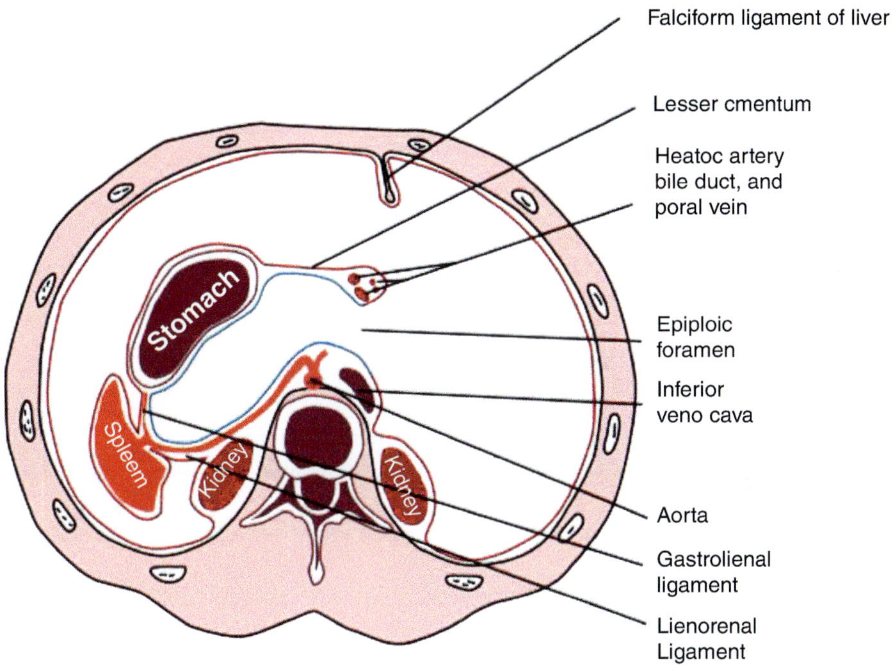

Fig. 6.2 Anatomy of lesser sac

6.4 Sphincter of Oddi

- The sphincter of Oddi (SO) is a group of smooth muscles encircling the terminal common bile duct.
- It is composed of three parts:
  - A common bile duct sphincter (sphincter choledocus)
  - A pancreatic duct sphincter (sphincter pancreaticus)
  - A common sphincter (sphincter papillae)
- The vasculature of the SO includes the anterior superior and posterior superior pancreatico-duodenal branches of the retro-duodenal artery. The gap between the retro-duodenal artery and the papillary orifice is usually more than 3 cm, and, in a small fraction, it could be lesser, which positions the artery within the field of sphincterotomy manipulations.
- Several neurotransmitters and hormones that affect an outer myenteric plexus and separate the submucosal plexus intrinsically regulate the SO. The extrinsic innervation includes parasympathetic vagal fibres and sympathetic nerves via the superior mesenteric ganglion.
- Cholecystokinin (CCK), secretin, and non-adrenergic non-cholinergic neurons via vasoactive intestinal peptide (VIP) and nitric oxide result in sphincter relaxation.

Table 6.1 SO—manometric pressures

Variable	Basal pressure (mmHg)	Amplitude (mmHg)	Frequency (n/min)	Antegrade (%)	Simultaneous (%)	Retrograde (%)	CCK 20 (ng/kg)
Normal	15	135	4	80	13	9	Inhibits
Median range	3–35	95–195	2–6	12–100	0–50	0–50	
Abnormal	>40	>300	>7			>50	Contracts

James Toouli, Alexander Craig, Sphincter of Oddi function and dysfunction. Can J Gastroenterol 2000;14(5):411–19

6.4.1 *Functions of SO*

- Regulate bile and pancreatic secretions into the duodenum.
- Avoid the reflux of enteric contents from the duodenum into the pancreatic and biliary system.
- Between meals, it redirects bile into the gallbladder during the resting state for concentration.
- In the fasting state, the myoelectric activity involves normal basal SO pressures of 3–35 mmHg (abnormal defined as greater than 40 mmHg), with episodic contractions (4 per min, each of 6-s duration) enhancing flow into the duodenum, normally 130 mmHg above basal pressures (Table 6.1).

6.4.2 *SO Dysfunction (SOD)*

- SOD is a benign, non-calculous obstructive disorder of the SO. It could be either a passive impediment due to fibrosis or inflammation or an active one due to sphincter muscle spasm; however, these may frequently occur simultaneously.

SOD is a possible cause of three clinical conditions:

1. Persistent or recurrent biliary pain following cholecystectomy, notwithstanding structural abnormalities.
2. Acute idiopathic pancreatitis.
3. Biliary pain in patients with intact gallbladders but without cholelithiasis (minimal evidence and an ambiguous clinical correlation).

Liver transplantation, acquired immunodeficiency syndrome, and hyperlipidaemia also increase the risk of SOD.

6.5 Pancreatic Stellate Cells

Pancreatic stellate cells (PaSCs) are found in the exocrine pancreas. It constitutes 8% of total pancreatic cells.

Table 6.2 Differences between quiescent and activated pancreatic stellate cells

	Quiescent	Activated (myofibroblastic state)
Morphologic features		
Vitamin A autofluorescence	+ (in fat droplets)	Absent or remnant
Endoplasmic reticulum	+	++
Nucleus size	No change	++
Molecular markers		
Vimentin	++	++
Desmin	+	+
Nestin	+	++
α-SMA	−	+
Properties		
ECM production	↔	Prominent
Cell proliferation	↔	↑
Cell migration	↔	↑

↔ limited, ↑ increased

Location
- Periacinar space
- Perivascular and periductal regions

Important role in
- Chronic pancreatitis
- Pancreatic cancer

Exist in two forms
- Quiescent state.
- Activated state (myofibroblastic state): it is the activated form that is responsible for the pathogenesis of chronic pancreatitis and pancreatic carcinoma (Table 6.2).

6.5.1 Activators of Pancreatic Stellate Cell

- Paracrine (initial event in activation)
  - Cytokines (IL-1, IL-6, IL-8, and TNF-α)
  - Growth factors (PDGF and TGF-β1)
  - Angiotensin II
  - Reactive oxygen species
- Autocrine (produced by the stellate cell itself and helps in perpetuating the cycle of activation)
  - PDGF, TGF-β1
  - Cytokines (IL-1 and IL-6)
  - COX2

6.5.2 Role of Stellate Cells in the Pathogenesis of Chronic Pancreatitis

- When the pancreas is exposed to ethanol and its metabolites or processes producing ROS PaSC are activated by autocrine and paracrine cellular factors derived from adjacent acinar cells, ductal cells, endothelial cells, and leukocytes.
- Activated PaSCs secrete certain products that may either repair by stimulating apoptosis or cause fibrosis due to deposition of excess ECM by sustained activation.

6.5.3 Role of Stellate Cells in the Pathogenesis of Pancreatic Cancer

- PaSC interacts with tumour cells, enhancing the formation of adhesions or fibrotic tissue or desmoplastic tissue prominently found in pancreatic cancer. These direct or indirect cell–cell interactions increase tumour invasion and angiogenesis susceptibility.

6.6 Pancreatic Stem Cells

- Stem cells are multipotent and may be present in the post-natal phase.
- However, the evidence for the presence of pancreatic stem cells in the post-natal phase is ambiguous. Available reports estimate it to be very scarce (0.02% of all pancreatic cells). Such clonogenic cells that can differentiate into ductal, acinar, and beta cells are rare and located in centri-acinar space (between acinus and terminal duct).
- It has been postulated that stem cells noted in the post-natal pancreas are probably derived from bone marrow or biliary epithelium.
- The use of pancreatic stem cells for the treatment of diabetes mellitus is a field worth further exploration. Two approaches are being studied:
  - In vivo stimulation of endogenous PSCs to differentiate into insulin-secreting beta cells using various growth factors and transcription factors (e.g. PDX-1, NGN-3, NeuroD, and nestin and clusterin).
  - Ex vivo expansion of pancreatic stem cells in culture medium and their subsequent transplantation into the host to stimulate insulin production.
- The role of pancreatic stem cells in recovery from the acute and chronic pancreas is also an area of intense research, as is its role in the initiation and perpetuation of pancreatic cancer.

6.7 Bile Production: Bile Acid-Dependent and Independent Pathways

- Every day 500–600 mL of hepatic bile is secreted.
- Bile acids form a large fraction of the organic components of bile. These are secreted actively across the canalicular membrane and induce secretion of other biliary components.

6.7.1 Bile Acid Synthesis

- Cholesterol is used to make bile acids in the pericentral hepatocyte of the hepatic acini. The *de-novo* synthesized bile acids include
  - Cholic acid (CA)
  - Chenodeoxycholic acid (CDCA)

6.7.2 Pathways of Bile Acid Synthesis

1. There are two pathways for bile acid synthesis, the first is the 'classic or major' or neutral pathway that uses cholesterol 7α-hydroxylase (CYP7A1) for CA biosynthesis, and the second is the 'alternative' or acidic pathway using oxysterol 7α-hydroxylase for CDCA biosynthesis.

6.7.3 Feedback Regulation of Bile Acid Synthesis

Excess bile acids indirectly provide negative feedback to suppress the expression of CYP7A1 as follows:

1. Bile acids act as ligands for the farnesoid X receptor (FXR), transcribing the gene for a small heterodimer partner (SHP). SHP is a liver receptor homolog-1 (LRH-1) antagonist necessary for the CYP7A1 gene expression.
2. Bile acids also suppress the CYP7A1 gene transcription via bile acid activation of the c-Jun NH2-terminal kinase (JNK) pathway.

   The enterohepatic circulation:
   The parts of the enterohepatic circulation include

1. Liver
2. Biliary tract
3. Small intestine

4. Portal venous circulation and, to a lesser extent, colon, systemic circulation, and kidney
    - Before secretion, conjugation of CA and CDCA to the amino group of taurine or glycine occurs to improve the hydrophilicity of the bile acid.
    - A large portion of the conjugated bile acids secreted into the small intestine are adequately absorbed.
    - Some unabsorbed bile acids reach the colon, and bacterial 7α-de-hydroxylase converts CA and CDCA to deoxycholic acid (DCA) and lithocholic acid (LCA), respectively.
    - Unconjugated bile acids form conjugated compounds again in the liver and are again secreted into bile.

6.7.4 Bile Secretion

Hepatocellular canalicular bile formation is usually divided into two components:

1. Bile salt-dependent [bile salt-dependent bile flow (BSDF)]
2. Bile salt-independent [bile salt-independent bile flow (BSIF)]

6.7.5 Bile Acid-Dependent Bile Flow

Primary Solutes

- Conjugated bile acids, conjugated bilirubin, glutathione, and heavy metals are some of the primary solutes pumped across the canalicular membrane to produce bile flow with the help of transporters.
- The choleretic activity refers to the quantity of bile flow induced per amount of solute secreted.

Secondary Solutes

- Certain solutes like electrolytes, calcium, and several others are secreted into the canaliculus due to the osmotic gradient.

6.7.6 Bile Acid-Independent Bile Flow

- **Reduced glutathione (GSH) and bicarbonate (HCO^{3-})** form the larger part of the bile acid-independent fraction of bile flow. The MRP2 conducts the ATP-dependent canalicular secretion of GSH. Intraluminal catabolism of GSH by gamma-glutamyl transpeptidase (GGTP) further increases the solute concentration and maintains the osmotic gradient for canalicular bile formation.

- Hepatic and biliary ATP-independent bicarbonate secretion occurs via the HCO^{3-}/Cl^{-} anion exchanger (AE2) contributing to the bile acid-independent bile flow.

6.7.7 *Hepatic Bile Acid Transport*

Approximately 95% of bile acids are obtained from the recirculation of bile.

- This vectorial transhepatocellular circulation of bile acids leads to the concentration of bile. It is steered by unique transport systems located at the sinusoidal and canalicular plasma membranes, namely primary (ATP-dependent), secondary (Na^{+} gradient-dependent), and tertiary (OH^{-} or HCO^{3-}-dependent anion exchange).
- Bile acid uptake differs between fed and fasting states, with periportal hepatocytes and liver acinar hepatocytes playing primary roles.
- Periportal hepatocytes mainly absorb and secrete recirculating bile acids. Perivenous cells primarily secrete fresh bile acids. A large part of the bile flow involves the recirculation of bile acids through periportal hepatocytes.

6.8 Neurohormonal Control of Pancreatic Secretion

Pancreatic secretion occurs in the interdigestive and digestive phases.

6.8.1 *Interdigestive Secretion*

Interdigestive secretion is cyclic and corresponds to the migratory motor complex (MMC), which occurs every 60–120 min. Increased secretion and bursts of motor activity occur in the duodenum (phases 2 and 3). The cholinergic system mediates pancreatic secretion during the interdigestive period. Motilin and pancreatic polypeptide control MMC activity.

6.8.2 *Digestive Secretion*

The digestive phase is divided into three phases: cephalic, gastric, and intestinal.

- Cephalic phase: Mediated through vagal mediated mechanism targeting acinar secretion. Acetylcholine is the principal mediator of the cephalic phase.

- Gastric phase: Gastric distension is the chief stimulus for pancreatic secretion. Pancreatic enzyme secretion occurs more than bicarbonate or water secretion during this phase. A variety of neural and hormonal mechanisms mediate this phase of pancreatic secretion.
- Intestinal phase: This phase begins when chyme enters the duodenum. It is mediated by enteropancreatic vagovagal reflex and multiple hormonal mediators.
  - The major mediator of bicarbonate secretion is intraluminal acid load. Secretin is the primary regulator of pancreatic bicarbonate secretion. Intraluminal fatty acids and proteins are major stimuli for pancreatic enzyme secretion via neurohormonal pathways.
  - Cholecystokinin is the primary humoral intermediate of pancreatic enzyme secretion. Fatty acids and peptides activate small intestinal mucosa to secrete cholecystokinin. Cholecystokinin acts on acinar cells through vasovagal reflexes (CCK receptors are rarely seen in the human pancreas, unlike other species),
  - Cyclic AMP-dependent and calcium-dependent mechanisms control the final common pathway for pancreatic secretion at the cellular level.

6.8.3 Feedback Regulation

- Luminal CCK-releasing factor acts on enteroendocrine cells in the duodenum to secrete CCK into the blood. Luminal CCK releasing factor is broken down by trypsin.
- During a meal, trypsin is bound to meal-based protein. Hence, trypsin is not available to catabolise the CCK-releasing factor. Thus, luminal concentrations of luminal CCK-releasing factor are high, and higher CCK levels are noted in blood, resulting in increased pancreatic enzyme secretion.
- After a meal, trypsin is free and catabolises the luminal CCK-releasing factor, resulting in lower secretion of CCK from enteroendocrine cells and lower secretion of pancreatic enzymes.

6.9 Neurohormonal Control of Biliary Secretion

Gallbladder motility and SO contraction are interlinked.

6.9.1 Gallbladder Contraction

- In a fasting (interdigestive) state, the gallbladder progressively fills up with bile without significantly increasing intraluminal pressure. Gallbladder volume decreases by 30–35% during 2/3 of the interdigestive phase. It corresponds to the migratory motor complex. During phase 1, the gallbladder progressively refills. This cycle helps mix the bile content in the gallbladder and helps prevent the formation of gallstones.
- After meals, gallbladder contraction is noted. This is mediated through vagal innervation during the cephalic phase. Cholecystokinin (CCK) is the main humoral mediator during the intestinal phase. Fat, amino acids, and intraluminal acid are potent stimulators of CCK secretion. Almost three-fourths of the gallbladder volume are expelled on stimulation by CCK.
- Other enteric hormones that control gallbladder motility are secretin and gastrin, which increase gallbladder contractility, while octreotide decreases gallbladder contraction.

6.9.2 SO Contraction

- The SO has a baseline pressure of 10 mmHg, which is masked by phasic contractions at 2–6 per min with an amplitude of 50–140 mmHg over duodenal pressure lasting 4–5 s. A small amount of bile flows in-between phasic contractions and helps clear the sphincter of any debris. When the SO pressure increases, bile is stored in the gallbladder, and when it relaxes, bile and pancreatic secretion flow into the duodenum.
- In the fasting period, the SO pressure corresponds to the migratory motor complex and increases during late phase 2/3. This aids in avoiding the reflux of intestinal contents into the bile duct. In the fed state, the SO pressures decrease and enhance the flow of biliary and pancreatic secretion into the duodenum.
- The contraction of the SO is controlled by neural and humoral mediators. Vagal innervation is excitatory to the SO. CCK is the major humoral mediator, which inhibits sphincter contraction. Secretin initially excites and subsequently inhibits sphincter contraction.
- Sex hormones and prostaglandins affect gallbladder motility and SO contraction.

6.10 Normal Pressures in Pancreatic and Bile Ducts (Table 6.3)

Table 6.3 Normal pressures in pancreatic and bile ducts

	Normal values[a]	Recommended cut-off for abnormal values (>2 S.D. above normal)[b]
Basal SO pressure (mmHg)	14.8 ± 6.3	>35
Ductal pressure (mmHg)		>13
Phasic contractions		
Amplitude (mmHg)	119 ± 32	>220
Duration (s)	4.7 ± 1	>8
Frequency (per min)	5.7 ± 1	>10

[a] Guelrud M, Mendoza S, Rossiter G, Villegas MI. Sphincter of Oddi manometry in healthy volunteers. *Dig Dis Sci.* 1990;35:38–46

[b] Cheon YK. How to interpret a functional or motility test—sphincter of Oddi manometry. J Neurogastroenterol Motil. 2012 Apr;18(2):211–7. https://doi.org/10.5056/jnm.2012.18.2.211

6.11 Pancreatic Function Tests

The pancreatic exocrine function tests are either direct tests that estimate the amount of pancreatic secretions in the duodenum after stimulation by meal or administration of secretagogue or indirect tests that determine the effect of pancreatic secretions on various nutrients (Table 6.4).

Table 6.4 Pancreatic function tests

Test	Description	Advantages	Disadvantages
Direct tests			
Secretin (+CCK)	IV secretin ± CCK Determine volume and HCO_3 secretion into duodenum ± enzyme concentrations	A highly sensitive and specific measurement of pancreatic function	Needs duodenal intubation, Requires IV hormones Uncommonly used and inadequately standardized
Lundh test meal	Oral ingestion of test meal Determine duodenal trypsin concentration (+amylase, lipase)	No IV hormones	Needs duodenal intubation and normal small intestinal mucosa Uncommonly used
Indirect tests			
Faecal fat	72-hour stool collection Fat measurement in stool Preferably compared to a known dietary fat intake, e.g. a 100 g/day of fat diet	Quantitative estimation of steatorrhea	Stool collection Tedious analysis Issues with adherence regarding diet/ recording fat intake
Triglyceride breath test	Ingestion of 13C- or 14C-labelled triglyceride meal, measurement of labelled CO_2 in exhaled breath	Cost-effective, non-invasive Estimate of steatorrhea	Time-consuming (5–6 h) Limited sensitivity for mild dysfunction
Faecal elastase-1	ELISA assay measuring elastase-1 in a pea-size stool sample	Simple one-time stool sample	Unable to detect mild or moderate dysfunction
Faecal chymotrypsin	Enzyme assay measuring chymotrypsin in a pea-size stool sample	Room temperature No IVs or tubes required	
NBT-PABA	Oral ingestion of NBT-PABA with a meal PABA in serum/urine	Simple measurement for severe pancreatic dysfunction	Do not detect mild or moderate dysfunction
Fluorescein dilaurate	Oral ingestion of fluorescein dilaurate with a meal Fluorescein in serum/urine		False positives with small bowel mucosal disease
Serum trypsin	Measure trypsin in the blood sample	Simple, cost-effective Commonly available No risk	Low sensitivity False positives reported in diabetes

CCK cholecystokinase, *IV* intravenous, *NBT-PABA N*-benzoyl-tyrosyl para-aminobenzoic acid
Vesterhus, Mette (2022). Monogenic diabetes and pancreatic exocrine dysfunction in mouse and man. PhD Dissertation. Accessed on 10/6/2022. https://www.researchgate.net/publication/265142356_Monogenic_diabetes_and_pancreatic_exocrine_dysfunction_in_mouse_and_man

6.12 Zymogens

- Digestive enzymes are released in inactive forms (pro-enzymes) called zymogens. This mechanism is helpful in the prevention of autodigestion of cells which secrete them. The enzymatic active site is obstructed by a peptide, which is cleaved later to release active enzymes. Some examples are pepsinogen and trypsinogen.
- Pepsinogen is released from the chief cells at the base of the gastric crypt. This is activated in the stomach's acidic milieu (pH < 5) to form a partially active pepsinogen molecule. A partially active pepsinogen molecule cleaves the peptide from another molecule to form active pepsin.
- Trypsinogen is secreted from pancreatic acinar cells. This is activated in an alkaline milieu in the duodenum by enterokinase secreted by intestinal brush border cells (microvilli). Pancreatic acinar cells also secrete trypsin inhibitors to prevent trypsinogen activation inside the pancreas. Trypsin also undergoes autolysis which acts as a negative feedback control over its activity. Another potential mechanism which protects the pancreas from inappropriate activation of trypsin is flushing out the enzymes from the pancreatic duct by pancreatic fluid secretion (mediated by chloride channels).
- Gene mutations which decrease trypsin inhibitor (SPINK gene mutation) or those mutations which make trypsin resistant to autolysis (PRSS gene mutation) or decrease the fluid secretion in the pancreas (CFTR gene in cystic fibrosis) result in pancreatitis.

6.13 Faecal Chymotrypsin and Elastase

- Faecal chymotrypsin and elastase are used as indirect tests for pancreatic exocrine insufficiency. These are non-invasive tests measuring the activity of pancreatic enzymes (chymotrypsin and elastase in stools) measured by ELISA on stool samples.
- Faecal chymotrypsin levels <24 U/g of stools were considered abnormal. The sensitivity of faecal chymotrypsin in definite and probable chronic pancreatitis was ~83% and 80%.
- Faecal elastase tests measure elastase-1. Faecal elastase-1 level <200 μg/g is considered diagnostic of pancreatic exocrine insufficiency. Compared to the secretin stimulation test as the gold standard, the faecal elastase test has a sensitivity and a specificity of 77% and 88%. Compared to quantitative faecal fat estimation, the faecal elastase test has a sensitivity and a specificity of 96% and 88%.
- Faecal chymotrypsin assays have a few limitations. Chymotrypsin is degraded during transit through the colon, and hence results are prone to variability. Present assays cannot differentiate native chymotrypsin from exogenously derived chymotrypsin (from pancreatic enzyme replacement therapy).

- Faecal elastase does not undergo degradation during transit and is hence more reliable. However, results may vary with the hydration status of stools.

6.14 Genetic Mutations Associated with Pancreatitis

There are two gene mutations associated with pancreatitis—those which increase susceptibility to pancreatitis on their own; others which are modifiers of the risk of pancreatitis.

A few important mutations are:

1. PRSS (serine protease—cationic trypsinogen)
2. SPINK-1 (serine protease inhibitors Kazal type 1)
3. CFTR (cystic fibrosis transmembrane conductance regulator)

Effect of genetic mutation on susceptibility to pancreatitis.

6.14.1 PRSS Gene Mutation

- PRSS gene codes for serine protease cationic trypsinogen. Trypsinogen is a proenzyme for trypsin. It is converted into active trypsin only after reaching the duodenal lumen by the effect of enterokinase. Trypsin can also activate trypsinogen (autoactivation) and also undergoes autolysis.
- Trypsinogen has an eight amino acid extension (trypsinogen activation peptide), which maintains its inactive from until it is split by enterokinase in the duodenum or by autoactivation by active trypsin. Trypsinogen has two calcium binding sites in the trypsinogen activation peptide segment. Calcium binding to the first site activates the conversion of trypsinogen to trypsin (autoactivation). Calcium binding to the second site prevents autolysis of trypsin. Thus, the ambient calcium levels control trypsin activity. High ambient calcium concentrations maintain trypsin in an active state, while low calcium concentration prevents trypsin activation and promotes autolysis.
- Maintaining low intracellular calcium levels inside acinar cells is key to preventing trypsinogen activation in the pancreas. Intracellular calcium in acinar cells rises with neurohormonal activation (activates basolateral calcium channel), bile acid reflux (activates apical calcium channel), and high extracellular calcium levels.
- There are 20 different mutations reported in the PRSS gene associated with recurrent acute pancreatitis, although R122H and N 29I are the most common. These mutations are associated with loci around calcium-dependent trypsin regulation sites, which add the gain of function characteristics by activating trypsin or attenuating trypsin autolysis.

6.14.2 SPINK 1 Gene

- Pancreatic secretory trypsin inhibitor (PSTI, serine protease inhibitor Kazal type 1) is a 56 amino acid peptide secreted along with trypsinogen and physically blocks the active site of trypsin. It colocalises along with trypsinogen within zymogen granules secreted by acinar cells. Thus, SPINK 1 gene product is the first-line of defence to avert premature activation of zymogens.
- Physiological concentrations of SPINK-1 are low as compared to trypsin. So, the inhibitory capacity of PSTI is low in normal physiological conditions. As SPINK-1 is an acute phase reactant, its concentration increases during active inflammation. Thus, SPINK-1 cannot regulate trypsin activity until acute inflammation is established. So, the effect of loss of SPINK-1 mutations on function would be apparent only in the later phase of acute pancreatitis.
- SPINK1 N34S is the most common mutation identified in the general population (1–4%). Pancreatitis associated with SPINK 1 gene mutation is early onset recurrent acute and chronic pancreatitis in paediatric patients, familial pancreatitis, and tropical pancreatitis, and is frequently a characteristic of polygenic pancreatitis-associated genotype.

6.14.3 Cystic Fibrosis Transmembrane Conductance Regulator (CFTR) Gene Mutations

- CFTR is expressed on pancreatic ductal epithelial cells and controls the secretion of chloride and bicarbonate. CFTR mutations are null mutations and result in thick inspissated secretions in the duct. This results in trypsin-mediated recurrent injury and results in fibrosis.

6.15 Cystic Fibrosis

- Cystic fibrosis is a multisystem disease characterised by a mutation in the CFTR gene. The CFTR gene is involved in the movement of chloride and bicarbonate past the apical membrane of ductal cells. It is expressed in lung alveoli, pancreatic ductal cells, intestinal epithelium, and male reproductive organs. Clinical manifestation reflects the effect on these organs.
- Effects of CFTR gene mutation are expressed in utero. The term cystic fibrosis was named after characteristic fibrocystic lesions noted in the pancreas in paediatric autopsy reports conducted by Dr. Dorothy Andersen. Pancreatic exocrine cells bear the major brunt of initial cell injury.
- CFTR mutation results in reduced chloride and bicarbonate content of pancreatic secretion and reduced volume of secretions. This results in viscid inspissated

secretions blocking ducts and acini, resulting in dilated ducts. Inappropriate activation of zymogen occurs in the pancreas, further perpetuation inflammation, fibrosis and fatty replacement of the pancreas. These changes are apparent as early as the 17th week of gestation.
- There are six different groups of patients depending on the type of CFTR mutations. Patients with classes I, II, III, IV, and VI have CFTR mutations which result in absent CFTR. Approximately 85% of patients fall within this category and have pancreatic insufficiency. Others have class V or mild class IV type mutations, which have functional CFTR with reduced ability without pancreatic insufficiency.
- Cystic fibrosis-related diabetes mellitus is noted in the 20–30 age group. Pancreatic exocrine insufficiency increases the risk of diabetes mellitus, a mix of types 1 and 2, since its characteristics include reduced islet cell mass and varying degrees of insulin resistance.

6.16 Schwachman Diamond Syndrome

It is an autosomal recessive disorder characterised by pancreatic exocrine insufficiency, bone marrow failure, and skeletal malformations.

- Pancreatic exocrine insufficiency: characterised by pancreatic acinar paucity and replacement with fatty tissue. Exocrine pancreatic dysfunction is universally present in infancy with or without nutrient malabsorption.
- Bone marrow failure: neutropenia is a common manifestation. Anaemia and thrombocytopenia are less common. Elevated HbF is noted in ~80% of patients. The bone marrow alterations elevate the risk of myelodysplasia and worsening to acute myeloid leukaemia.
- Skeletal malformations: skeletal dysplasia involving metaphyses involving long bones and costochondral junction are noted. Mild shortening of ribs and distal rib flaring are the earliest radiological findings.
- No specific treatment exists apart from pancreatic enzyme supplementation. Hematopoietic stem cell transplantation is a potential treatment for bone marrow failure.

6.17 Pancreas–Endocrine Function

- Islet cells of Langerhans are the key endocrine cells distributed among the acinar cells in the pancreas. About two million islet cells constitute 2–4% of the total weight of the pancreas.
- The islets contain four different types of cells: α, β, δ, and F, distributed in a unique pattern. The α and δ cells are found in the peripheral sections, while β

cells are in the centre. α and β cells constitute 20% and 75% of islet cell mass, respectively.
- Each cell secretes a different hormone, α cells—glucagon; β cells—insulin, δ cells—somatostatin; and F cells—pancreatic polypeptide.
- In a day, the healthy human pancreas secretes about 30 units of insulin. The fasting insulin concentration is 10 μU/mL, and post-prandial levels seldom cross 100 μU/mL in otherwise healthy subjects. The insulin levels in islets increase in response to glucose and FFA, and simultaneously glucagon levels are inhibited, indicating nutrient storage, while the reverse indicates the release of nutrients.
- Glucagon and glucagon-like peptide, GLP-1 and GLP-2, are produced due to a single gene-regulated proglucagon protein. The GLPs are produced by entero-endocrine cells in the intestine and, along with glucagon, have metabolic functions.

6.18 Biliary Microlithiasis

- Biliary sludge is a heterogenous mixture of solid particulate matter in the bile with partly larger particles of about 2–3 mm called microliths. Their presence in bile may result in clinical manifestations such as colicky pain, cholecystitis, cholangitis, and acute pancreatitis.
- Biliary sludge: biliary sludge is an ultrasound finding in the gallbladder characterized by
  - Layering
  - Occurs in the dependent portions
  - Low echo pattern
  - Produces no acoustic shadow

6.18.1 Sludge: Chemical Composition

1. Microcrystals—cholesterol monohydrate crystals (broken window panes appearance and birefringent on polarising microscopy), calcium bilirubinate crystals, and calcium microspheroliths (phosphates and carbonates)
2. Mucin

6.18.2 Pathogenesis of Sludge and Microlithiasis

The biliary sludge and gallstones have a similar pathogenesis and include supersaturation of bile, stasis and gall bladder hypo-motility, and pro-nucleating factors in bile.

Risk factors for microlithiasis

1. Constitutional
2. Haemolysis
3. Diet
4. Age
5. Obesity
6. Pregnancy
7. Gastric bypass surgery and liver transplantation
8. Oestrogen, clofibrate, and octreotide
9. Cirrhosis
10. TPN

6.18.3 Clinical Features

- The symptoms of biliary sludge and microlithiasis are not clear currently. However, both manifest as typical biliary colic, primarily via cystic or common bile duct blockage with acute cholangitis or pancreatitis. These may also cause biliary pain and cholecystitis in the absence of gallstones.
- Among all cases of acute pancreatitis, 30% are idiopathic. Studies have shown that up to 70% of these patients have biliary microlithiasis on evaluation, while only less than 7% of healthy controls have microlithiasis.

6.18.4 Natural History

- A prospective follow-up study on 96 patients: On follow up for 3 years showed that about 18% of cases of biliary sludge disappeared but recurred during observation. Asymptomatic gallstones developed in 8% of the patients and 6% developed symptomatic gallstones, while another 6% developed recurrent biliary colics.
- Pregnancy progressively increases the risk of sludge or stone formation with >5% in the second trimester, ~8% in the third trimester, and 10% 4–6 weeks post-partum. Although sludge and stones are asymptomatic, the latter is associated with biliary pain. Post-partum gallbladder motility is restored in 60–70% of the affected females.

6.18.5 Diagnosis

- Given the higher sensitivity compared to transabdominal ultrasound (55%), endoscopic ultrasound (92%) is superior in diagnosing biliary sludge. However, the diagnosis of biliary microlithiasis requires sampling of gall bladder bile, which can be done by

  - Direct percutaneous puncture
  - During surgery
  - During ERCP
  - From duodenum after gall bladder stimulation, using a Bilbao Dotter tube, after injection of cholecystokinin or cerulin
  - The bile thus obtained is examined under the microscope for crystals, as described above

6.18.6 Treatment

- Only symptomatic cases of biliary sludge or microlithiasis need treatment. A conservative approach is suggested in incidentally identified biliary sludge cases until symptoms manifest.
- In patients who need treatment, the four major treatment options include
  - Stopping the offending agent or controlling the clinical situation leading to microlithiasis, which should be done in all cases if possible.
  - Therapy to dissolve the microliths, which involves therapy with ursodeoxy cholic acid (UDCA) which is given in a dose of 10–15 mg/kg/day for 6–12 months[2].
  - Widening the sphincter, by ERCP and sphincterotomy.
  - Removing the source of microliths by cholecystectomy.

6.19 Gallstone Formation

Globally, gallstones are seen in up to 10% of the population. The main pathogenetic mechanisms are:

- Supersaturation of bile
- Nucleation
- Gallbladder dysmotility

6.19.1 Supersaturation of Bile

- Cholesterol, bile acids, and phospholipids are the major lipid constituents of bile. Cholesterol is insoluble and is maintained in the micelle state by the detergent effect of bile acids and phospholipids.
- Cholesterol is synthesised from acetyl CoA by HMG CoA reductase and is stored in the liver after esterification by acyl coA cholesterol transferase (ACAT). If HMG CoA reductase activity is elevated and esterification by ACAT is dimin-

ished, cholesterol is not stored in the liver and is excreted in bile. Thus, there is an increased concentration of cholesterol in bile.

- The solubility of cholesterol measured as cholesterol saturation index (CSI) can be measured by plotting it on a triangular phase diagram involving cholesterol, lecithin, and bile salts. If CSI > 1, bile is saturated with cholesterol and likely to precipitate and form crystals.
- Normally, cholesterol is transported as micelles, but when cholesterol saturation rises, it gets carried in the form of vesicles, which are initially unilamellar, and then multilamellar, resulting in crystal formation.
- Bile acid composition is altered in people with gallstones. Primary bile acids (cholic acid and chenodeoxy cholic acid) normally constitute 70% of the bile acid pool. Secondary bile acids (deoxycholic acid and lithocholic acid) constitute 25%. Tertiary bile acids (ursodeoxycholic acid) constitute ~4% of the bile acid pool. Patients with gallstones have reduced primary bile acid (cholic acid) and increased secondary bile acids (deoxycholic acid). Deoxycholic acid increases CSI and favours precipitation.
- Some receptors participate in bile acid secretion like ABCB11, phospholipids like ABCB4 and cholesterol like ABCG5 and ABCG8, which alter the normal composition of bile. Patients with ABCB4 mutation are noted to have gallstones at a younger age.

6.19.2 Nucleation

- Nucleation involves merging microscopic crystals formed from supersaturated bile followed by crystallisation, forming cholesterol monohydrate crystals that can aggregate to form gallstones.
- The amount of protein in bile with cholesterol crystals is higher than in bile without crystals. Mucin glycoprotein is the most important pronuclear protein to be identified. The core of mucin glycoproteins contains hydrophobic regions that can bind cholesterol, phospholipids, and bilirubin. Mucin and bilirubin are common components of cholesterol gallstones and together may serve as a seedbed for developing gallstones.
- The calcium levels in the bile correlate with bilirubin precipitation and gallstone formation, with higher concentrations noted in patients with gallstones. Excess calcium compounds, such as carbonate, bilirubinate, and phosphate, act as seedbeds for cholesterol crystallisation.

6.19.3 Gallbladder Dysmotility

- The gallbladder concentrates bile by absorbing water content and acidifying it by absorbing bicarbonate and secreting hydrogen ions. Acidic milieu enhances the dissolution of calcium salts in bile and avoids precipitation.

- Patients with gallstones have elevated fasting and post-prandial gallbladder volume with reduced emptying after meals. The contractile response of the gallbladder with gallstones to CCK is reduced. This leads to stasis and biliary sludge formation (composed largely of cholesterol monohydrate crystals, calcium bilirubinate granules, and mucus).

6.19.4 Pathogenesis of Pigment Stones

The pathogenetic mechanisms of pigment stones differ from those of other types. Black pigment stones are seen in haemolytic anemias and result from the hypersecretion of bilirubin conjugates into bile, which co-precipitates with calcium due to supersaturation. Anaerobic bacteria infect the bile causing brown pigment stones, which is facilitated by gallbladder stasis.

6.20 Types and Composition of Gallstones

There are three types of gallstones based on composition:

(a) Cholesterol stones
(b) Pigment stones
(c) Mixed stones

6.20.1 Cholesterol Stone

- The most common type of gallstone
- Primarily contains cholesterol
- Pure cholesterol stones usually appear sizeable and yellow-white. Microscopic examination reveals long, thin cholesterol monohydrate crystals interwoven by a matrix of mucin glycoproteins with a black centre made of an unconjugated bilirubin calcium salt

6.20.2 Pigment Stones

- Pigment stones lack a crystalline structure and are either black or brown
- Black pigment stones, generally found in cirrhotics and haemolytic states, constitute calcium bilirubinate with different quantities of mucin glycoprotein, calcium, and copper polymer complex

- Brown pigment stones, usually associated with infection, comprise calcium salts of unconjugated bilirubin, cholesterol, and protein

6.20.3 *Mixed Stones*

- Mixed cholesterol gallstones are often found as multiples, are smaller, commoner than pure cholesterol stones, and more than half of their content is cholesterol

6.21 Primary Biliary Cholangitis (Cirrhosis)

- Autoimmune disease with no specific HLA predisposition.
- Affects middle-aged females (F:M = 9:1).
- Characterised by autoimmune injury to epithelial cells of interlobular bile ducts.
- One-half of patients are identified in the asymptomatic phase based on the cholestatic pattern of LFT; fatigue and itching are the commonest symptoms; occasionally, patients present with decompensation of liver cirrhosis.
- The most common signs are hyperpigmentation, hepatosplenomegaly, xanthelasma, and sometimes jaundice if the disease worsens.
- Autoimmune thyroiditis, Sjogren's syndrome, scleroderma, rheumatoid arthritis, and renal tubular acidosis are commonly associated conditions.
- Diagnosis is established by the high alkaline phosphatase with or without increases in bilirubin in liver function tests and the presence of an antimitochondrial antibody (AMA).
- Antimitochondrial antibody (AMA) is directed against the M2 component of
  - Pyruvate dehydrogenase complex (PDC-E2)—most frequent
  - Branched-chain 2-oxo-acid dehydrogenase complex (BCOADC-E2)
  - 2-Oxo-glutarate dehydrogenase complex (OGDC-E2)

    Predominantly of the IgG1 and IgG3 classes
    Non-cytotoxic
    Continue to be present post-liver transplantation in the absence of disease recurrence
    Disease severity does not correlate to antibody titre
    Not necessarily present in PBC

- Liver biopsy shows injury to small interlobular bile ducts and non-caseating granuloma formation. Ludwig's staging has been widely used to stage primary biliary cirrhosis (Fig. 6.3).

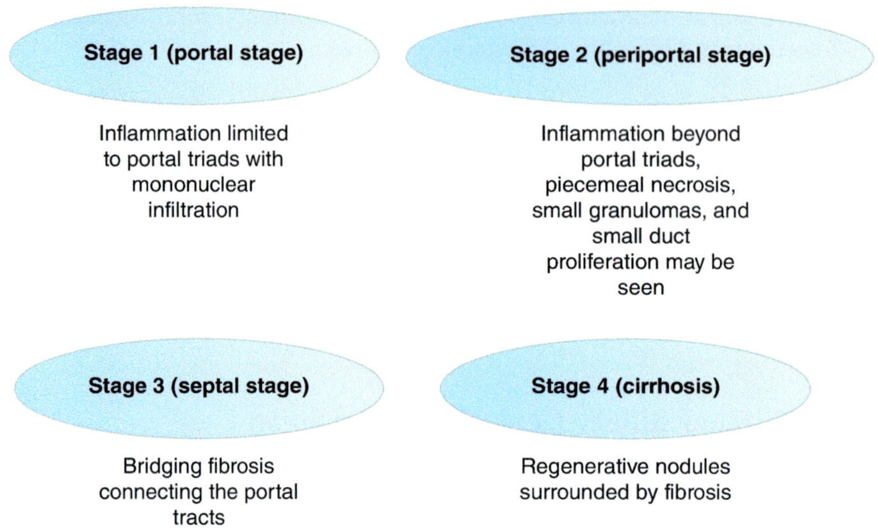

Fig. 6.3 Ludwig's staging for primary biliary cirrhosis

6.21.1 Complications of Chronic Cholestasis

(a) Osteopenic bone disease

- The risk of osteopenia (t-score between −1 and −2.5) and osteoporosis (t-score < −2.5) is very high (4–8 fold) compared to the age-matched general population, with only 20% having normal bone mineral density and more than 25% suffering osteoporosis.

(b) Fat-soluble vitamin deficiency
(c) Hypercholesterolemia
(d) Steatorrhea

- Using survival models helps estimate the probable prognosis and optimal time of liver transplantation.
- Mayo risk score cross-validated.

6.21.2 Treatment

- Ursodeoxycholic acid (UDCA) 10–15 mg/kg is the only medication with proven benefit.

  It acts through the following mechanisms:

  (a) Blocks the absorption of toxic, hydrophobic, and endogenous bile salts
  (b) Stabilises the hepatocyte membranes against toxic bile salts

  (c) Replaces endogenous hepatotoxic bile acids
  (d) Reduces expression of MHC class I and class II antigens

- Treatment with UDCA leads to

  (a) Quick improvement in hepatic biochemical abnormalities.
  (b) Attenuation of the histologic severity.
  (c) Decrease in the risk of gastroesophageal varices formation.
  (d) Delay in the progression to cirrhosis.

- Liver transplantation improves the probability of survival in patients with the end-stage PBC.

6.22 Primary Sclerosing Cholangitis (PSC)

- Chronic cholestatic disorder of unrecognised aetiology typically displays patchy inflammation, destruction, and fibrosis of intra and extra-hepatic bile ducts.
- It usually affects men in the age group 25–45 years.
- More than three-fourths of the affected patients have co-existing IBD, of which the majority have UC and some have Crohn's colitis.
- Approximately 2–4% of patients with chronic UC have PSC, while ~1–3% of patients with CD have PSC.
- It is inversely associated with smoking status.
- Pathogenesis is incompletely understood. Genetic factors like HLA predisposition are not consistent across studies. The role of autoimmunity is supported by the presence of perinuclear-anti neutrophil cytoplasmic antibody (p-ANCA) in 65–88% and ANA in 25–50%, and type 1 diabetes mellitus, Grave's disease, and autoimmune hepatitis in few patients. Non-immune factors like ischemia and toxin-mediated injury are also postulated.
- The presentation may be asymptomatic (incidental findings on imaging), with a biochemical abnormality on liver function test (increased ALP and GGT) or symptomatic with pruritus, fatigue, jaundice, or recurrent cholangitis. A few patients may present in the decompensated state.
- Typical cholangiographic abnormalities: these include diffuse multi-focal strictures, which are short, annular, with a band-like or beaded appearance, usually at duct bifurcations, as well as ectasia of bile ducts. The gallbladder and cystic duct are involved in 15–20%. Diffuse irregularity, structuring, and beading of the main pancreatic duct may also occur. Biliary strictures are intrahepatic alone in 15–30%, extrahepatic in 10% and involve both in most cases. The commonest site of dominant stricture is the bifurcation of hepatic ducts (75%).
- A liver biopsy is required if the cholangiogram is unremarkable, but clinical signs and symptoms indicate PSC. It can also provide prognostic information. The findings include periductal concentric fibrosis, called onion skin fibrosis (50%), and obliterative fibrous cholangitis, which is pathognomonic but seen in 10% only. Histologically there are four stages, ranging from 1 to 4, which include portal inflammation, portal fibrosis, bridging fibrosis, and cirrhosis.

Table 6.5 Models to predict outcomes in PSC

Mayo Model	King's College	Multicenter (Revised Mayo Model)	Swedish	New Mayo Model
1989	1991	1992	1996	2000
Age	Age	Age	Age	Age
Bilirubin	Hepatomegaly	Bilirubin	Bilirubin	Bilirubin
Histologic stage	Histologic stage	Histologic stage	Histologic stage	Albumin
Hemoglobin	Splenomegaly	Splenomegaly		AST
IBD	ALP			Variceal hemorrhage

- The clinically used prognostic models are the King's College, the Mayo Clinic (original and revised), and the multicentric and Swedish models (Table 6.5).
- The broad lines of management include
  - Medical therapy: symptomatic therapy for pruritus and fatigue, and disease-modifying, mainly UDCA. This has been tried in smaller doses, 13–15 mg/kg/day, and in higher doses, up to 30 mg/kg/day, but results have been conflicting. It predominantly leads to improved liver function tests, but survival advantage is controversial.
  - Cancer surveillance: for cholangiocarcinima and colonic malignancies.
  - Antibiotics for cholangitis.
  - Dilate symptomatic strictures, especially if dominant. Stenting is not usually preferred due to the risk of cholangitis.
  - Assess for osteoporosis and other vitamin deficiencies.
  - Management of cirrhosis.
  - Liver transplantation: indications include jaundice not alleviable endoscopically, jaundice which cannot be treated medically, severe recurrent cholangitis, cirrhosis with reduced liver function, and if HCC is demonstrated or CCA is suspected.

6.23 Recurrent Pyogenic Cholangitis (RPC)

- Characterised by recurrent pyogenic infection with intrahepatic pigment stones.
- First described in Hong Kong in the 1930s, also known as oriental cholangitis, Hong Kong fever.

- Malnutrition and helminthic infection may be involved in recurrent pyogenic cholangitis. Malnutrition alters the enzyme expression involved in the deconjugation of bilirubin. Increased deconjugation of bilirubin leads to the availability of unconjugated bilirubin in the bile, along with calcium, which forms calcium bilirubinate and the crux for forming intrahepatic ductal calculi.
- Helminthic infections like ascaris and Clonorchis trigger biliary epithelial inflammation, modulate host immune response, and elevate host vulnerability to upward bacterial translocation to enteric bacterial pathogens like *E. coli*, *Klebsiella*, and *Proteus*. Vicious cycles of recurrent biliary obstruction with bacterial cholangitis due to hepatolithiasis with biliary epithelial inflammation and stricture formation primarily underlies recurrent pyogenic cholangitis pathogenesis.
- Patients present with recurrent episodes of cholangitis and often appear sick with jaundice and right hypochondriac tenderness. Liver function tests reveal elevations in liver enzymes (ALP and GGT). Imaging features include intrahepatic and extrahepatic biliary tree dilatation (central > peripheral zones); unenhanced CTs also show hyperdense intraductal calculi, thickening of the biliary wall, biliary stricture, and, in infrequently, pneumobilia. In the advanced disease stage, atrophy of the liver parenchyma and cholangiocarcinoma occur. Among the various assessment tools, MRCP offers the optimal assessment of RPC.
- Apart from broad-spectrum antibiotics, medical management includes biliary decompression with ERCP. ERCP provides a roadmap of the disease process along with therapeutic procedures like sphincterotomy, stone removal and placement of biliary stents across strictures. Decompressive procedures may be required multiple times.
- Unilobar RPC disease should be managed with surgical excision with biliary reconstruction and access loop formation, while in bilobar, diffuse type, liver transplantation should be considered.

6.24 Secondary Sclerosing Cholangitis

- Secondary sclerosing cholangitis (SSC) is a chronic cholestatic biliary disease with classic signs of hepatic inflammation, obliterative fibrosis of the bile ducts, formation of strictures, and worsening damage to the biliary tree causing cirrhosis.
- The most frequently described causes of SSC are the following:
  - longstanding biliary obstruction
  - surgical trauma to the bile duct
  - ischemic injury to the biliary tree in liver allografts
  - intra-arterial chemotherapy
  - inadvertent biliary spillage of scolicidal agents in percutaneous aspiration, injection, and re-aspiration (PAIR) treatment for hydatid cysts

- Critical illness sclerosing cholangiopathy is a recently described aetiology. Possibly related to ischemic bile duct injury due to hypotension and the use of vasopressors. Biliary cast formation is the hallmark of critical illness associated with sclerosing cholangiopathy.
- The majority of patients with secondary sclerosing cholangitis progress to cirrhosis and decompensation. Limited therapeutic options exist apart from liver transplantation.

6.25 Liver Fluke Infestation

- The parasitic infestation due to the consumption of infested-raw fish causes inflammation of the bile ducts and is a chronic liver fluke disease. The commonly occurring species that cause human infestation are Clonorchis sinensis, Opisthorchis viverrine, and Opisthorchis felineus. The adult flukes inhabit the small intrahepatic bile ducts for 20–30 years.
- The prolonged infestation causes sustained chronic inflammation of the bile ducts leading to epithelial hyperplasia, periductal fibrosis, and bile duct dilatation. Although many individuals remain asymptomatic, those with excessive infection tend to experience fatigue and non-specific abdominal complaints. Such infections may increase the risk of complications such as stone formation, recurrent pyogenic cholangitis, and cholangiocarcinoma.

6.25.1 Management

- Screening tests for liver fluke infection with stool ova test are recommended in areas with a high prevalence of raw or freshwater fish consumers or individuals with a history of such consumption and whose CT images reveal dilatation of the peripheral intrahepatic bile ducts in the absence of obstruction.
- In patients diagnosed with this infection, anthelmintics are advised.
- Individuals over the age of 50 years with a history of clonorchiasis, whether treatment-naive or experienced, should be periodically screened for cholangiocarcinoma.

6.25.2 Hepatobiliary Ascariasis

Individuals with hepatobiliary ascariasis usually develop acute abdominal pain and jaundice. Frequently reported in the Southeast Asian population, HBA leads to recurrent pyogenic cholangiopathy.

6.25.3 Clinical Features

The frequency of HBA is three-fold higher in females than males, who also have been reported to experience ascariasis more frequently. Although HBA can occur in a wide range of age groups, it is predominantly reported in the third decade of life. Paediatric patients are predisposed to develop intestinal instead of biliary obstruction owing to the smaller size of their biliary system. The following factors may increase the susceptibility of individuals to developing HBA:

- History of biliary surgery (cholecystectomy, choledocholithotomy, sphincteroplasty, and endoscopic sphincterotomy).
- Pregnancy due to hormonal modulation of the ampulla.
- Use of certain medicinal agents, such as anaesthetics or tetrachloroethylene, which modifies the internal milieu surrounding the worm.

6.25.4 Modes of Presentation

1. The sudden onset of right hypochondrial pain, episodic or sustained for some days, correlates with the worm's entry into the ampullary orifice from the duodenum. Along with biliary colic other clinical characteristics seldom reported include chills, fever, and mild jaundice.
2. Acute cholangitis in HBA manifests as high-grade fever, chill, icterus, and upper abdominal pain and is a medical emergency. Other features include hypotension, tender hepatomegaly, leucocytosis, elevated bilirubin, serum alanine aminotransferase, and alkaline phosphatase. Exacerbation of acute cholangitis into pyogenic cholangitis involves pus formation at the ampullary orifice, which is aspirated by ERCP.
3. Acute cholecystitis also involves right hypochondrial pain and guarding due to tenderness, low-grade febrile illness, and emesis. The pain radiates to the interscapular area or the right shoulder extremity. The walls of the gallbladder are thickened, and it is distended. Biliary sludge is a frequent finding.
4. Common in children, hepatic abscesses are pus-filled dead ova that migrate upward in the CBD and appear as single or multiple lesions. Its clinical features include high-grade fever, right hypochondrial pain due to hepatomegaly, intercostal tenderness, and edema. These ova induce a granulomatous inflammatory reaction with eosinophil infiltration.
5. Hemobilia is a rare occurrence in individuals with biliary ascariasis.

6.25.5 Diagnosis

- Stool examination showing eggs and vomit-containing worms are diagnostic.
- Pus-containing aspirates from hepatic abscesses show *Ascaris* ova since these produce inflammation leading to granulomatous necrosis.
- Larval invasion of vasculature causes peripheral eosinophilia.
- Cross-reactivity of antibodies with several parasitic antigens hinders its use in the immuno-diagnosis of an infested individual.

6.25.6 Treatment of HBA

The principles of treatment of biliary ascariasis are the following:

1. Cholangitis or cholecystitis should be treated conservatively.
2. Oral anthelmintics facilitate the excretion of paralyzed worms via normal intestinal activity.
3. Endoscopy or ERCP may be alternative options in cases refractory to pharmacological therapy or those with biliary obstruction.

Conservative therapy comprises broad-spectrum antimicrobials, painkillers, and fluid and electrolyte replacement. Infestation in the majority of patients with acute acalculous cholecystitis resolves without sequelae. Although oral antihelminth is effective, the larval forms remain unaffected.

6.26 AIDS Cholangiopathy

- Biliary signs and symptoms of HIV infection are found in patients with low CD4 counts.
- Opportunistic pathogens (*Cryptosporidium parvum*, CMV, isospora, and cyclospora) are considered to be aetiological agents for AIDS cholangiopathy.
- The mechanism of AIDS cholangiopathy is not clear. Cryptosporidium and HIV work synergistically and are postulated to cause biliary epithelial apoptosis and stricture formation. CMV has been associated with ischemic injury to the biliary epithelium.
- Clinical presentation varies across a large spectrum from absence of symptoms to elevated ALP and GGT to biliary pain associated with biliary obstruction.
- Imaging studies show dilated CBD or IHBRD or the characteristic 'beaded' appearance of sclerosing cholangitis.
- Four patterns of cholangiopathy have been described: papillary stenosis is the commonest (30%). Papillary stenosis in association with intra-and extra-hepatic strictures (50%), intrahepatic 'beaded' sclerosing cholangitis-like pattern (10%), and segmental extrahepatic stricture (10%) are the other patterns noted.

- Biliary sphincterotomy alleviates right hypochondrial pain. Treatment of opportunistic infections and antiretroviral therapy are the main treatment modalities for treating AIDs cholangiopathy.
- Rarely, cholangiocarcinoma can result in long-standing AIDS cholangiopathy.

6.27 Vanishing Bile Duct Syndrome

- Vanishing bile duct syndrome is a severe form of chronic cholestatic liver disease involving severe damage to the intrahepatic bile ducts.
- Pathogenesis is not well understood. Immunological injury is postulated. T cell-mediated immunological reaction leading to apoptosis has been implicated.

6.27.1 Causes of Ductopenia

- Drugs: amoxicillin clavulanate, macrolides (azithromycin), quinolones, sulfonamides, antifungals (itraconazole), anti-epileptics (carbamazepine, lamotrigine), anti-psychotics (phenothiazines), and NSAIDS (Ibuprofen).
- Infections: CMV, EBC, HBV, HCV (fibrosing cholestatic hepatitis).
- Malignancy: Hodgkin's lymphoma.
- Sarcoidosis.
- Macrophage activation syndrome.
- PSC, PBC, and graft-versus-host disease are differentials.

6.27.2 Clinical Presentation and Course

Variable: Drug-related vanishing bile duct syndrome usually presents 1–6 months after initiating offending drugs. Some are asymptomatic with elevated ALP. Symptoms include pruritus, jaundice, and fatigue. A few individuals progress relentlessly to cirrhosis with decompensation.

6.27.3 Diagnosis

- Diagnosis is established by the presence of ductopenia in liver biopsy specimens after exclusion of all other possible etiologies. Ductopenia, the pathological hallmark of this disease, is defined as the loss of interlobular bile ducts in more than half of the portal area (histology specimen should contain >10 portal tracts).

6.27.4 Management

Troublesome symptoms, including pruritus, fatigue, and hypercholesterolemia with xanthomata, need medical treatment with antihistaminics, bile acid binding resins, and statins. Occasionally, plasmapheresis may be required to correct dyslipidemia. Nutritional supplements are required, including fat-soluble vitamin supplementation (vitamins A, D, E, and K). For patients progressing to cirrhosis, liver transplantation is required.

6.28 Mirizzi's Syndrome

- Mirizzi's syndrome is associated with acute or chronic inflammation due to solitary large or multiple small gallstones accumulated in the Hartmann's pouch or the gallbladder infundibulum and cystic duct.
- Extended cystic duct parallel to the bile duct and low penetration of the cystic duct into the bile duct are probable risk factors.
- During recurrent cholecystitis episodes, the gallbladder wall, which is initially edematous, becomes fibrosed and thinned out. Due to repeated inflammation, the gallbladder with impacted stone in cystic duct or Hartman's pouch adheres to the bile duct. The cystic duct becomes fibrosed and obliterated. The impacted stone forms a pressure ulcer over the gallbladder wall and subsequently fistulises into the bile duct (cholecystobiliary fistula).
- Mirizzi's syndrome has been classified into five types (Csendes classification)
  - Type I: lodging of the gallstone in the gallbladder infundibulum or cystic duct results in external constriction of the bile duct.
  - Type II: formation of cholecystobiliary fistula due to erosion of the bile duct wall by a gallstone involving less than one-third of the circumference of the bile duct.
  - Type III: cholecystobiliary fistula involving up to two-thirds of the bile duct circumference.
  - Type IV: cholecystobiliary fistula with totally damaged bile duct wall with the gallbladder entirely merged with the bile duct forming a single structure with no discernible segregating planes between these structures.
  - Type V: includes the presence of a cholecystoenteric fistula and any other type of Mirizzi.

    Type Va: absence of gallstone ileus
    Type Vb: complicated by gallstone ileus
- It presents as recurrent biliary pain with obstructive jaundice. Jaundice may be attributed to coexisting choledocholithiasis or extrinsic compression of the bile duct by the impacted calculi in the cystic duct.

- Ultrasonography shows gallstones with intrahepatic biliary dilation with normal calibre lower CBD. MRCP can demonstrate extrinsic compression of proximal CBD with an impacted gallstone at the neck of the gallbladder. ERCP can identify Mirizzi's syndrome and provide therapeutic benefit with the insertion of the biliary stent across the compression level and help alleviate jaundice.
- Surgical management depends on the type of Mirizzi's syndrome. Cholecystectomy with the repair of the bile duct is preferable. Due to dense adhesions, partial cholecystectomy with hepaticojejunostomy may be needed in type III or IV Mirizzi's syndrome.

6.29 Indeterminate Biliary Stricture

- Indeterminate biliary strictures are those whose aetiology remains uncertain after cross-sectional imaging and conventional ERCP-based tissue acquisition techniques. Due to the limited sensitivity of biliary brush cytology and biliary biopsies (30–45%), many strictures remain indeterminate.
- Attempts to obtain tissue from biliary stricture are made using additional modalities like EUS-guided biopsies and cholangioscopy with biopsies. The sensitivity and specificity of the different cholangioscopy-aided biopsies are listed in Table 6.6.
- The visual impression of malignancy on cholangioscopy is more sensitive than biopsies. Irregular vascular patterns, villous mass, finger and papillary projections, concentric stenosis and infiltrative stricture, and irregular granular mucosa are a few characteristic patterns noted on cholangioscopy.

Table 6.6 Performance characteristics of cholangioscopy visual impression and biopsies

	Visual impression			Targeted biopsy		
	Sensitivity (%)	Specificity (%)	Accuracy (%)	Sensitivity (%)	Specificity (%)	Accuracy (%)
All types of scope	67–100	49–100	50–94	38–100	75–100	61–100
Fiberoptic SOC	78–100	77–93	80–94	49–100	94–100	73–100
Digital SOC	83–90[a] 94[b] (89–97)	89–96[a] 95[b] (90–98)	87–93[a] 94[b] (90–98)	80–85	100	93
DPOC	78–89	73–91	75–90	80–100	75–100	79–93
DOC	67–100	49–100	50–97	38–100	100	61–100
History of prior negative or unconfirmed diagnosis	90–100	79–96	89–97	38–88	94–100	61–93

DOC dual operator cholangioscopy, *DPOC* direct per-oral cholangioscopy, *SOC* single operator cholangioscopy,

[a] Data from a systematic review and presented in range

[b] Data from meta-analysis with a 95% confidence interval

- Intraductal EUS is useful in identifying the degree of bile duct association. Irregular wall thickening and perilesional hypoechoic nodes or infiltration into adjacent vessels indicate malignancy.
- EUS with FNB is useful in mass-forming types of biliary neoplasm and has a better yield in lower bile duct neoplasms than hilar neoplasm.
- Surgical excision is recommended if tissue diagnosis remains inconclusive after all attempts and clinical suspicion of malignancy remains high.

6.30 Difficult Bile Duct Stone

- Bile duct calculi are noted in up to 10% of patients who undergo cholecystectomy. ERCP with biliary sphincterotomy, balloon, or basket extraction are standard techniques for extracting bile duct stones. These standard techniques may not succeed in up to 10% of patients.
- There are different characteristics described for 'difficult CBD stone' in literature. They may be related to CBD calculi or the bile duct.
- CBD stone-related features:
  - Size >20 mm
  - Multiple CBD stones (>10)
  - The shape of the stone (barrel or sharp edges)
  - Location of stone (intrahepatic, above stricture, Mirrizi's disease)
- Bile duct-related features
  - Acute angulation of distal CBD from duodenum (<135°)
  - Post Billroth II gastrectomy status
  - Sigmoid-shaped CBD
  - Periampullary diverticulum
  - Impacted stones
- Additional procedures used in the management of 'difficult' bile duct stones include:
  - Endoscopic papillary large balloon dilatation (sphincteroplasty)
  - Mechanical lithotripsy
  - Cholangioscope-assisted laser lithotripsy
  - Cholangioscope-assisted electrohydraulic lithotripsy
  - Placement of biliary stents and reattempt
  - External shockwave lithotripsy

Surgical choledochotomy with CBD stone extraction with/without bilio-enteric anastomosis should also be considered in the treatment option for difficult CBD calculi (Table 6.7).

Table 6.7 Results of advanced techniques used for 'difficult' CBD calculi

	Technical success	Adverse events
Cholangioscopy-assisted laser lithotripsy	95%	9%
Cholangioscopy-assisted electrohydraulic therapy	88%	13%
Extracorporeal shock wave lithotripsy	84%	8%

6.31 Bile Duct Injuries

Post-operative complications, especially gallbladder, pancreas, and gastric surgery, may include bile duct injuries. Of these, more than 80% are attributed to laparoscopic cholecystectomy.

Classification systems:

- **Bismuth–Corlette**: It considers the complete section of the common bile duct and the length of the proximal bile duct stump (Table 6.8)
- **Strasberg's classification** (Table 6.9 and Fig. 6.4)

Table 6.8 Bismuth–Corlette

Type	Level of injury	Biliary stump length (cm)	Hepatic confluence
I	Low	>2	Preserved
II	Middle	<2	Preserved
III	High		Preserved
IV			Lost

Table 6.9 Strasberg's classification

Class	Leakage from	Continuity with bile duct or confluence
A	Cystic duct or an accessory duct	Yes
B	Occlusion of accessory duct	No
C	Accessory duct	No
D	Partial section of bile duct	Incomplete loss of continuity
E (1–5)	A complete section of bile duct with varying stump length	Complete loss

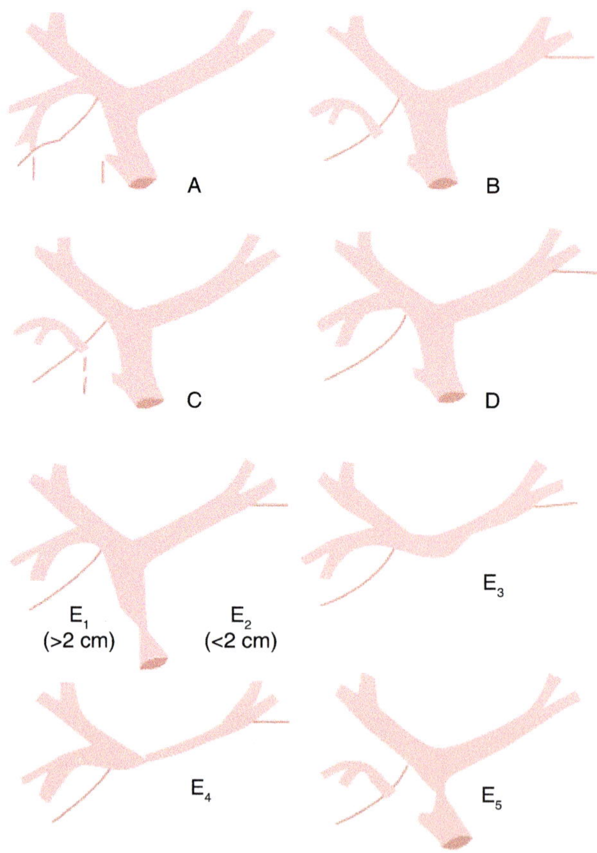

Fig. 6.4 Strasberg's classification of bile duct injuries

6.32 EUS in the Management of Biliary Diseases

EUS is not only an important diagnostic but a therapeutic modality in managing biliary diseases.

- Its sensitivity is better than MRCP in identifying biliary microlithiasis and smaller CBD calculi. Dilated CBD with no obvious etiology on CT or MRCP and abnormal LFT is an important indication for EUS.
- EUS-guided FNA/B is useful in identifying the aetiology of CBD stricture/mass or head of pancreas mass causing obstructive jaundice.
- Interventional EUS has progressed in the last two decades. Obstructions of malignant types in the biliary tract can be drained using EUS instead of PTBD in patients whose ERCP is unsuccessful. Recent literature even supports using upfront EUS-guided biliary drainage (choledochoduodenostomy, hepaticogastrostomy, or antegrade stent placement) in these patients due to its equivalent efficacy and better tolerability than ERCP. Customised stents for hepaticogastrostomy are available, as are single puncture devices (Hot Axios), which have made the procedure safer.
- EUS-guided biliary drainage is technically successful in ~90–95% and has ~10–15% adverse events, including bleeding, cholangitis, and bile leak.
- EUS-guided drainage is an option in patients with benign conditions who fail to achieve deep cannulation of the bile duct by conventional ERCP.

6.33 Portal Biliopathy

- Portal cavernoma cholangiopathy elucidates abnormalities in the extrahepatic biliary ducts with or without the involvement of the first- and second-generation intrahepatic ducts in patients with portal cavernoma.
- Venous drainage of the bile duct is through an extensive network of veins. Subepithelial plexus of veins communicate through perforators with epicholedochocal plexus (of Saint) of veins, which are fine reticular meshwork seen on the surface of the bile duct. These, in turn, communicate with paracholedochocal plexus (of Petren), which form two (or three) marginal veins which run at 3-o'clock and 9-o'clock positions. Marginal veins communicate with the hilar plexus of veins and drain into the portal vein superiorly. These veins communicate with gastric, pancreaticoduodenal, and gastrocolic veins at the inferior end.
- In EHPVO, multiple porta-portal collaterals form with the hepatoportal flow to carry blood across the segment of the thrombosed portal vein. The bile duct is extrinsically compressed by engorged paracholedochal plexus and epicholedochal plexus. Increased pressure is passed through perforators to the intraepithelial plexus, which bulges into the lumen of the bile duct.

- Biliary changes are attributed to compression exerted by the porto-portal collateral veins or ischemia due to inadequate portal circulation. The former cause is reversible after shunt surgery, while the latter is not reversible. Based on the findings of cholangiography, the severity of biliary tree abnormalities in PCC is graded as follows:
  - Grade I: irregularities or angulations of the biliary tree
  - Grade II: indentations or strictures without upstream biliary dilation
  - Grade III: strictures with upstream biliary dilation

Dilatation of ducts means an increase of ductal diameter of more than 7 mm for extra-hepatic ducts and/or more than 4 mm for intra-hepatic ducts

- Only 20% of patients are symptomatic with obstructive jaundice or cholangitis due to CBD calculi/stricture. The majority have an asymptomatic cholestatic pattern of LFT abnormality. MRCP with portal venography is the most comprehensive non-invasive modality to study portal cavernoma cholangiopathy.
- PCC is a progressive condition, but symptoms develop only later in the disease stage when protracted exposure to portal hypertension and sustained obstruction occurs. Treatment is not advised in the absence of symptoms. Patients with symptoms of PCC should receive treatment in a gradual staggered manner with endoscopic decompression, biliary sphincterotomy, and stone removal for patients with biliary colic or cholangitis. Shunt surgery is the next step for patients with persistent biliary obstruction due to strictures. If the biliary obstruction persists, proceed to the bilio-enteric anastomosis.

6.34 Etiopathogenesis of Gallbladder Cancer

Gallbladder cancer (GBC) is the most frequent biliary tract cancer in northern India. The fact that this cancer shows noticeable differences based on ethnicity and geographic regions suggests an influence of genes or heredity and environment in its pathogenesis. However, these mechanisms are yet hypothetical and need more evidence. The proposed etiopathogenetic factors include the following:

- Age
- Sex
- BMI
- Family h/o gallbladder malignancy
- Cholelithiasis
- Porcelain gallbladder
- Salmonella carrier
- Anomalous pancreaticobiliary duct junction
- Gallbladder polyp

6.34.1 Gallstones

- The occurrence of gallstones in Indian patients with GBC is 40–75%, as reported in certain studies. It has been suggested that cancer development follows a chronic inflammation—metaplasia—dysplasia—carcinoma sequence. The gallstone size was found to correlate with the probability of developing gallbladder cancer in some series. Stone more than 3 cm in size and porcelain gall bladder increase the susceptibility to GBC in older studies.
- However, prospective studies have shown that only 1% of all gallstones result in GBC. Although the association of gallstones with gallbladder cancer suggests a common antecedent, clinical data are insufficient to establish a causative role. So, it is obvious that some other risk factors must trigger gallbladder carcinogenesis.

6.34.2 Bacterial Infections

Anaerobic bacteria → Deconjugate bile acids → Induce tumorigenesis

- GBC patients have positive bile cultures more often than controls. Compared to healthy subjects, patients with GBC have been shown to have considerably greater levels of secondary bile acids owing to bacterial infections of bile.

6.34.3 Typhoid Carrier State

- It has been shown that the progression of a chronic typhoid carrier state and not *Salmonella typhi* infection per se correlates with elevated GBC risk. Indian studies have shown that compared to non-carriers, there is an eight-fold greater risk of GBC in those with culture-positive typhoid carriers.

6.34.4 Genetic Factors

- p53 and ras gene mutations have been most frequently associated with GBC. Studies have shown that p53 expression was greatly enhanced in about 60% of GBCs, while kras mutations at codon 12 were reported in 40–60% of cases.
- The precise molecular mechanisms of the development and progression of GBC are yet to be elucidated. Although more evidence is required to confirm, initial data indicate a high incidence of LOH at several chromosomes in patients with GBC.

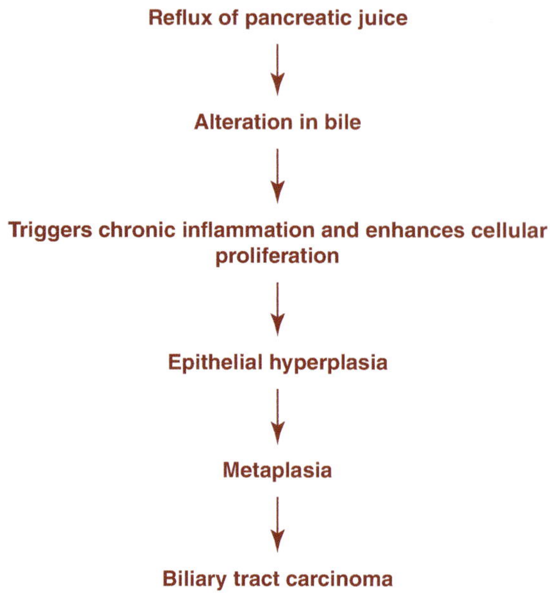

Fig. 6.5 Effect of anomalous pancreatico-biliary duct junction

- Single-nucleotide polymorphisms in certain genes like the apolipoprotein B 100 gene involved in cholesterol metabolism, the CYP1A1 gene involved in carcinogen metabolism, and glutathione-S transferase gene polymorphisms have been associated with GBC in a few studies.

6.34.5 Congenital Anomalies of the Biliary System

Pancreatic juice reflux into the biliary tract due to anomalous pancreatico-biliary duct junction may underlie biliary tract malignancy (Fig. 6.5).

A specific K-ras mutation may be involved in GBC associated with anomalous pancreaticobiliary duct junction.

6.34.6 Gallbladder Adenoma

- Gallbladder adenoma is a frequently found benign neoplasm. Evidence of adenoma–carcinoma progression has been well documented in early studies. All adenomas showing malignant changes in these studies were >12 mm in diameter.

6.34.7 Dietary and Lifestyle Factors

- Some studies suggest an association between GBC and a cholesterol-rich diet, but it is yet to be proven. Other less important risk factors include obesity, endogenous, and exogenous estrogens which heighten the risk of cholelithiasis.
- Occupation exposure to oil, paper, chemical, shoes, and textile components in industries may increase risk.

6.35 Staging of Gallbladder Cancer (Table 6.10)

Table 6.10 Staging of gallbladder cancer

<table>
<tr><th>AJCC 8th</th><th>AJCC 7th</th><th>TNM 8th</th><th colspan="3">Changes between AJCC 8th and 7th editions</th></tr>
<tr><td>I</td><td>I</td><td>T1N0M0</td><td colspan="3">Same criteria</td></tr>
<tr><td>IIA</td><td>II</td><td>T2aN0M0</td><td colspan="3">IIA: Peritoneal side perimuscular connective tissue invasion</td></tr>
<tr><td>IIB</td><td>II</td><td>T2bN0M0</td><td colspan="3">IIB: Hepatic side perimuscluar connection tissue invasion</td></tr>
<tr><td>IIIA</td><td>IIIA</td><td>T3N0M0</td><td colspan="3" rowspan="4">N staging changed from a location based to a number-based approach
N1: 1–3 regional nodes
N2: ≥ 4 regional nodes</td></tr>
<tr><td>IIIB</td><td>IIIB</td><td>T1–3N1M0</td></tr>
<tr><td>IVA</td><td>IVA</td><td>T4N0–1M0</td></tr>
<tr><td>IVB</td><td>IVB</td><td>TxN2M0
TxNxM1</td></tr>
<tr><td>T</td><td>T criteria</td><td>N</td><td colspan="3">N criteria (at least ≥5 LNs)</td></tr>
<tr><td>TX</td><td>Primary tumour cannot be assessed</td><td>NX</td><td colspan="3">Regional LC cannot be assessed</td></tr>
<tr><td>T0</td><td>No evidence of primary tumor</td><td>N0</td><td colspan="3">No regional LN metastasis</td></tr>
<tr><td>Tis</td><td>Carcinoma in situ</td><td>N1</td><td colspan="3">Metastases to 1–3 regional LNs</td></tr>
<tr><td>T1</td><td>Tumour invades the lamina propria or muscular layer</td><td>N2</td><td colspan="3">Metastases to ≥4 regional LNs</td></tr>
<tr><td>T1a</td><td>Tumor invades the lamina propria</td><td>M</td><td colspan="3">M criteria</td></tr>
<tr><td>T1b</td><td>Tumor invades the muscular layer</td><td>M0</td><td colspan="3">No distant metastasis</td></tr>
<tr><td rowspan="3">T2</td><td rowspan="3">Tumor invades the permuscular connective tissue on the peritoneal side, without the involvement of the serosa (visceral peritoneum) or tumour invades the perimuscular connective tissue on the hepatic side, with no extension into the liver</td><td>M1</td><td colspan="3">Distant metastasis present</td></tr>
<tr><td>T</td><td>N</td><td>M</td><td>Stage</td></tr>
<tr><td>Tis</td><td>N0</td><td>M0</td><td>0</td></tr>
</table>

(continued)

Table 6.10 (continued)

AJCC 8th	AJCC 7th	TNM 8th	Changes between AJCC 8th and 7th editions		
T2a	Tumour invades the perimuscular connective tissue on the peritoneal side, without involvement of the serosal (visceral peritoneum)	T1	N0	M0	I
		T2a	N0	M0	IIA
T2b	Tumour invades the perimuscular connective tissue on the hepatic side, with no extension into the liver	T2b	N0	M0	IIB
T3	Tumor perforates the serosa (visceral peritoneum) and/or directly invades the liver and/or other adjacent organs or structures, such as the stomach, duodenum, colon, pancreas, omentum, or extrahepatic bile ducts	T3	N0	M0	IIIA
		T1–3	N1	M0	IIIB
		T4	N0–1	M0	IVA
T4	Tumor invades the main portal vein or hepatic artery or invades two or more extrahepatic organs or structures	Any T	N2	M0	IVB
		Any T	Any N	M1	IVB

6.36 Cholangiocarcinoma (Classification/Palliative options/ Status of Liver Transplant)

Cholangiocarcinoma (CCA) is the commonest biliary tract cancer and the second most frequent primary hepatic malignancy. It is categorised into intrahepatic (iCCA), perihilar (pCCA), and distal (dCCA) subtypes, pCCA being the commonest subtype (Fig. 6.6).

6.36.1 Classification

Classification of cholangiocarcinoma based on its anatomic location within the biliary tree is shown in Table 6.11.

6.36.2 Management

Surgical resection: Surgical resection is the preferred treatment for CCA.

- Liver transplantation is the preferred alternative for PSC patients with pCCA due to field defects and common advanced fibrosis. Regional lymph node metastases are not considered an absolute contraindication to resection, although N1 disease is an independent prognostic factor for worse outcomes.

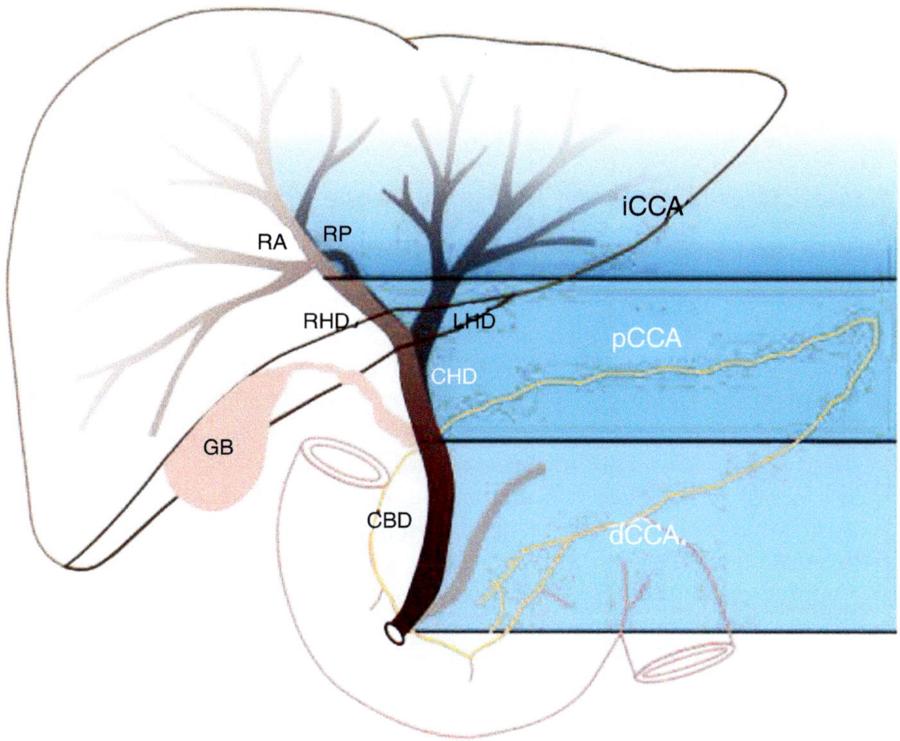

Fig. 6.6 cholangiocarcinoma—classification

Table 6.11 Classification of cholangiocarcinoma based on its anatomic location

Subtype	Location
Intrahepatic cholangiocarcinomas (iCCA)	Proximal to the secondary branches of the left and right hepatic ducts
Perihilar cholangiocarcinoma (pCCA)	Between the secondary branches of the right and left hepatic ducts and the common hepatic duct near the cystic duct origin
Distal cholangiocarcinoma (dCCA)	Tumours of the common bile duct near the ampulla Vateri do not invade it. The dCCA within the intrapancreatic portion of the common bile duct is indistinct from pancreatic head carcinomas and, therefore, challenging to identify.

- Based on the dimensions of the tumour spread, iCCA is resected by segmentectomy. More than three-fourths of the patients need major segmentectomy (>3 segments). When the low volume of the balance hepatic lobe discourages resection, preoperative portal vein embolisation of the ipsilateral hepatic lobe is performed, which causes compensatory hypertrophy of the contralateral lobe, thus facilitating successive hemi-hepatectomy.

- Surgical resection must not be performed in patients with bilateral, multifocal disease, distant metastases, and comorbidities, which are associated with operative risks and outweigh their benefits.

6.36.3 Orthotopic Liver Transplantation

- High recurrence rates and poor long-term survival with OLT in CCA are reasons for recommendations against it.
- Neoadjuvant chemoradiation preceding OLT has demonstrated good efficacy in pCCA. Recurrence rates are lower (20%), and recurrence-free 5-year survival is higher (68%). Unfortunately, patient-selection criteria are too narrow, and disease progression occurs in more than 25% of patients while awaiting OLT, leading to exclusion from this treatment protocol.
- The lack of large sample sizes and study designs (retrospective) and heterogeneity in disease characteristics render the benefit of OLT with or without adjuvant treatment for iCCA demonstrated in these studies questionable. Besides, recurrence rates were as high (35–75%) and 5-year survival low (34–51%). Therefore, OLT is currently not recommended for iCCA.

6.36.4 Nonsurgical Therapies or Palliative Options

- A phase III randomized controlled (ABC-02) trial (RCT) comparing gemcitabine/cisplatin-combination therapy versus gemcitabine monotherapy demonstrated a 6-month survival benefit in patients with CCA.
- Another phase III RCT comparing EGFR-inhibitor erlotinib-gemcitabine/cisplatin to gemcitabine/cisplatin demonstrated two-fold greater objective response rates and twice the duration of progression-free survival.
- Locoregional therapies such as radiofrequency ablation, transarterial chemoembolisation (TACE), drug-eluting bead-TACE, and selective intra-arterial radiotherapy with microspheres or external beam radiation therapy form a part of palliative care. Although some studies found some benefit in tumor regression and survival, data are derived from a retrospective nature, small sample size, and heterogenous inclusion criteria. Additionally, high rates of grade III/IV toxicity have been reported.
- Similarly, stenting with photodynamic therapy (PDT) has shown survival benefits, improved biliary drainage, and enhanced quality of life in smaller settings.
- When resection is not an option in patients with CCA and cholestasis, drainage of more than half of the liver parenchyma provides survival benefits. Since the means of stent placement, whether percutaneous or endoscopy, does not affect the outcomes, the route should be guided by accessibility and comorbidities.

6.37 Pathophysiology of Pancreatitis

6.37.1 Pathophysiology of Chronic Alcoholic Pancreatitis

- Many patients with alcoholic CP have a period of recurrent episodes of acute pancreatitis, lasting 5–6 years, followed by a phase of chronic pain, exocrine and endocrine insufficiency, classically referred to as CP. Abstinence from alcohol after the onset of the disease slows its progression but does not necessarily halt the same.
- Various hypothetical bases of alcoholic CP pathophysiology have been suggested.

Ductal Obstruction Hypothesis (Fig. 6.7)

Toxic Metabolic Hypothesis

(a) Alcohol, or one of its metabolites, has detrimental effects on pancreatic ductal and acinar cells. Elevated membrane lipid peroxidation and free radical formation occur in CP.
(b) Alcohol stimulates the release of cholecystokinin from I cells. CCK stimulation results in the activation of proinflammatory transcription factors.
(c) Alcohol and its metabolites also stimulate the pancreatic stellate cells increasing extracellular matrix deposition.

Necrosis Fibrosis Hypothesis

- Frequent acute pancreatitis with necrosis increases the risk of developing CP. Several natural history studies, as well as animal models, serve as support for this hypothesis.

Chronic Alcoholism

↓

Secretion of pancreatic juice (protein-rich, low in volume and bicarbonate)

↓

Favour formation of protein precipitates, plugs and stones

↓

Obstruction of small ductules, damaging duct and parenchyma

Fig. 6.7 Ductal obstruction hypothesis

Genetic Predisposition

- It is also thought that certain combinations of genetic mutations or polymorphisms would make an individual more susceptible to the detrimental effects of alcohol. Alcohol consumption in such genetically predisposed individuals results in cellular and metabolic derangements triggering a sustained inflammatory response, eventually causing chronic pancreatitis.

6.37.2 Pathophysiology in Acute Pancreatitis

- **Premature zymogen activation and colocalisation theory**: The pancreas normally synthesises a variety of digestive enzymes which have a propensity to damage the gland. These enzymes are secreted as inactive zymogens, activated only when they reach the duodenum. Premature activation of zymogens within the acinar cells is believed to be the pivotal pathophysiological factor in acute pancreatitis.
- The zymogens secreted by the acinar cells initially pass through the Golgi apparatus, where they are packaged within condensing vacuoles and are then taken to the luminal area of the cell for excretion into the ducts. However, the acinar cells also contain lysosomal hydrolases like cathepsin B, which can activate the zymogens within the cells. Normally, these hydrolases are segregated from the zymogens. Colocalisation of digestive enzymes and lysosomal hydrolases within the acinar cells is one of the important hypotheses regarding the pathophysiology of acute pancreatitis. Experimental studies support these data.
- The second important factor in the pathogenesis of acute pancreatitis is a severe inflammatory process. Once colocalisation and zymogen activation has occurred, acinar cell necrosis occurs, which triggers a cascade of inflammation in the pancreas. Several kinase-activating cascades involving p38-MAPK, ERK 1/2, and JNK are upregulated. There is also activation of transcription factors like NF-κB worsening the inflammation. Activation of the complement system, recruitment of macrophages and polymorphs, and release of TNF alpha, IL-1, 6, and 8 all perpetuate the injury. Activated phospholipase A plays a role in the occurrence of ARDS. In mild cases, inflammation is confined to the pancreas, but in severe cases, it becomes systemic and results in multi-organ dysfunction.

Genetic Factors

In conditions like hereditary pancreatitis and cystic fibrosis, the underlying genetic factors also play a major role in the pathophysiology of acute pancreatitis. The majority of these mutations result in premature trypsin activation.

Gallstone Pancreatitis

Suggested mechanisms include:

1. Bile reflux into the pancreatic duct
2. Ampullary obstruction due to stones
3. Edema resulting from a stone passage

   - Studies exploring the pathophysiology of acute pancreatitis are usually done in animal models, including cerulin-induced pancreatitis in rats and bile duct-ligated opossum models.

6.38 Revised Atlanta Classification for Acute Pancreatitis

Revised definitions of morphological features of acute pancreatitis

			CECT	
Type	Sections of the pancreas with acute inflammation	Tissue necrosis	Pancreatic parenchyma enhancement	Pancreatic necrosis
Interstitial oedematous pancreatitis	Pancreatic parenchyma and peripancreatic tissues	No	Yes	No
Necrotising pancreatitis	Pancreatic parenchymal	Yes	No	Yes
			Fluid density	Appearance
Acute peripancreatic fluid collection (no pseudocyst)	Peripancreatic fluid associated with interstitial oedematous pancreatitis	No	Homogenous	Confined by normal peripancreatic fascial planes No definable wall encapsulating the collection
Pancreatic pseudocyst	Well-demarcated inflammatory wall, completely encapsulated, usually outside the pancreas	No or minimal	Homogenous, non-liquid component	Well-circumscribed, usually round or oval
Acute necrotic collection	Variable amounts of both fluid and necrosis in pancreatic parenchyma and/or the peripancreatic tissues	Yes	Heterogeneous and non-liquid density	No definable wall encapsulating the collection
Walled-off necrosis	Well-defined inflammatory wall, mature, encapsulated collection of pancreatic and/or peripancreatic necrosis	Yes	Heterogeneous with liquid and non-liquid density	Well-defined wall encapsulating the collection

6.39 Enteral Nutrition in Acute Pancreatitis

- Considering the fact that acute pancreatitis is a catabolic condition, patients may be at moderate–high risk of malnutrition.
- In mild acute pancreatitis, low fat, soft solid diet should be initiated orally as soon as clinically tolerated.
- In patients with acute pancreatitis incapable of tolerating oral feeds, enteral nutrition should be initiated rather than parenteral. Enteral nutrition protects the gut mucosal barrier, stimulates intestinal motility, decreases intestinal bacterial overgrowth, and improves splanchnic blood flow. Randomised trials have shown decreased mortality and multi-organ dysfunction rates with enteral nutrition. Enteral nutrition should be initiated within 24–72 h of hospital admission.
- A standard polymeric diet is the preferred diet in acute pancreatitis. Compared with a semi-elemental diet, a standard polymeric diet is equally tolerated and provides similar clinical benefits. If patients develop malabsorption with a polymeric diet, a semi-elemental diet may be considered.
- Enteral nutrition should be administered through a nasogastric tube in acute pancreatitis. If the nasogastric tube is not tolerated due to gastroparesis or gastric outlet obstruction, nasojejunal tube feed may be initiated. Randomised trials have shown similar tolerance, complication rates, and mortality with nasogastric and nasojejunal feeds.
- Post minimally invasive necrosectomy, it is considered safe to initiate oral intake within 24 h. If unable to tolerate oral feeds, nasojejunal feeding may be initiated.
- In patients with inta-abdominal compartment syndrome, a nasojejunal tube may be initiated.
- Parenteral nutrition is preferred for patients who cannot match caloric requirements with enteral feeding, tolerate either form of enteral nutrition, or have contraindications to enteral feeding.

6.40 Scoring Systems in Acute Pancreatitis

Various methods are used to score the severity grade of acute pancreatitis. Similarly, scoring systems are available for radiological assessment of local complications (Tables 6.12, 6.13, 6.14, and 6.15).

Clinical scoring systems:

1. Ranson's score at hospitalisation and subsequently at the end of 2 days
2. Glasgow score
3. APACHE score
4. BISAP score
5. Marshall's score

Radiological scores:

1. CT severity indices (Table 6.16)

Table 6.12 Variables used in Ransons. Glasgow and APACHE scores for classifying acute pancreatitis

Variable	Ranson (at 0 and 48 h)	Glasgow (within the first 48 h)	Apache II[a] (on admission, then daily)
Age	>55 years	–	+
	–	–	Premorbid state
WBC count (×10⁹/L)	>16 (on admission)	>15	+
Blood glucose (mmol/L)	>11.1 (on admission)	>10	–
AST (U/L)	>250 (on admission)	>200	–
LDH (U/L)	>350 (on admission)	>600	–
Serum urea (mmol/L)	>1.8 increase (at 48 h)	>16	Creatinine
Serum calcium (mmol/L)		<2	–
Serum albumin (g/L)	<2 (at 48 h)	<32	–
PaO_2 (mmHg)		<60	Arterial pH
Base deficit	<60 (at 48 h)		–
Fluid sequestration	>4 (at 48 h)		+
Packed cell volume (%)	>6000 mL (at 48 h)		+
Serum sodium	≥10 fall (at 48 h)		+
Serum potassium	–		+
Temperature	–		+
Mean arterial BP			+
Heart rate			+
Respiration rate			+
Glasgow coma scale suggested a cut-off number	11 criteria: <3 criteria indicate mild AP	8 criteria: ≥3 criteria indicate severe AP	14 criteria: 8 points[a] indicate severe AP

AP acute pancreatitis, *AST* aspartate transaminase, *BP* blood pressure, *LDH* lactate dehydrogenase, *WBC* white blood cell

[a] Points system per variable: ranges from 0 (normal) to +4 (very abnormal); minimum total score: 0, maximum score = 71

Table 6.13 BISAP score

BUN	BUN >25 mg/dL (8.9 mmol/L) (1 point)
Impaired mental status	Impaired mental status abnormal mental status with a Glasgow coma score <15 (1 point)
SIRS	SIRS evidence of SIRS (systemic inflammatory response syndrome) (1 point)
Age	Age >60 years old (1 point)
Pleural effusion	Imaging study reveals pleural effusion (1 point)

0–2 Points: lower mortality (<2%); 3–5 points: higher mortality (>15%)

Table 6.14 Revised Atlanta criteria for severity of acute pancreatitis

Severity of AP	Definitions
Mild	Absence of organ failure and absence of local complications
Moderately severe	Local complications[a] and/or transient organ failure (<48 h)
Severe	Persistent organ failure (>48 h)

[a] Local complications defined by one or more of the following: peripancreatic fluid collections, pancreatic and peripancreatic necrosis (sterile or infected), pseudocyst and walled-off necrosis (sterile or infected), gastric outlet dysfunction, splenic and portal vein thrombosis, and colonic necrosis

Table 6.15 Modified Marshall scoring system for acute pancreatitis

Organ system	Score 0	Score 1	Score 2	Score 3	Score 4
Respiratory[a]	>400	301–400	201–300	101–200	<101
Renal (mg/dL)[b]	≤1.5	>1.5 to 1.9	>1.9 to ≤3.5	>3.5 to ≤5.0	>5.0
Cardiovascular: BP (mmHg)[c]	>90	<90, fluid responsive	<90, not fluid responsive	<90, pH < 7.3	<90, pH < 7.2

Organ failure is defined as a score = 2 for at least one of the three organ systems. Duration of organ failure is defined as transient (=48 h from the time of presentation) or persistent (>48 h from the time of presentation). Persistent multiorgan failure is defined as two or more organs failing during the same 3-day period

[a] Ratio of partial pressure of arterial oxygen to fraction of inspired oxygen

[b] Serum creatinine level. To convert to Système International units (μmol/L), multiply by 88.4

[c] Systolic blood pressure

Table 6.16 CT severity indices for acute pancreatitis

CT severity indices	
Prognostic indicator	
Pancreatic inflammation	
• Normal pancreas	0
• Enlargement of pancreas	1
Pancreatic abnormalities with peripancreatic inflammation	2
Single fluid collection	3
Two or more collections or gas	4
Pancreatic necrosis	
None	0
<30%	2
30–50%	4
>50%	6
Pancreatic inflammation	
Normal pancreas	0
Pancreatic abnormalities with or without peripancreatic inflammation	2
Pancreatic or peripancreatic fluid collection or fat necrosis	4

CT severity indices	
Prognostic indicator	
Pancreatic necrosis	
None	0
<30%	2
>30%	4
Extrapancreatic complications (pleural effusion, ascites, parenchymal complications, GI tract involvement)	6

6.41 Endoscopic Management of Gallstone Pancreatitis

Multiple RCTs have been performed to evaluate the timing of ERCP in acute biliary pancreatitis. Based on a meta-analysis of these RCTs, the following conclusions have been achieved.

- There is no benefit of routine ERCP in acute biliary pancreatitis, irrespective of the severity.
- Early ERCP (<72 h of admission) benefits patients with cholangitis or biliary obstruction due to CBD stones (Table 6.17).

Table 6.17 Summary of a few randomised controlled trials

Author, year	Place of study	Inclusion criteria	Timing of early ERCP	*n*	Primary outcome
Chen et al., 2010	Single centre, China	Severe biliary pancreatitis (APACHE >11)	<72 h of admission	55	In hospital mortality
Fan et al., 1993	Single centre, Hongkong	Acute pancreatitis, any etiology	<24 h of admission	Of 192 patients, 127 had gallstones	In hospital mortality
Folsch et al., 1997	Multiple centres (22), Germany	Suspected biliary pancreatitis without cholangitis or biliary obstruction	<72 h of the onset of symptoms	120 patients enrolled	Plan to enrol 190 patients. The study stopped after the second interim analysis due to higher mortality in the early ERCP group
Neoptolomos et al., 1988	Single centre, UK	Patients with suspected biliary pancreatitis with or without cholangitis or biliary obstruction	<72 h of admission	121 patients	In hospital mortality
Oria et al., 2007	Single centre, Argentina	Acute biliary pancreatitis with biliary obstruction; clinically excluded	<72 h of admission	103	Change in SOFA and CT severity indices
Tang et al., 2010	Single centre, China	Severe biliary pancreatitis	<48 h of admission	90	In hospital mortality
Zhou et al., 2002	Single centre, China	Acute biliary pancreatitis with or without cholangitis/ biliary obstruction	<24 h of admission	45	Complications
Dutch Pancreatitis Study Group, APEC trial, 2021	Multicentre, Netherland	Severe gallstone pancreatitis (APACHE >3)	<24 h of admission	232	Composite of mortality and major complications

6.42 Step-Up Approach in Management of WOPN

- WOPN refers to the peripancreatic collection with a well-demarcated wall (usually >4 weeks after the onset of acute pancreatitis) containing solid contents within.
- In case of unyielding abdominal pain, emesis or early feeling of fullness or infection, treatment for WOPN is indicated.
- A few decades earlier, laparotomy and necrosectomy were the standard walled-off necrosis treatments. It has been shown that minimally invasive procedures are associated with better clinical outcomes. A minimally invasive step-up procedure involves placing a percutaneous drainage catheter, after which video-assisted retroperitoneal debridement of necrosis is performed. More than 30% of patients improved only with a single percutaneous drainage procedure in the PANTER trial. This trial found that a minimally invasive step-up surgical approach is associated with significantly fewer major complications and mortality than a primary open necrosectomy.
- EUS-guided cystogastrostomy has become the preferred one for pancreatic pseudocysts. Direct endoscopic necrosectomy (DEN) involves placing transgastric stents across the gastric wall into the cavity in the lesser sac and passing the endoscope across into the cavity after maturation of the tract and doing endoscopic debridement. The endoscopic step-up approach for WOPN involves two steps: the first is a drainage procedure involving EUS-guided/endoscopic transluminal stent placement (plastic/lumen apposing metal). After 48–72 h, if the patient continues to be symptomatic, the endoscope can be channelled into the necrotic cavity via the transluminal tract, and necrosectomy can be achieved.
- Multiple RCTs have compared endoscopic step-up with a surgical step-up approach for WOPN. The PENGUIN trial showed fewer inflammatory markers (IL-6) in the endoscopic step-up approach group than in the surgical step-up approach. Two RCTs with clinical outcomes as primary endpoints have been reported to date: the TENSION and MISER trials. The TENSION trial from the Dutch pancreatitis group showed non-inferiority of endoscopic step-up compared to the surgical step-up group regarding composite endpoint of mortality and major complications. MISER trial found comparatively fewer major complications with the endoscopic step-up approach regarding the external fistula tract than with surgical step-up.

6.43 Post-ERCP Acute Pancreatitis

Acute pancreatitis is the most frequent complication after ERCP (2–10%)

- Previously defined mild post-ERCP pancreatitis includes abdominal pain necessitating new hospitalisation or prolonged hospital stay for 2–3 days and a serum amylase ≥3× ULN 24 h post-procedure. Subsequent amendments include lipase instead of amylase levels with new onset or exacerbation of existing abdominal pain present before ERCP.

Table 6.18 Risk factors for post-ERCP pancreatitis

Patient related	Provider related	Procedure related
Young	Experience	Difficult cannulation
Female		Frequency of cannulation efforts
h/o post-ERCP pancreatitis		Pancreatic duct cannulation
Suspected SODD		Pancreatic duct contrast injection
Normal bilirubin		Precut sphincterotomy
Non-dilated CBD		Ampullectomy

- Prolongation of 4–10 days of hospitalisation is classified as acute pancreatitis of moderate severity, while more than 10 days or development of necrosis or pseudocyst or need for intervention (drainage or surgery) post-ERCP is classified as severe acute pancreatitis (Table 6.18).

6.43.1 Prevention of Post-ERCP Pancreatitis

Pancreatic duct stent placement and rectal NSAID attenuate the risk of post-ERCP pancreatitis. In each patient undergoing ERCP, ESGE recommends routine placement of rectal indomethacin or diclofenac unless there is a contraindication. ESGE recommends the placement of a pancreatic stent in patients susceptible to acute pancreatitis, as detailed above.

6.44 Recurrent Acute Pancreatitis

- Acute pancreatitis recurs in ~22% of patients after an index episode.
- Up to 75% of recurrent episodes of pancreatitis are due to biliary calculi or alcohol. In 25% of patients, other aetiologies must be evaluated for repeat episodes.
- The risk of progression to chronic pancreatitis is 10% (95% CI: 6–15) after a single episode of acute pancreatitis. The risk of progression to chronic pancreatitis is 36% (95% CI: 20–53%) with recurrent acute pancreatitis.
- Other potential causes for recurrent pancreatitis include:
  - Biliary microlithiasis
  - Hypertriglyceridemia (>1000 mg/dL), hypercalcaemia
  - Drugs
  - Drug-induced pancreatitis may be caused due to immunological reactions (6-mercaptopurine, aminosalicylates, and sulphonamides), build-up of toxic metabolites of certain drugs like valproate, pentamidine, tetracycline, ischaemia due to diuretics, azathioprine, and intravascular thrombosis due to estrogen
  - Pancreas divisum and choledochocele

Pancreas divisum occurs when the dorsal and ventral ducts do not undergo fusion during embryonic development. Pancreas divisum, as identified on MRCP, was prevalent in similar proportions regardless of recurrent acute pancreatitis. Moreover, up to 7% of the general population have pancreas divisum noted on MRCP. Most of these patients do not report any pancreatic symptoms. Hence, the pancreas divisum alone does not cause recurrent acute pancreatitis

In patients with genetic mutations (CFTR gene), the prevalence of pancreas divisum was higher and associated with recurrent acute pancreatitis than those without. So, it is postulated that pancreas divisum, associated with genetic predisposition, may result in repeat episodes of pancreatitis

A small proportion of patients with minor duct dilatation due to obstruction at minor papilla may benefit from minor papilla endotherapy. No large RCTs in patients with pancreas divisum presenting with recurrent acute pancreatitis have assessed the role of endotherapy

Choledochocele: type 3 choledochal cyst increases the frequency of recurrent pancreatitis due to additional biliary sludge. Deroofing of choledochal cysts with sphincterotomy is beneficial

– SO dysfunction

High pressures in the biliary and pancreatic duct due to raised sphincter pressures have been hypothesised to cause impediments in outflow and result in recurrent acute pancreatitis. The role of SOD in recurrent pancreatitis has been debated. The clinical significance of raised SOD pressures in patients with recurrent acute pancreatitis compared to controls is not evident. The risk of post-ERCP pancreatitis in this subset of patients is greater than normal

– Ampullary and pancreatic tumours

Pancreatic tumours and ampullary tumours can cause outflow obstruction and result in pancreatitis. EUS is recommended in patients above 40 yrs. of age who present with idiopathic pancreatitis to exclude small ampullary or pancreatic tumours

– Autoimmune pancreatitis

Autoimmune pancreatitis is a long-standing fibro-inflammatory condition of the pancreas. It exists in two forms

Type 1 is the pancreatic manifestation of systemic IgG4-related disorder. It usually presents in patients >60 years, predominantly males, as obstructive jaundice, with 70% having elevated serum IgG4 levels. Histological features include lymphoplasmacytic acinar infiltration and storiform fibrosis

Type 2 autoimmune pancreatitis is distinct in young adults with no gender predisposition. It may present as acute pancreatitis in 50% of patients, though recurrent acute pancreatitis is rare. Serum IgG4 may not be elevated. Histologically it is characterised by granulocyte epithelial lesions and is termed duct-centric pancreatitis

AIP responds to steroids; type 1 tends to recur in up to 30–60%, while recurrence is rare in type 2 (<5%)

- Genetic

  In patients younger than 30 years of age with recurrent acute pancreatitis, ~30% had genetic mutations in PRSS, SPINK 1, and CFTR genes

  Genetic mutations on their own may predispose to pancreatitis or behave as modifiers. PRSS gene is associated with hereditary pancreatitis, recurrent acute pancreatitis, and pancreatic malignancy. The presence of other genes like CFTR and SPINK 1 gene correlates with recurrent acute episodes

6.45 Pathophysiology of Pain in Chronic Pancreatitis

6.45.1 Introduction

The most common and most significant clinical problem in patients with chronic pancreatitis 50–90% of patients experience pain during the disease.

6.45.2 Neural Pathways (Fig. 6.8)

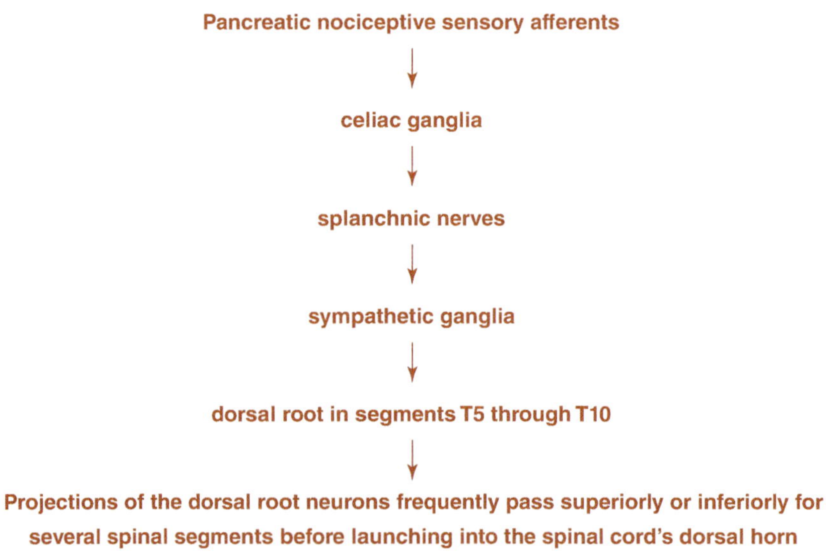

Fig. 6.8 Neural pathways in pancreatitis pain

6.45.3 Patterns of Pain

Two different pain patterns have been elucidated:

Type A pain—repeat episodes of short-term, relapsing pain
Type B pain—extended and sustained; it is associated with the large-duct CP, elevated intraductal pressure, or subsequent complications of CP such as the development of pseudocysts or biliary blockages

6.45.4 Causes

1. Acute pancreatic inflammation
2. Retroperitoneal inflammation
3. Increased intrapancreatic pressure
4. Alterations in pancreatic nerves: damage of the perineural sheath enabling inflammatory mediators entry into the neural components
5. Altered neurotransmitters: rise in substance P and calcitonin gene-related peptide in interlobular and intralobular nerve bundles
6. Central sensitisation of pain state similar to any other state of chronic pain

6.45.5 Other Causes of Pain: Local Complications

- Pancreatic pseudocyst
- Mechanical obstruction of the common bile duct or duodenum
- Development of pancreatic adenocarcinoma
- Gastroparesis

6.45.6 Significance

- Decreases the appetite and limits food intake resulting in weight reduction and malnutrition
- Reduction in quality of life and loss of social functioning
- Potential for addiction to narcotic analgesics
- Intractable pain is the commonest reason for opting for surgical intervention in patients with chronic pancreatitis

6.45.7 Treatment

- Medical therapy
  - Analgesics:

    Always start with acetaminophen and/or tramadol with morphine as a last resort
    Risk of narcotic addiction
  - Pancreatic enzymes:

    Degrade CCK preleasing factor in the duodenum → reduce CCK levels → reduce pain
    Non-enteric-coated preparations are better than enteric-coated preparations
    Effective in patients with 'small-duct' or 'minimal change' CP
- Endoscopic therapy
  - Celiac ganglion nerve block under endoscopic ultrasound guidance
  - Dilation of pancreatic duct strictures
  - Pancreatic duct stenting
  - Removal of obstructing pancreatic duct stones
  - Extracorporeal shock wave lithotripsy (ESWL)
- Surgical therapy
  - Modified Puestow or lateral pancreaticojejunostomy (widely used)
  - The pancreatic duct should be dilated enough to be detected at surgery (generally a lower limit of 5–6 mm)
  - Short-term pain relief is very good (80%)
  - Long-term pain relief (50%)
  - Several modifications have been explored, including

    Traditional pancreaticoduodenectomy (Whipple operation)
    Limited resection of the pancreatic head (Beger operation)
    Combined opening of the pancreatic duct and partially revealing pancreatic head (Frey operation)
- Two randomised trials compared endoscopic therapy with surgery
- Results are better with surgical therapy, particularly in long term

6.46 IgG4-Related Diseases: GI and Extraintestinal Manifestations

IgG4-related disease (IgG4-RD) is a long-standing relapsing multi-organ fibroinflammatory syndrome, a supposed autoimmune aetiology. Its features include elevated serum levels of IgG4 and tissue infiltration by IgG4-positive cells. It typically presents as mass-forming synchronous or metachronous lesions in nearly any organ throughout the body. Histologically it reveals a dense lymphoplasmacytic infiltrate rich in IgG4-positive plasma cells, storiform fibrosis, and obliterative phlebitis, which progresses and eventually leads to permanent tissue damage and insufficiency. It responds to corticosteroids and B-cell-depleting drugs.

- Has a 3:1 male predominance, except for head and neck signs with similar frequencies in both genders.
- Typically present in the elderly or at an advanced age and rarely seen in children.
- Its risk factors are poorly understood, although some studies suggest occupational exposure to extreme and extended exposure to solvents, industrial dust, industrial oil, or polymers in vulnerable people.

6.46.1 Pathogenesis

The exact pathogenesis is still unclear; however, the likely mechanism is sustained exposure of predisposed individuals to unknown antigens that trigger immunological reactions via processes described below:

(a) A process injuring tissues carried out by activated lymphocytes, macrophages, and fibroblasts—tissue infiltration by CD4 + T lymphocytes and subsequent shift towards T-Helper 2 (Th2) cells.
(b) IgG4 and regulatory T cells mediated the anti-inflammatory response.

Nevertheless, the pathognomonic antigens likely vary in displayed signs and symptoms, and dissimilar antigens or autoantibodies produce similar immune reactions in different organs.

6.46.2 Clinical and Investigation Findings

The five defining features of IgG4-RD are included in the mnemonic HISORt and elaborated as histopathology and immunostaining, imaging, serology, organ involvement, and response to therapy.

Histopathology and Immunostaining

Dense lymphoplasmacytic infiltrate, fibrosis laid out in a woven storiform arrangement, and obliterative phlebitis are typical features. These characteristics differ in different organs but are highly evident in the pancreas and bile duct on resection. Tissue samples obtained with resection enable detection of IgG4-RD, while it is challenging to reach a diagnosis using samples from fine-needle aspiration.

When affected tissues are immune-stained, diffuse and abundant infiltration by IgG4-positive plasma cells is noted, usually more than 50. A ratio of IgG4-positive/IgG-positive plasma cells >0.4 indicates IgG4-RD and can be used to distinguish it from other conditions that may falsely show high IgG4-positive cells.

Serology

The pooled sensitivity and the specificity are 85% and 93%, respectively, for detection of IgG4-RD with serum level of IgG4. Since IgG4 levels are not useful in prognosis, seronegativity is not the aim of therapy. Search for better serological markers has led to a few new markers, such as the serum ratio of IgG4/IgG1, which provides considerably enhanced positive predictive value and specificity for differentiating IgG4-SC from PSC with a cutoff value of 0.24.

Organ Involvement

IgG4-RD can involve any organ, with most evidence indicating an involvement of at least two.

Gastrointestinal Manifestations

Most commonly involved are the pancreas and bile ducts, and seldom affected are the liver, gallbladder, stomach, small intestine, and large intestine.

Pancreas

The most common manifestation is autoimmune pancreatitis (AIP), in which patients experience jaundice develop a focal pancreatic mass or enlargement without jaundice, pancreatic insufficiency manifesting as new or worsened hyperglycemia and steatorrhea, and infrequently acute pancreatitis.

- Even in severe inflammation, there is no pain in AIP, so the manifestation of pain suggests a different pathology. AIP may be an incidental finding during the investigation of symptoms experienced due to impact on other organs.
- AIP display characteristic traits in nearly 40% of patients undergoing CT and MRI. These typically include a sausage-shaped pancreas with capsule-like rim. Some unusual appearances include focal or multifocal growth of the gland. A low-density mass, dilation of the main pancreatic duct, and distal atrophy should be evaluated for other causes since these are not features of AIP.

Bile Ducts

Biliary signs and symptoms, which include obstructive jaundice or asymptomatic elevation in liver tests of IgG4-RD, are grouped under IgG4 sclerosing cholangitis (IgG4-SC).

- Proximal bile ducts resemble those of PSC and cholangiocarcinoma features on imaging; isolated distal bile duct strictures should raise a suspicion of pancreatic cancer. The most important marker for IgG4-SC diagnosis is concomitant AIP. AIP males' absence makes IgG4-SC from PSC or cholangiocarcinoma challenging. Concomitant IBD often exists with PSC but rarely with IgG4-SC.
- It is necessary to rule out malignancy in patients presenting with dominant stricture or mass using the following techniques: endoscopic biliary brushing for cytology analysis, fluoroscopy-directed intrabiliary or ampullary biopsies for histology analysis, bile fluid samples, and cholangioscopy if required.
- Type 1 IgG4-SC mimics malignancy but is infrequently reported. Patients with both IgG4-SC and AIP increase the relapse risk and, therefore, require long-term maintenance therapy. In the absence of treatment, it worsens secondary biliary cirrhosis. Unlike in PSC, cholangiocarcinoma may complicate IgG4-SC frequently.

Other Hepatobiliary and Gastrointestinal Manifestations

IgG4-hepatopathy comprises secondary lesions in hepatic parenchyma in patients with AIP and IgG4-SC. However, mass-like lesions indicating peri-portal and lobular hepatitis and portal inflammation are hepatic indications of IgG4-RD. These also display biliary damage and eosinophilic infiltrates, but parenchymal necro-inflammatory abnormalities are absent. IgG4-related cholecystitis as diffuse wall thickening due to IgG4-positive plasma cell infiltration and transmural fibrosis was reported in more than one-fourth of patients with AIP.

IgG4-RD in the alimentary tract primarily involves glandular tissues or the vasculature of the GIT, causing the development of ulcers, polypoid lesions, submucosal masses, and wall thickening. As per a study, Eosinophilic Esophagitis (EoE), at least in adults, is linked with IgG4 and is not an IgE-induced allergy. The radiologic and pathologic features of submucosa or wall thickening can be characteristic signs of IgG4-related gastric lesions, which display typical features of IgG4-RD.

Non-gastrointestinal Tract Manifestations

Although it affects multiple organs, systemic or constitutional symptoms are seldom found.

IgG4-related kidney disease manifests as tubulointerstitial nephritis, membranous glomerulonephritis, and obstructive nephropathy due to IgG4-related retroperitoneal fibrosis or ureteral IgG4-RD.

Salivary and lacrimal gland signs include persistent, painless, and firm bilateral enlargement of ≥2 sets of glands. The salivary function is unaffected, and it is necessary to differentiate it from Sjögren's syndrome and lymphoma.

The **IgG4-related ophthalmic disease** manifests as exophthalmos and ptosis from infiltration of extraocular muscles and orbital pseudo-tumor. The effect on the optic nerve can result in blindness.

Retroperitoneal fibrosis involves the development of peri-aortic or peri-iliac fibroinflammatory tissue, which often swamps the ureter, indicated by obstructive uropathy or hydronephrosis. IgG4-RD has been reported in nearly 60% of patients with idiopathic retroperitoneal fibrosis. Since most patients lack symptoms, the diagnosis is usually incidental during imaging, while some may manifest a few systemic or obstructive symptoms due to mass lesions.

Pulmonary manifestations: Around 8–30% of patients lack symptoms and are incidentally diagnosed on routine imaging or during the evaluation of nonspecific symptoms such as cough, chest pain, or shortness of breath. Imaging features, except paravertebral band-like soft-tissue thickening, are not definite indicators and may suggest any pulmonary disease. Lymphadenopathy, prostate, thyroid, skin, sinuses, or breast signs are indicators of organ involvement.

6.46.3 Diagnosis

IgG4-RD resembles many common diseases, including chronic inflammatory diseases, malignancies, infections, other autoimmune disorders, and various forms of vasculitis. The diagnostic evaluation is influenced by the affected organ. Based on fundamental clinical manifestations of AIP and IgG4-RD, modified HISORt criteria for AIP may act as a basis for diagnosing IgG4-RD (Table 6.19).

6.46.4 Treatment

- Although the inflammatory component of IgG4-RD consistently and rapidly responds to steroids, changes due to fibrosis (such as pancreatic atrophy, intrahepatic bile duct strictures, and retroperitoneal fibrosis) result in irreversible end-organ damage.
- Broad treatment goals include induction of remission and prevention of relapse, which can be achieved with a combination of steroids, immunomodulators (azathioprine, mycophenolate mofetil), and B-cell depletion therapy with an antibody against CD20. Maintenance therapy may immediately follow remission induction or may be delayed until the first or the second disease relapse.
- High-dose steroid treatment (40 mg/day for 4 weeks, followed by taper by 5 mg/week, until patients are off the drugs) is an acceptable regimen for prednisolone rates of response to steroid therapy are 79–97%, whereas rates of relapse to the maintenance therapy are 31–60%.

Table 6.19 HISORT criteria for autoimmune pancreatitis

Criterion H-histology (at least one of the following)	1. Periductal lymphoplasmacytic infiltrate, obliterative phlebitis, storiform fibrosis 2. Lymphoplasmacytic infiltrate, storiform, abundant IgG4+ cells (≥10 HPF)
Criterion I—Imaging of the pancreas	1. Typical-diffusely enlarged gland with delayed (rim) enhancement; diffusely irregular, attenuated main pancreatic duct 2. Others: Focal pancreatic mass/enlargement, focal pancreatic duct stricture, pancreatic atrophy, pancreatic calcification, pancreatitis
Criterion S—Serology	Elevated serum IgG4 (normal: 8–140 mg/dL)
Criterion other organ involvement (can be confirmed by biopsy or resolution/improvement with steroid therapy)	Hilar/intrahepatic biliary strictures, persistent distal biliary stricture, parotid/lacrimal gland involvement, mediastinal lymphadenopathy, retroperitoneal fibrosis
Criterion R—Response to steroid therapy	Resolution or marked improvement of pancreatic/extrapancreatic manifestation with steroid therapy
Diagnostic of autoimmune pancreatitis when any of the following is fulfilled	1. Criterion H 2. Criterion I + S 3. Strong clinical suspicion of autoimmune pancreatitis (idiopathic pancreatic disease + criterion S and/or O) + criterion R

- Among the maintenance regimens, relapses occur in the lowest proportion of patients (10%) receiving rituximab maintenance (every 6 months). Relapse occurs in 25–35% of patients who receive low-dose steroid and immunomodulator regimens for maintenance.

6.47 Role of EUS in Chronic Pancreatitis

- Chronic pancreatitis is a syndrome which involves inflammation and fibrosis of the pancreas in susceptible individuals developing sustained responses to parenchymal injury or stress.
- Confirmed and advanced cases of CP display pancreatic atrophy, fibrosis, pain syndromes, duct distortion and strictures, calcifications, pancreatic exocrine dysfunction, pancreatic endocrine dysfunction, and dysplasia.
- Abdominal imaging with ultrasound, CT, or MRI abdomen helps identify advanced chronic pancreatitis when pancreatic calcifications, ductal dilatation, stricture, or atrophy is established. These imaging modalities are not sensitive to identify early chronic pancreatitis.
- Endoscopic ultrasound has been studied to characterise early chronic pancreatitis. Various criteria have been proposed to define chronic pancreatitis based on endoscopic ultrasound imaging features, including ductal and parenchymal characteristics (Table 6.20).

Table 6.20 Rosemont criteria for diagnosing chronic pancreatitis based on EUS imaging

Parenchymal features	Ductal features
1. **Hyperechoic foci with shadowing (major A)**: They have been defined as echogenic structures ≥2 mm in length and width that shadow. They histologically correlate with pancreatic parenchymal calcification. 2. **Lobularity with honeycombing (major B)**: Well-circumscribed, ≥5 mm structures with hyperechoic rims compared to the echogenicity of its central areas labelled as lobules. ≥3 adjacent lobules in the body or tail of the pancreas are labelled as "honeycombing" lobularity. The histological correlate of lobularity is unknown. 3. **Lobularity without honeycombing (minor)**: As defined above, lobules are non-contiguous and present in the body and tail. 4. **Hyperechoic foci without shadowing (minor)**: Defined as above but without shadowing. 5. **Cysts (minor)**: Anechoic, rounded/elliptic structures ≥2 mm in short axis. 6. **Stranding (minor)**: Hyperechoic lines ≥3 mm in length noted in ≥2 different directions with reference to the imaged plane. ≥3 strands are essential.	Excluding ductal calculi, all other features should be looked at in the body and tail of the pancreas only. 1. **Main pancreatic duct (MPD) calculi (major A)**: Echogenic structures within MPD with acoustic shadowing located in any part of the pancreas. 2. **Irregular MPD contour (minor)**: MPD with a non-uniform outline in the body and tail of the pancreas only. 3. **Dilated side branches (minor)**: ≥3 tubular anechoic structures, each measuring ≥1 mm in width and communicating with MPD. These should be demonstrated in the body and tail of the pancreas only. 4. **Main pancreatic duct dilatation (minor)**: MPD diameter ≥3.5 mm in the pancreatic body or ≥1.5 mm in the pancreatic tail. 5. **Hyperechoic duct margin (minor)**: Relatively hyperechoic duct wall found in >50% of the entire MPD of the body and tail.
Endoscopic ultrasound diagnosis of chronic pancreatitis based on Rosemont criteria 1. Consistent with CP: • One major A feature with ≥3 minor features or • One major A feature and major B feature or • Two major A features 2. Suggestive of CP: • One major A feature with <3 minor features or • One major B feature with ≥3 minor features or • ≥5 minor features 3. Indeterminate for CP: • Three or four minor features with no major features or • Major B feature alone or with <3 minor features 4. Normal: <3 minor features, no major features	

6.48 Secretin-Enhanced MRCP

- Secretin is an endogenous hormone comprising 27 amino acids, secreted from the duodenum when acid chyme comes in contact with duodenal mucosa. It is a short-acting hormone which increases pancreatic ductal pressure and causes complete relaxation within 1 and 5 min, respectively.

Table 6.21 Reporting of secretin-enhanced MRCP

Imaging before secretin administration	Dynamic secretin-enhanced imaging	Duodenal fluid
Pancreatic parenchyma: • Signal intensity • Any surrounding free or organized fluid • Any solid or cystic lesion	Maximum diameter of main pancreatic duct: • Reported in tenths of mm, along with the location of measurement at baseline, 3 min after secretin administration and on most delayed (≥8 min) sequence after secretin administration	Grade 0: ↔ fluid Grade 1: ↑ fluid in duodenal bulb only Grade 2: ↑ fluid to proximal third duodenal segment Grade 3: ↑ fluid in the distal third duodenal segment or beyond
Main pancreatic duct: • Morphology • Any dilated or ecstatic side branches • Presence or absence of pancreas divisum	Changes after secretin: • Qualitative presence or absence of dynamic changes to main pancreatic duct dilation after secretin administration. • ↑ in dilation or ectasia of side branches • Any indicators of ductal communication of cystic lesions or side branches, such as an increase in size or signal on heavily T2-weighted MRCP	Peripancreatic fluid collections after secretin: • Presence or absence • If present, any communication with the main duct or side branches

- FDA has approved synthetic secretin for use during ERCP. The standard dose is 0.2 mcg/kg of body weight. Secretin administration results in the secretion of bicarbonate-rich fluid from the pancreas. Nausea and abdominal pain are noted in 0.5% of patients.

6.48.1 Protocol for Secretin-Enhanced MRCP

- Subsequent to secretin administration, 2D heavily T2-weighted coronal slab images should be obtained in a plane centred on the pancreaticobiliary ductal system for assessment (Table 6.21).
- The position of the coronal plane at an angle determined based on the radial slab offering the best view of the main pancreatic duct on earlier MRCP should be selected and maintained till the end of the procedure. This ensures improved imaging facilitating the identification of changes in the ducts.
- Quick snapshots can be obtained in sequence as secretin effects disseminate. A breath hold time of 30–60 s for imaging is advised since it permits a window for titration, providing dynamic analysis. Over 9 min, this method provides optimal images.

Table 6.22 Epidemiology of chronic pancreatitis in India

Parameter	Geevarghese, 1968	Balakrishnan et al., 2006	Balakrishnan et al., 2006	Midha et al., 2010	Bhasin et al., 2009
Number of patients		220	244	242	155
Age at onset	Childhood	20.7	30.6	24.7	33
BMI	<18.5	15.9	20.4	19.6	20.2
Etiology					
Idiopathic	100%	98%	67%	59%	42%
Alcohol	–	2%	33%	38%	38%
Diabetes	90%	77%	60%	35%	23%

6.48.2 Indications for Secretin-Enhanced MRCP

- Established indications for S-MRCP include anomalous pancreaticobiliary junction, pancreas divisum, Wirsungocele, Santorinicele, chronic pancreatitis, main pancreatic duct stenosis, and evaluation of complex postoperative anatomy.
- There is a certain level of ambiguity in other indications such as acute or recurrent acute pancreatitis, pancreatic exocrine function, sphincter of Oddi dysfunction, and pancreatic neoplasms.

6.49 Chronic Pancreatitis in India: Changing Epidemiology and Genetic Associations

The epidemiology of chronic pancreatitis has been dynamic over the years in India, and several genetic mutations have been reported, as depicted in Tables 6.22 and 6.23.

6.50 Pancreatic Cancer: Molecular Mechanism, Clinical Relevance, and Therapeutic Implication

- Pancreatic cancer is associated with poor survival. Resection is associated with the best chance of survival in pancreatic malignancy (up to 30%, 5-year survival rate). However, most pancreatic malignancies are diagnosed in an advanced stage, when curative resection is not feasible. Even in resectable pancreatic malignancies, adjuvant chemotherapy has limited efficacy.
- There is increasing interest in understanding the molecular mechanism of pancreatic malignancy with its resultant utility in prognostication and treatment.
- Pancreatic cancer classically surfaces as a precancerous lesion (pancreatic intraepithelial neoplasia, PanIn), which undergoes genetic mutations to develop into pancreatic adenocarcinoma.

Table 6.23 Gene mutations in chronic pancreatitis—Indian data

	SPINK gene		CFTR gene		CTRC gene		MORD4/CLDN2 locus	
	Patients	Controls	Patients	Controls	Patient	Controls	Patients	Controls
Chandak et al.	32%	2%	–	–	–	–	–	–
Bhatia et al.	44%	2%	–	–	–	–	–	–
Midha et al.	42%	4%	9%	0	–	–	–	–
Derilx et al.	31%	4%	–	–	5%	0.7%	–	–
Singh et al.	41%	2%	32%	10%	–	–	–	–
Paliwal et al.	–	–	–	–	12%	3.7%	–	–
Sundaresan et al.	38%	4%	–	–	–	–	–	–
Giri et al.	–	–	–	–	–	–	OR MOD4: 1.62 OR CLDN2: 1.57	

- K-ras mutation has been reported in many pancreatic malignancies and is associated with poor survival. About ~50–70% have mutations in TP53, 40–60% in cyclic-dependent kinase CDKN2A, and 30–40% in SMAD4 genes.
- Epigenetic alterations modify histone or DNA modification underlie the pathogenesis of pancreatic cancer.
- Classical/pancreatic progenitor and the basal-like/squamous/quasi-mesenchymal (QM-PDA) are subtypes with different features. Basal-like tumours predominantly have TP53 mutations, while the classical subtype features include GATA binding protein 6 (GATA6) expression and KRAS dependency. Patients with basal-like tumours have a severe-grade disease associated with poor outcomes, as shown by the COMPASS trial with the first-line chemotherapy. However, the treatment is not tailored to any specific subtype. QM-PDA subtype responded better to gemcitabine while classical subtype to erlotinib and mFOLFIRINOX.
- Morphological features can aid in predicting pancreatic cancer molecular subtypes. Squamous patterns indicate basal-like, while glandular patterns indicate classical subtypes, respectively.
- Metabolic reprogramming is necessary to maintain and enhance neoplastic growth in cancer cells, which KRAS sustains in the pancreas. Based on the metabolism, pancreatic cancer can be sub-grouped as quiescent, glycolytic, cholesterogenic, and mixed metabolic type. Glycolytic tumours are associated with poor prognosis, while cholesterogenic tumours correlate with better survival due to higher energy expenditure.
- Liquid biopsies based on circulating tumour cells (CTC), circulating tumour DNA (ct DNA), and circulating exosomes are being evaluated in pancreatic cancer. A high CTC burden has been associated with reduced survival. Circulating tumour DNA has been correlated with pancreatic cancer progression during or after treatment. For early detection of pancreatic cancer, circulating tumour-

derived exosomes (miRNAs) are useful since these demonstrate 100% sensitivity and specificity.

- Current guidelines recommend FOLFIRINOX, mFOLFIRINOX, gemcitabine, or gemcitabine plus nab-paclitaxel as neoadjuvant therapy. Gemcitabine, although the initial therapy in advanced stages, was replaced with FOLFIRINOX due to an improved survival advantage. Subsequently, gemcitabine, combined with nab-paclitaxel, was recommended as the first-line combination since it proved comparable to FOLFIRINOX therapy.
- For those patients with MMR deficient genotypes, pembrolizumab immunotherapy has been approved as the second-line therapy. 5 FU with liposomal irinotecan is the approved second-line treatment for the rest.

6.51 Cystic Neoplasms of the Pancreas

- Pancreatic cystic lesions have been reported in 0.21–24.3%, with minimal identifications through ultrasound and the maximum via autopsies. Due to the evolving novel diagnostic alternatives, more cases have been diagnosed in the last few years (Table 6.24).
- Pseudocysts constitute >70% of pancreatic lesions. These collections lack epithelial lining, noted in or around the pancreas as a sequelae of acute or chronic pancreatitis. True cysts (Van Hippel Lindau, cystic fibrosis, and lymphoepithelial cysts) are rare.
- Cystic neoplasms are indolent neoplasms with slow progression over many years. Mucinous neoplasms have malignant potential (mucinous cystadenoma and intraductal mucinous) neoplasms. Solid pseudopapillary neoplasms are potentially malignant. Some cystic neoplasms are due to cystic degeneration of pancreatic adenocarcinoma or neuroendocrine tumour. Serous cystadenomas rarely undergo malignant transformation.

Table 6.24 Clinical features and morphological characteristics of cystic neoplasms

	Pseudocyst	Serous cystadenoma	Mucinous cystadenoma	IPMN
Age	Any age	60–70 y	40–50 y	60–80 y
Gender	M > F	75% F	>90% F	Equal
Location in pancreas			95% in the tail and body of the pancreas	
Morphology	Thin wall; can have thin septa; solid debris was seen in walled-off pancreatic necrosis (WOPN)	Lobulated microcystic	Macrocystic	Main duct IPMN—dilated MPD; 'fish mouth papilla' Branch duct IPMN—a bunch of grapes communication to PD
Mural nodules		–	+	+
Wall calcification		–	+ (25%)	
Central calcification		18–20% central scar with calcification	–	
Clinical presentation	Preceded by abdominal pain episodes	Asymptomatic-noted on imaging	Asymptomatic-noted on imaging	Asymptomatic-noted on imaging Rarely pancreatitis due to thick viscid mucin causing obstruction
Fluid analysis	High fluid amylase; low fluid CEA	Low amylase, low CEA	Low amylase; high fluid CEA; low fluid glucose	High fluid CEA and amylase; low fluid glucose
			String sign + (viscous fluid)	String sign + (viscous fluid)
DNA markers in cyst fluid		Van Hippel Lindau gene	K ras mutation	GNAS mutation
Risk of malignant transformation	–	Minimal	15% have invasive carcinoma at resection if size >3 cm	Highest risk (>45%) with main duct IPMN >10 mm PD diameter; 30% risk with 5–9 mm MPD diameter Branch duct IPMN—15–30% invasive malignancy on resected cyst >3–4 cm

(continued)

Table 6.24 (continued)

	Pseudocyst	Serous cystadenoma	Mucinous cystadenoma	IPMN
Management	Drainage—endoscopic cystogastrostomy if symptomatic	Usually follow up; resection if symptomatic	Resection if cyst size >3 cm or mural nodules (to follow up rest); option of resection to discuss with all	Resection if MPD >10 mm or enhancing mural nodules

6.51.1 Management of Incidentally Identified Cystic Neoplasms of the Pancreas

- MRI with MRCP or MDCT is recommended for all pancreatic cysts >1 cm
- To look for high-risk or worrisome features
  - 'High-risk features'—obstructive jaundice in a patient with a mass in the head region of pancreas, enhancing solid component; MPD >10 mm; if high-risk features—surgical resection.
  - 'Worrisome features'—cyst measuring greater than 3 cm, thickening enhanced cyst walls, 5–9 mm MPD size, non-enhanced mural nodules, sudden alteration in the MPD calibre with distal pancreatic atrophy, and lymphadenopathy. The presence of worrisome characteristics should prompt evaluation with EUS. Resection should be opted for if EUS evaluation shows a definite mural nodule (no mobility or flow on Doppler, FNA showed tumour tissue) or main pancreatic duct is affected, or cytology confirms malignancy.

6.52 Pancreatic Neuroendocrine Neoplasm

- Pancreatic neuroendocrine tumours (PNET) comprise 1–2% of all pancreatic tumours. They are being identified early by imaging, and, hence, frequency is increasing.
- Most pancreatic NETs occur sporadically. Association with tuberous sclerosis, neurofibromatosis, multiple endocrine neoplasia (MEN), Von Hippel Lindau disease, and Cowden disease form a fraction of syndromic pancreatic NETs.
- Most pancreatic NETs are non-functional (60–90%). The rest of pancreatic NETs secrete hormones and are named accordingly (Table 6.25). Due to the absence of symptoms, non-functional NETs present late and are often advanced/metastatic at presentation.
- Pancreatic NETs are graded based on proliferation index (Ki-67 index <3, 3–20, >20 for G1, G2, and G3 tumours). TNM staging has been used for staging these tumours. DOTATOC scan uses radiolabelled octreotide which binds to soma-

Table 6.25 Symptomatic neuroendocrine tumours

	Malignancy	Pancreatic involvement	Symptoms
Insulinoma	10%	99%	Hypoglycaemic symptoms
Gastrinoma	60–90%	25%	Zollinger Ellison syndrome (peptic ulcer, diarrhea)
VIPoma	40–70%	90%	Watery diarrhea, hypokalemia, dehydration, and achlorhydria
Glucagonoma	50–80%	100%	Rash, migratory erythema, cachexia, and diabetes mellitus
Somatostatinoma	70%	55%	Diabetes mellitus, cholelithiasis, and diarrhea
Carcinoid	60–90%	1%	Diarrhea, flushing, asthma, and heart disease

tostatin receptors expressed on pancreatic neuroendocrine tumour cells and is useful for staging well-differentiated NETs. PET CT is useful for staging poorly differentiated neuroendocrine tumours or carcinoma.

- Surgery remains the only curative option for the resectable disease. Apart from curative resection, palliative resection to decrease symptoms and improve survival is also performed for locally advanced NETs. Partial hepatectomy and non-radical resections are also performed for metastatic NETs to achieve cytoreduction and improve symptoms.
- In patients with metastatic disease, systemic pharmacotherapy (long-standing octreotide, mTOR inhibitors everolimus, multitarget tyrosine kinase inhibitors sunitinib, chemotherapeutic agents 5 FU, and streptozocin) are used. Peptide receptor radiation therapy (PPRT) using radiolabelled somatostatin analogues ^{177}Lu-DOTOTATE is approved for advanced PNETs.

6.53 Pancreatic Enzyme Supplements

- Steatorrhea is the main consequence of pancreatic exocrine insufficiency. Although the pancreas secretes various enzymes (lipase, amylase, and various protease), lipase deficiency is the prominent aspect of pancreatic exocrine insufficiency displayed as abnormal fat digestion.
- Lipase is irreversibly inactivated at acidic pH. Low bicarbonate secretion in chronic pancreatitis leads to loss of neutralisation of acidic chyme in the duodenum and jejunum, leading to the inactivation of lipase.
- The insufficiency of amylase and protease is countered by replacement with salivary amylase, intestinal glycosidase, gastric pepsin, and intestinal peptidases. Nevertheless, these mechanisms cannot tackle lipase insufficiency. However, due to the enormous reserve of the pancreas, steatorrhea is deferred until levels fall below 10% of normal.

6.53.1 Indications for Enzyme Replacement

(a) Chronic pancreatitis patients with steatorrhea
(b) Post-pancreatic resection (Whipple's surgery)

6.53.2 Critical Aspects of Enzyme Replacement Therapy

- Consumption of a normal meal stimulates the release of lipase up to a maximum 3000–6000 IU/min or an average of 2000–4000 IU/min physiologically.
- Enzyme substitution therapy should be able to mimic the physiological meal-associated lipase secretion in patients with pancreatic exocrine insufficiency and deliver a minimum of 30,000 IU to avoid steatorrhea.
- To gain optimal benefits from digestion and assimilation, pancreatic enzyme preparations should be taken with meals.
- Enteric-coated preparations, especially mini-microspheres (1–1.2 mm), are preferred over conventional ones to resist inactivation due to gastric acid and favourable gastric emptying effect.

6.53.3 How to Improve the Efficacy of Enzyme Therapy

- Confirm adherence of the patient to therapy.
- The dose of enzymes can be escalated up to 80,000 USP U of lipase per meal.
- Use PPIs for inhibition of gastric acid secretion.
- Non-responders should be evaluated for bacterial overgrowth and treated.

6.54 Treatment for Pruritus in Intrahepatic Cholestasis

The pharmacological treatment of intrahepatic cholestasis-associated pruritus includes the agent listed in Table 6.26.

Table 6.26 Treatment for pruritus

	Drug	Dose	Interaction/precautions
1	Cholestyramine	4–16 g/day	Constipation Interference with intestinal absorption
2	Rifampicin	150–600 mg/day	Hepatic enzyme induction, hepatotoxicity
3	Benzfibrate/fenofibrate	200–400 mg/day	Renal and hepatotoxicity; myopathy and rhabdomyolysis
4	Naltrexone (mu receptor antagonist)	25–50 mg/day	Opioid withdrawal reaction (pain, confusion)—to start with low dose
5	Sertraline	75–100 mg/day	QTc prolongation, malignant neuroleptic syndrome
	Newer approaches		
1	k-Receptor agonist (Nalfurafine)	2.5–5 μg/day	Used in Japan
2	Ileal bile acid transporter inhibitor (linerixibat, maralixibat, odevixibat)		Bile acid diarrhea
3	Other experimental therapies: Albumin dialysis UVB light		

Further Reading

Bishr Omary M, Lugea A, Lowe AW, Pandol SJ. The pancreatic stellate cell: a star on the rise in pancreatic diseases. J Clin Invest. 2007;117(1):50–9.

Blechacz B. Cholangiocarcinoma: current knowledge and new developments. Gut Liver. 2017;11(1):13–26.

Chari ST, Smyrk TC, Levy MJ, Topazian MD, Takahashi N, Zhang L, Clain JE, et al. Diagnosis of autoimmune pancreatitis: the Mayo Clinic experience. Clin Gastroenterol Hepatol. 2006;4:1010–6; quiz 934.

Chignard N, Mergey M, Veissiere D, et al. Bile acid transport and regulating functions in the human biliary epithelium. Hepatology. 2001;33:496.

Choudhuri G, Srivastava A. Biliary microlithiasis. Trop Gastroenterol. 1998;19(1):11–4.

Di Sebastino P. Mechanisms of pain in chronic pancreatitis. In: Beger HG, editor. Diseases of the pancreas, current surgical therapy. Springer; 2008.

Domínguez-Muñoz JE. Pancreatic enzyme therapy for pancreatic exocrine insufficiency. Curr Gastroenterol Rep. 2007;9:116–22.

Elta GH. Sphincter of Oddi dysfunction and bile duct micro lithiasis in acute idiopathic pancreatitis. World J Gastroenterol. 2008;14(7):1023–6.

Forsmark CE. Chronic pancreatitis. In: Feldman M, editor. Sleisenger & Fordtran's gastrointestinal and liver disease. 8th ed. Saunders; 2006.

Garg PK, Narayana D. Changing phenotype and disease behaviour of chronic pancreatitis in India: evidence for gene-environment interactions. Glob Health Epidemiol Genom. 2016;1:e17. https://doi.org/10.1017/gheg.2016.13. PMID: 29868209

Jaster R. Molecular regulation of pancreatic stellate cell function. Mol Cancer. 2004;6(3):26.

Jüngst C, Kullak-Ublick GA, Jüngst D. Gallstone disease: microlithiasis and sludge. Best Pract Res Clin Gastroenterol. 2006;20(6):1053–62.

Kamisawa T, Zen Y, Pillai S, et al. IgG4-related disease. Lancet. 2015;385:1460–71.

Lans JI, Geenen JE, Johanson JF. Endoscopic therapy in patients with pancreas divisum and acute pancreatitis: a prospective, randomized, controlled clinical trial. Gastrointest Endosc. 1992;38:430.

Lee SP, Maher K, Nicholls JF. Origin and fate of biliary sludge. Gastroenterology. 1988;94:170–6.

Madura JA II, Madura JA. Diagnosis and management of sphincter of Oddi dysfunction and pancreas divisum. Surg Clin N Am. 2007;87:1417–29.

McMahon AP, Ingham PW, Tabin CJ. Developmental roles and clinical significance of hedgehog signalling. Curr Top Dev Biol. 2003;53:1.

Miyabe K, Zen Y, Cornell LD, Rajagopalan G, Chowdhary VR, Roberts LR, Chari ST. Gastrointestinal and extra-intestinal manifestations of IgG4–related disease. Gastroenterology. 2018;155(4):990–1003.

Muller M, Jansen PL. Molecular aspects of hepatobiliary transport. Am J Phys. 1997;272:G1285.

Russell DW. The enzymes, regulation, and genetics of bile acid synthesis. Annu Rev Biochem. 2003;72:137.

Sahel J, Cros RC, Bourry J, Sarles H. Clinico-pathological conditions associated with pancreas divisum. Digestion. 1982;23:1.

Salonen IS. Congenital duodenal obstruction—a review of the literature and a clinical study of 66 patients, including a histopathological study of the annular pancreas and a follow-up of 36 survivors. Acta Paediatr Scand Suppl. 1978;272:1.

Shaffer EA. Gallbladder sludge: what is its clinical significance? Curr Gastroenterol Rep. 2001;3(2):166–73.

Sharma BC, Agarwal DK, Dhiman RK, Baijal SS, Choudhuri G, Saraswat VA. Bile lithogenicity and gallbladder emptying in patients with microlithiasis: effect of bile acid therapy. Gastroenterology. 1998;115(1):124–8.

Sherman S, Lehman GA. Sphincter of Oddi dysfunction: diagnosis and treatment. JOP. 2001;2(6):382–400.

Swensson J, Zaheer A, Conwell D, Sandrasegaran K, Manfredi R, Tirkes T. Secretin-Enhanced MRCP: How and Why-AJR Expert Panel Narrative Review. AJR Am J Roentgenol. 2021;216(5):1139–1149. https://doi.org/10.2214/AJR.20.24857

Skandalakis LJ, Rowe JS Jr, Gray SW, Skandalakis JE. Surgical embryology and anatomy of the pancreas. Surg Clin North Am. 1993;73:661.

Toouli J, Craig A. Sphincter of Oddi function and dysfunction. Can J Gastroenterol. 2000;14(5):411–9.

Vlahcevic ZR, Pandak WM, Stravitz RT. Regulation of bile acid biosynthesis. Gastroenterol Clin N Am. 1999;28:1.

Vonlaufen A, Phillips PA, Xu Z, Goldstein D, Pirola RC, Wilson JS, Apte MV. Pancreatic stellate cells and pancreatic cancer cells: an unholy alliance. Cancer Res. 2008;68(19):7707–10.

Xu WL, Ling YC, Wang ZK, et al. Diagnostic performance of serum IgG4 level for IgG4-related disease: a meta-analysis. Sci Rep. 2016;6:32035.

Chapter 7
Nutrition

Gourdas Choudhuri

7.1 FODMAPs: Intervention Approach

FODMAP is an acronym for fermentable oligo-, di, monosaccharide and polyol. FODMAP diets are associated with increased symptoms of bloating in IBS patients. Bloating in IBS patients may be related to small intestinal bacterial overgrowth or a diet rich in FODMAP.

7.1.1 FODMAP and IBS

Diets rich in FODMAP increase intraluminal water content due to the osmotic effect. Bacterial breakdown of FODMAP content increases intraluminal hydrogen and carbon dioxide content. Both these factors increase distension of the intestine and stimulate mechanoreceptors. The afferent signals from these mechanoreceptors are perceived as bloating. This causes reflex alterations in gut motility, abdominal muscle and diaphragmatic movements resulting in discomfort.

Diets rich in FODMAP have also been associated with mast cell activation in the intestine with increased histamine production resulting in visceral hypersensitivity. It has also been associated with significantly altering the colon's short-chain fatty acid production.

G. Choudhuri (✉)
Department of Gastroenterology and Hepatobiliary Sciences, Fortis Memorial Research Institute, Gurugram, India

G. Choudhuri et al. (eds.), *Gastroenterology and Hepatology*,
https://doi.org/10.1007/978-981-99-9287-4_7

7.1.2 Low FODMAP Diet in IBS

Randomized controlled trials have shown a low FODMAP diet to be associated with significant improvement in symptoms in patients with IBS compared to standard dietary advice or no dietary modification. Low FODMAP diet intervention involves different stages—elimination of FODMAP, challenge and subsequent substitution with food containing low FODMAP content.

A low FODMAP diet is also under evaluation for non-celiac gluten hypersensitivity and IBD (ulcerative colitis).

7.1.3 Challenges in the Implementation of a Low FODMAP Diet

A low FODMAP diet requires close monitoring and implementation under a trained dietician. FODMAP content of common food items is not listed in the contents of the package. There have been concerns regarding the alteration in gut microbiome with the institution of a low FODMAP diet. There is concern that excessive limitation of FODMAP content may cause substantial alteration in dietary composition, leading to the ingestion of an unbalanced or nutritionally deficient diet. The sustainability of low FODMAP intervention in the long term has to be studied.

7.2 Glutamine

Glutamine is the most abundant amino acid in the human body. The liver, skeletal muscle, adipose tissue, and liver have glutamine synthetase activity and can synthesize glutamine. Enterocytes, leucocytes and renal tubule cells are primary glutamine-utilizing tissue. The balance between production and consumption results in stable plasma levels of glutamine.

Glutamine is key amino acid useful in the synthesis of purines, pyrimidines. It also plays a key role in interorgan ammonia transport, NADPH synthesis, and other intermediary pathways involved in maintaining cell integrity. Although glucose is the key nutrient for neutrophils, lymphocytes and macrophages, during catabolic phases like sepsis, and postoperative stress, these cells utilize glutamine in preference to glucose. For this reason, glutamine is also considered as "fuel of the immune system."

Glutamine though classified as non-essential amino acid, due to the ability of our body to synthesize it, becomes conditionally essential during periods of stress or critical illness. Glutamine is supplemented in either amino acid or dipeptide form up to 25–40 g/day. It may be administered enterally or parenterally. Multiple RCTs and systematic reviews have concluded that glutamine supplementation reduces infectious complications, hospital stays, and mortality in critically ill patients. It has also been shown to reduce post-chemotherapy mucositis.

There have been concerns about the ability of cancer cells to utilize glutamine as a nutrient source. Vitro studies have shown that certain lung cancer cells have the abundant capacity to metabolize glutamine, though, in vivo, they utilize glucose as the primary energy source.

Plasma glutamine concentration may independently predict mortality in sick, hospitalized patients. There is a poor correlation between body stores of glutamine and plasma concentration level in critically ill patients. It has been shown that macrophages have intact phagocytic activity even in low plasma glutamine levels. So, supplementation with glutamine should be based on nutritional assessment and immune-inflammatory mediators rather than on plasma glutamine concentration alone.

7.3 Immunonutrition

Immunonutrition is the modulation of the immunological system by nutrients or specific food items fed in amounts above those normally encountered in the diet. The common food substances included in immunonutrition are omega-3 fatty acids, glutamine, arginine, nucleotides, and micronutrients like selenium.

Patients undergoing surgery or in ICU must be assessed for malnutrition. Patients with cachexia due to catabolic stress have impaired immune responses. They are prone to more wound infections and prolonged hospital stay after surgery.

Omega-3 fatty acids (polyunsaturated) downregulate eicosapentaenoic acid production and reverse immunosuppression. Sulphur-containing amino acids maintain glutathione concentration in cells and thus exert antioxidant activity. Glutamine is an important nutrient for producing nucleic acids in rapidly dividing cells like leucocytes. Glutamine supplementation increases glutathione production and exerts an anti-inflammatory effect. Arginine increases nitric oxide synthesis and plays a key role in T cell production. Certain enzymes like superoxide dismutase, catalase and glutathione peroxidase mitigate oxidant injury by converting oxidants to harmless byproducts. These enzymes require micronutrients like copper, selenium, zinc and manganese.

Enteral supplementation of immunonutrition is beneficial as compared to parenteral supplementation. There have varied results with immunonutrition supplementation in critically ill patients. A recent systematic review has shown perioperative immunonutrition to benefit GI cancer patients undergoing surgery by reducing the length of hospital stay and surgical site wound infection without altering mortality.

7.4 Arginine

Arginine is a key amino acid involved in cellular growth. It plays a vital role in the urea cycle. It is also an important nitrogen donor for synthesizing nitric oxide, which has vasodilatory and immunological properties. It plays a key role in the synthesis of creatine in muscle.

Arginine is a conditionally essential amino acid whereby basal needs are met by endogenous production. Arginine is synthesized through two independent mechanisms—"intestine–renal axis" and "citrulline–nitric oxide cycle". In the intestine–renal axis, dietary amino acids are converted to citrulline in enterocytes and transported via the portal vein to the liver. Citrulline is not metabolized in the liver and reaches systemic circulation. Citrulline is subsequently converted to arginine in the kidney. In the second pathway, immune cells convert arginine to citrulline and nitric oxide. Cytosolic arginine synthetase again converts citrulline to arginine, and this cycle ensures the self-sustaining production of nitric oxide, which is required for T-cell function.

During periods of stress (surgery, infection etc.), the increased demands may not be met by endogenous production, and dietary supplementation may be necessary.

As arginine is important in cellular growth and immunological response, various studies have examined if arginine supplementation in critically ill patients as part of immunonutrition improves outcomes. Similarly, arginine deprivation (using PEGylated arginine deaminase) is being evaluated in clinical trials as part of treating malignancies.

7.5 Quality of Life Assessment

Health-related quality of life (HRQOL) measures the impact of chronic diseases on a patient's health. HRQOL measurement is an attempt to predict a patient's daily living based on the patient's subjective assessment of physical, social and psychological well-being. This tool has become an important part of assessing responses to treatment strategies. FDA has suggested including patient-reported outcome measures in trial designs to assess new treatment strategies.

7.5.1 Types of QOL

Quality of life assessments are of three types—global, generic and disease-specific. Global QOL summarises overall well-being using a visual analogue or graded scale. Though easily administered, reasons and domains of dysfunction cannot be assessed in global QOL scales.

Generic systems use multiple questions addressing various aspects of health and well-being. Psychological well-being (PGWB) and short form 36 (SF-36) are common generic HRQOL systems used in GI diseases. They are complex to administer. This questionnaire may not include certain disease-specific symptoms or social dysfunction.

Disease-specific QOL is more specific to the disease. They can be used to study the effect of disease over time or change with treatment. They are complex to administer and time-consuming. There are many disease-specific QOLs like IBS-QOL for IBS, IBDQ for IBD and GI cancer-specific QOLs.

7.5.2 Development of QOL Questionnaire

Key steps in developing generic or disease-specific HRQOL are item generation, item reduction, pretesting, psychometric assessment and cross-cultural adaptation. Item generation identifies all possible consequences of a given disorder using a literature review, patient forum and core group. Item reduction involves reducing items to be studied to a manageable number. Pretesting ensures clear wording and patient understanding of the questionnaire. The three keys psychometric attributes of robust HRQOL are Validity, reliability and responsiveness. Cross-cultural adaptation involves appropriate forward and backward translation among languages, harmonization, and pretesting.

7.6 Food Allergy vs. Intolerance

Food allergy involves immunological reactions (IgE and non-IgE mediated response) to the protein in the offending food substance and may result in a systemic reaction involving multiple organs in the body. Severe food allergies can lead to life-threatening anaphylactic reactions. About 40% of patients with food allergies develop atopic dermatitis, which resolves by avoiding offending food. Perioral dermatitis and urticaria are other skin manifestations seen in food allergies. Up to 25% of IgE-mediated food allergy patients can develop wheeze after offending food ingestion.

Some common food substances associated with food allergy are crustaceans, fish, eggs, nuts (like peanuts, almonds, cashew, walnut), soya and sesame. About 6% of children have food allergies-mainly to cow milk, egg and peanut. Most children develop tolerance and outgrow milk and egg allergies by school age while peanut, nut and seafood allergies remain for a long.

Food protein-induced enterocolitis, proctocolitis and enteropathy are non-IgE-mediated hypersensitivity to food products (commonly cow's milk and soya milk) affecting children around the age of weaning. Children present with blood in stools, diarrhea, abdominal distension, and failure to thrive. Celiac disease is a chronic form of food allergy with intestinal and extraintestinal manifestations triggered by exposure to gluten in wheat products.

Diagnosis of food allergy relies on eliciting detailed history. Skin prick tests and serum food-specific IgE antibodies (RAST) are ancillary tests to diagnose food allergy. A Double-blind, placebo-controlled oral provocative challenge is the gold standard for diagnosing food allergy.

Food intolerance does not cause a systemic reaction and is not life-threatening. Lactose intolerance is one common example. Lactose intolerance becomes more common in adults as intestinal brush border enzyme lactase activity diminishes with age. Food intolerance may be dose-related, like lactose intolerance. Often, symptoms are limited to the gut-bloating, diarrhea and abdominal cramps/discomfort. Occasionally, food (chocolates) can trigger migraine headaches.

7.7 Enteral vs. Parenteral Nutrition

7.7.1 Enteral Nutrition

Enteral nutrition is a way of providing diet directly into the stomach or intestine. It can be used exclusively to provide nutrition to meet dietary requirements or as an adjunct to oral feeding. The diet is fed through special tubes placed through the nose into the stomach (nasogastric tube) or small intestines (nasojejunal tube). If patients are at high risk of aspiration or need prolonged enteral feeding (>4 weeks), percutaneous tube placement into the stomach (percutaneous gastrostomy) or into the jejunum (percutaneous jejunostomy or percutaneous gastrostomy-jejunostomy extension) may be used.

Indications for Enteral Feeding

- Stroke or dementia or CNS conditions in whom swallowing reflexes are compromised (motor neuron disease).
- Acute pancreatitis.
- Critically ill patients.
- Pre or postoperative enteral support.
- Patients with esophageal or pharyngeal cancer without obstruction planned for radiotherapy.

Contraindications for Enteral Feeding

- Bowel obstruction
- Bowel ischemia

- Bowel perforation
- Major upper GI bleeding
- Intractable vomiting
- Ileus

Starting early enteral nutrition within 24–48 h in critically ill patients is beneficial. It is suggested to start with hypocaloric feeds (~10 kcal/kg or ~60% of caloric requirement) in the first week, followed by an increase in required calories (25–30 kcal/kg) in the subsequent week. Protein requirement is calculated at 1.2–1.5 g/kg body weight. Randomized trials have shown no difference in mortality or ICU stay with initial hypocaloric feeds compared to standard caloric feeds with less GI intolerance.

The current body weight should be considered the initial dosing weight in normal or underweight patients. If obese, 110 × ideal body weight may be considered initial dosing weight.

Advantages of Enteral Feeding

- Maintenance of gut mucosal barrier.
- Less risk of hospital-acquired infections like pneumonia or sepsis.

Risks of Enteral Feeding

- Aspiration.
- GI intolerance—diarrhea.
- Metabolic abnormalities—refeeding syndrome.
- Access site infections/mechanical complication-PEG site infection, bleeding, buried bumper.
- Change of enteral tubes—nasogastric or nasojejunal tubes in 4 weeks; PEG tubes in 3 months.

7.7.2 Parenteral Nutrition

Parenteral nutrition is a mode of providing nutrition through an intravenous route. It is done through peripheral venous access or central venous access. Peripheral veins can only give dilute solutions with osmolality <900 mOsm (Peripheral parenteral nutrition). The higher osmotic load caused inflammation of the veins with pain and occlusion. Central venous access is preferred if patients require exclusive parenteral nutrition as the infusion volume is likely high and needs higher caloric content with high osmolality.

If parenteral nutrition is expected to be required for less than 30 days, ASPEN guidelines recommend using the term short-term parenteral nutrition. Peripherally inserted central venous catheters may be preferred for short-term parenteral nutrition. Subclavian or internal jugular or femoral venous access may also be used for central venous access if parenteral nutrition is required for a short duration (<1 week). If long-term home parenteral nutrition is required, tunnelled central venous catheters (Hickmann line) are preferred.

Indications

- Short gut syndrome.
- Intestinal failure due to any etiology—radiation enteritis, motility disorder or Crohn's or extensive mesenteric thrombosis.
- High output small intestinal fistulas.
- Postoperative bowel leak.

Contraindications

- Sepsis
- Disseminated intravascular coagulation
- Hemodynamic instability

Risks

- Catheter-related sepsis.
- Thrombosis.
- Insertion-related complications—pneumothorax, hematoma central line displacement.
- Metabolic complications—refeeding syndrome.

7.8 Common Sources of Micro-nutrients in India

Water and fat-soluble vitamins and minerals constitute micro-nutrients. The common sources of micronutrients in India are tabulated in Table 7.1.

Table 7.1 Common sources of micro-nutrients and minerals in India

	Micronutrient	RDA (includes requirement and safety margins)	Common source (plant-based)	Common source (animal-based)
1.	Vitamin A	600 μg/day	Green leafy vegetables (coriander leaves, curry leaves, ponanganni, agathi, molaikeerai or amaranth, radish leaves, spinach), carrots, tomatoes, sweet potatoes, papaya, pumpkin, mango	Butter, ghee and edible oil
2.	Vitamin D	400–800 IU/day	Mushroom	Cod liver oil, fish, egg yolk
3.	Vitamin E	15 mg/day (alpha tocoferol)	Almonds, spinach, kiwifruit, broccoli, vegetable oil	Shrimp
4.	Vitamin K	0.1 mg/day	Green leafy vegetables	Animal liver
5.	Vitamin B1 (Thiamine)	1.2 mg/day	Cereals (wheat, rice), legumes (most dhals)	
6.	Vitamin B2 (riboflavin)	1.4 mg/day	Cereals (wheat, ragi), legumes (blackgram, Bengal gram), leafy vegetables (molaikeerai, radish leaves, spinach), nuts (almond, walnut)	Egg, sheep's liver
7.	Vitamin B3 (Niacin)	16 mg/day	Cereals (wheat), legumes (dhal)	Chicken and fish
8.	Folate	0.2–0.5 mg/day	Green leafy vegetables (amaranth or molaikeerai, mint, spinach), legumes (Bengal gram, blackgram), nuts (gingelly and soybean)	Liver
9.	Vitamin B12	1 μg/day	No natural plant products; Only fortified cereals	Fish, meat, milk and milk products
10.	Vitamin C	40 mg/day	Fresh amla, citrus fruits, guava, banana and certain vegetables such as tomatoes, agathikeerai, drumstick leaves, cabbage	

	Minerals	RDA	Common source (plant-based)	Common source (animal-based)
1.	Iron	15–20 mg/day	Green leafy vegetables (molaikeerai or amaranth, cauliflower greens, radish leaves), legumes, dry fruits	Meat, fish and poultry
2.	Calcium	600 mg/day	Nuts (almond, dry coconut), whole Bengal gram, blackgram, ragi, soybean and green leafy vegetable (amaranth or molaikeerai, agathi)	Milk, curd, fish

7.9 Intermittent Fasting

Intermittent fasting is a form of eating or diet that focuses on timing (Time restricted diet) in contrast to most other diets, which focus on contents (Calorie restricted,

Mediterranean diets).

The scientific basis of IF started with the initial observation that rodents, kept lean by once-a-day feeding, were healthier and lived longer than their counterparts who had unrestricted access to food throughout the day.

7.9.1 Basic Physiology

Feeding increases insulin secretion, an anabolic hormone that causes extra glucose deposition as glycogen and fat as triglycerides in adipose tissues. Excess energy accumulation stimulates mTOR activity, enhancing protein synthesis and cell growth.

Energy restriction (fasting) for around 16 h or more results in the depletion of liver glycogen stores and fat mobilization from adipose tissue as the body switches from glucose-based energy extraction to fat-based energy extraction (**popularly called the metabolic switch**). Triglycerides are hydrolyzed to free fatty acids (FFAs) in fat cells (adipocytes) and are transported to the liver, where they are converted to the ketone bodies (KBs), acetoacetate and β-hydroxybutyrate (β-HB).

1. β-HB and acetoacetate can be utilized in the Krebs cycle to generate energy. β-HB also has intracellular signalling functions for activating various transcription factors in neurons.
2. FFAs also activate transcription factors with widespread effects on cells throughout the body and brain.
3. Reduced levels of glucose and amino acids during fasting result in reduced activity of the mTOR pathway and up-regulation of autophagy.

Depletion of excess energy stimulates the body to preserve the maintenance and repair systems, enhancing stress resistance, recycling damaged molecules, stimulating mitochondrial biogenesis, and promoting cell survival. All these effects support improvements in health and disease resistance.

Types of Intermittent Fasting

1. 16:8 Fasting. In this method, one eats during an 8-h window and fasts for 16 h at a stretch. This permits two meals a day and skipping one.
2. Alternate-day (AD type) fasting: complete fasting (water only) every other day. It appears to be the toughest regime. A less severe form of this regimen allows 25% of the average daily consumption of Calories on the 'fast' day.
3. 5:2 Intermittent Fasting (i.e. fasting two non-consecutive days each week. One may take up to 500 Cal with black coffee and unrestricted water on the fast days. The other 5 days are for normal food (minus refined carbohydrates).

The benefits of IF have been proven in many studies and are summarised in Table 7.2. The benefits are as strong as you would get by permanently cutting down

Table 7.2 Proven benefits of intermittent fasting

Muscle	Intestine	Liver	Blood
• Improved insulin sensitivity • Enhanced efficiency and endurance • Reduced inflammation	• Improved motility • Reduced inflammation • Healthy microbiota	• Glycogen depletion • Increased ketone production • Improved insulin sensitivity • Reduced liver fat	• Elevated ketone body levels • Reduced insulin, glucose and leptin • Elevated adiponectin • Reduced inflammatory cytokine level • Reduced marker of oxidative stress
Heart and vascular system	*Brain*	*Adipose tissue*	*Overall benefit*
• Reduced blood pressure • Reduced resting heart rate • Increased parasympathetic tone • Stress resistance	• Improved cognition • Resistance to injury and disease • Synaptic plasticity • Mitochondrial biogenesis • Neurotrophic factor production	• Lipolysis • Reduced leptin production • Reduced inflammation	• Weight loss • Improvement in diabetes • Improvement in metabolic syndrome • Improvement in NAFLD

your calorie intake by ~30% in all meals, which remains an option. These benefits include weight loss, improvements in glucose regulation, blood pressure, heart rate, endurance training efficacy, and abdominal fat loss.

A recent article in N England J of Medicine (December 2019) summarised most health benefits. Further, several studies have demonstrated impressive weight loss, liver fat reduction and improved glycaemic status.

The adverse effects are primarily

1. Hunger, as the body has been accustomed to three meals per day since childhood. Those persisting with the regime, however, report declining hunger within a week or two.
2. Hypoglycaemia: especially patients with diabetes tightly controlled with insulin or sulphonylureas. They are advised to try it only under medical supervision.

7.9.2 Conclusion

The concept and benefit of IF seem to resonate with most cultural and traditional practices worldwide and across time, where faith-based intermittent fasting was an integral part and possibly accounted for the low prevalence of metabolic disorders caused by energy excess. It offers a simple strategy to harness several metabolic abnormalities resulting from the long-term imbalance between excess calorie consumption and decreased expenditure. It could help stem type 2 diabetes mellitus, obesity, fatty liver, hypertension and coronary heart disease, and contribute to health, decreasing the incidence of cancers and longevity.

7.10 Refeeding Syndrome

- Refeeding syndrome is a common metabolic consequence of parenteral nutrition. This syndrome results from the sudden provision of a large amount of calories, especially carbohydrates, to a previously malnourished patient. On parenteral nutrition infusion, the metabolism of these patients attempts to become rapidly anabolic.
- Insulin production is increased, pushing potassium, phosphorus, and magnesium into intracellular compartments. This increases the risk of hypokalemia, hypophosphatemia, and hypomagnesemia and may lead to fatal arrhythmias.
- Sodium retention and large fluid shifts can also occur and may place the patient at risk of developing congestive heart failure.
- The metabolic and physiologic consequences of the depletion, repletion, compartmental shifts, and inter-relationships of the following nutrients have been defined:
  - Phosphorus (<1 mg/dL, death within hours)
  - Potassium
  - Magnesium
  - Glucose metabolism
  - Vitamin deficiency
  - Fluid resuscitation

7.10.1 Metabolic Response to Refeeding

- A shift from body fat to carbohydrates as a major fuel source.
- Glycogenolysis, gluconeogenesis, and fatty acid mobilization from adipose tissue is inhibited.
- Cellular uptake of glucose, potassium, phosphorus, and magnesium is enhanced by insulin.
- Anti-natriuretic effect (sodium retention and extracellular fluid expansion).

7.10.2 Patients at Risk of Refeeding Syndrome

- Chronic alcoholism.
- Anorexia nervosa.

- Classic marasmus.
- Classic kwashiorkor.
- Chronic undernourishment.
- Morbid obesity with massive weight loss.
- Prolonged hypocaloric intravenous hydration.
- Nil per os (NPO) for >7–10 days.
- Cardiac and cancer cachexia.

7.10.3 Recommendations to Avoid Refeeding Syndrome

- Be aware of the syndrome.
- Recognize the patient at risk.
- Correct electrolyte imbalance before initiating nutritional support, whether by the oral, enteral, or parenteral route.
- Judiciously restore circulatory volume, and monitor heart rate, intake, and output.
- Increase caloric delivery slowly.
- Administer vitamins routinely.
- Closely monitor electrolytes over the first week: Serum phosphorus, potassium, magnesium, sugar, and urine electrolytes.